The Regularization Cookbook

Explore practical recipes to improve the functionality
of your ML models

Vincent Vandenbussche

<packt>

BIRMINGHAM—MUMBAI

The Regularization Cookbook

Copyright © 2023 Packt Publishing

Group Product Manager: Ali Abidi

Associate Publishing Product Manager: Aditya Datar

Senior Editor: Tiksha Lad

Technical Editor: Sweety Pagaria

Copy Editor: Safis Editing

Project Coordinator: Farheen Fathima

Proofreader: Safis Editing

Indexer: Manju Arasan

Production Designer: Vijay Kamble

Marketing Coordinator: Vinishka Kalra

First published: July 2023

Production reference: 1200723

Published by Packt Publishing Ltd.

Grosvenor House

11 St Paul's Square

Birmingham

B3 1R.

ISBN 978-1-83763-408-8

www.packtpub.com

Writing a book mirrors the kaleidoscope of life, encompassing its ups and downs, joys and sorrows, the quest for purpose, fulfillment, and the haunting whispers of inner emptiness. In this swirling storm of intense emotions, Amandine's unwavering support served as my guiding light, keeping me on course, and it is with deep appreciation that I dedicate this book to her.

– Vincent Vandenbussche

Foreword

In the ever-evolving landscape of machine learning, it can often feel like you're trying to navigate a complex labyrinth of algorithms, models, and methods. Among the multitude of concepts that make up this domain, regularization plays an essential role – it keeps the balance between bias and variance, helping models to generalize well rather than just memorizing the training data.

As the art and science of machine learning continue to mature, it becomes imperative that we make these technical concepts accessible, intuitive, and practical to practitioners. In this realm, Vincent, with his depth of experience in machine learning, has embarked on a journey to unravel the complexity of regularization in machine learning models. In this cookbook, he offers a structured approach, grounded in practicality, that meticulously covers everything from the basics to advanced regularization techniques.

Beginning with an introduction to the topic, the book lays the groundwork by offering a refresher on machine learning best practices. It then moves on to tackle regularization techniques for both linear and tree-based models, elegantly bridging the gap between theory and practice.

But Vincent's cookbook does not stop there. It ventures into regularization with data, covering feature aggregation and the handling of imbalanced datasets, followed by insights into regularization for deep learning and recurrent neural networks.

Recognizing the significance of application-specific regularization techniques, Vincent dedicates several chapters to regularization in natural language processing and computer vision. These chapters explore regularization methods specific to these domains, including the use of embeddings, data augmentation, and synthetic image generation.

The final chapter, on synthetic image generation for regularization, is an intriguing intersection of creativity and technology, offering a testament to Vincent's expertise in cutting-edge techniques.

Throughout this cookbook, Vincent's deft blend of theory, practice, and cutting-edge knowledge is evident. This book serves not just as a guide, but as a toolbox, filled with practical recipes that you can directly implement and adapt as per your needs.

Whether you are a beginner finding your feet in the world of machine learning or an experienced practitioner looking to deepen your understanding of regularization, this cookbook offers a comprehensive and hands-on approach. It speaks volumes to Vincent's expertise and his ability to make complex concepts accessible and practical.

This cookbook promises to be an indispensable resource for every machine learning enthusiast's journey. It is a culmination of Vincent's vast experience, ingenuity, and desire to share his knowledge with the world. As you embark on this journey with him, it is my hope that this book will guide, inspire, and empower you to harness the true power of regularization in machine learning.

Happy reading and cooking in the world of machine learning!

–Akin Osman Kazakci

Head of Data Innovation Lab at MINES ParisTech and Entrepreneur-in-Residence at Caltech

Contributors

About the author

Since gaining a Ph.D. in physics, **Vincent Vandenbussche** has worked for a decade in diverse companies, deploying ML solutions at scale. He has worked at numerous companies, such as Renault, L'Oréal, General Electric, Jellysmack, Chanel, and CERN.

He also has a passion for teaching; he co-founded a data science boot camp, was an ML lecturer at MINES ParisTech engineering school and EDHEC Business School, and trained numerous professionals at companies such as ArcelorMittal and Orange.

About the reviewer

Rajat Agrawal is an accomplished data scientist with a keen focus on deep learning and a profound interest in generative AI and language models. With four years of experience in research work and two years in the field, he has established himself as a valuable asset in the realm of data science. Rajat's expertise in statistical modeling and predictive analytics, coupled with his groundbreaking publication on *Lane Detection and Collision Prevention System for Automated Vehicles* in Springer, demonstrates his innovative approach to computer vision algorithms. He is dedicated to a noble mission of democratizing AI, ensuring its accessibility to every individual.

My parents' unwavering support has been instrumental in shaping my career.

Table of Contents

3

Regularization with Linear Models   47

4

Regularization with Tree-Based Models   73

5

Regularization with Data   113

6

Deep Learning Reminders   141

7

Deep Learning Regularization   181

8

Regularization with Recurrent Neural Networks   221

9

Advanced Regularization in Natural Language Processing   259

Preface

Have you ever wondered why so many machine learning projects fail in production?

In many cases, this is because of a lack of generalization of models, leading to unexpected predictions when facing new, unseen data. This is what regularization is about: making sure a model provides the expected predictions, even when facing new data.

In this book, we will explore many forms of regularization. To accomplish this, we will explore two primary avenues for regularization solutions, depending on the recipes in the chapter:

- When given a machine learning model, how do we regularize it? Regularizing is most suited in applications where the model is already imposed (whether there is a legacy solution to be updated or strong requirements) and the training data is fixed, so the only solution is to regularize the model.

- Given a machine learning task, how do we get a robust, well-generalizing solution? This approach is most suited in applications where only the problem is defined, but no strong constraints have been provided yet, so more solutions can be explored.

Hopefully, these recipes will provide you with the necessary tools and techniques to solve most of the machine learning problems you may face that require regularization, as well as a solid practical understanding of the underlying concepts.

Who this book is for

This book is suitable for anyone with prior Python knowledge: the only strong requirement to fully understand the proposed solutions is that you can read and run simple Python code. For each new method or model that is introduced, some context and practical explanations are provided, so that anyone with a computer science background can fully understand what they are doing.

Although any Python practitioner is able to follow this book, the main target audience is the following:

- Machine learning practitioners, such as machine learning engineers, applied scientists, and data scientists, who want ready-to-use methods and code to pick from whenever they face a new problem or task. Using this book, hopefully they can handle many situations by slightly adapting the code to their own problems.

- Machine learning enthusiasts who want to gain a deeper knowledge and understanding of machine learning, with concrete examples and working code. Using this book, they can get deep knowledge with hands-on examples and build a solid portfolio of projects.

What this book covers

Chapter 1, An Overview of Regularization, provides a high-level introduction to what regularization is, as well as all the fundamental knowledge and vocabulary to fully understand the remaining chapters of this book.

Chapter 2, Machine Learning Refresher, guides you through a typical machine learning workflow and best practices, from data loading and splitting to model training and evaluation.

Chapter 3, Regularization with Linear Models, covers regularization with common linear models: linear regression and logistic regression. Regularization with L1 and L2 penalization is covered, as well as some practical tips for how to choose the right regularization method.

Chapter 4, Regularization with Tree-Based Models, provides reminders about decision trees for both classification and regression, as well as how to regularize them. Ensemble methods, such as Random Forest and Gradient Boosting, and their regularization methods are then covered.

Chapter 5, Regularization with Data, introduces regularization with data, using hashing and its features and feature aggregation. Resampling methods for imbalanced datasets are then covered.

Chapter 6, Deep Learning Reminders, provides reminders about deep learning, both conceptually and practically. Starting with a Perceptron, we then train models for regression and classification.

Chapter 7, Deep Learning Regularization, covers regularization for deep learning models. Several techniques are explored and explained: L2 penalization, early stopping, network architecture, and dropout.

Chapter 8, Regularization with Recurrent Neural Networks, dives into **Recurrent Neural Networks** (**RNNs**) and **Gated Recurrent Units** (**GRUs**). It starts by explaining what they are and how to train such models. Regularization techniques are then covered, such as dropout and maximum sequence length.

Chapter 9, Advanced Regularization in Natural Language Processing, explores regularization methods specific to **Natural Language Processing** (**NLP**). Regularization using word2vec embeddings and BERT embeddings is covered. Data augmentation with word2vec and GPT-3 is explored. Zero-shot inference solutions are also proposed.

Chapter 10, Regularization in Computer Vision, dives into regularization for computer vision and **Convolutional Neural Networks** (**CNNs**). After explaining CNNs conceptually and practically on classification, recipes with regularization for object detection and semantic segmentation are provided.

Chapter 11, Regularization in Computer Vision – Synthetic Image Generation, dives deeper into synthetic image generation for regularization. Simple data augmentation is first explored. Then, a QR code object detection mechanism is built with only synthetic training data. Finally, we explore a real-time style transfer whose training is based on Stable Diffusion data, as well as explain how to work with such a dataset by yourself.

To get the most out of this book

You will need a version of Python installed. All the code has been tested with Python 3.10 on Ubuntu 22.04, with CUDA version 12.1. However, it should work with Python 3.9 and later versions, on any OS, and with CUDA version 11 and later.

Software/hardware covered in the book	OS requirements
Python 3.9	Windows, macOS, or Linux (any)

For the deep learning chapters, especially from *Chapter 8* onward, the use of a **Graphics Processing Unit (GPU)** is recommended. The code was tested on an Nvidia GeForce RTX 3090 with 24 GB of memory. Depending on your hardware specifications, the code may need to be adjusted accordingly.

Recipes begin from the second chapter. If you are using the digital version of this book, we advise you to type the code yourself or access the code via the GitHub repository (link available in the next section). Doing so will help you avoid any potential errors related to the copying and pasting of code.

Download the example code files

You can download the example code files for this book from GitHub at `https://github.com/PacktPublishing/The-Regularization-Cookbook`. If there's an update to the code, it will be updated on the existing GitHub repository.

We also have other code bundles from our rich catalog of books and videos available at `https://github.com/PacktPublishing/`. Check them out!

Conventions used

There are a number of text conventions used throughout this book.

`Code in text`: Indicates code words in text, database table names, folder names, filenames, file extensions, pathnames, dummy URLs, user input, and Twitter handles. Here is an example: "Those two lines should download a `.zip` file, and then unzip its content, so that a file named `Tweets.csv` is available."

A block of code is set as follows:

```
# Load data
data = pd.read_csv('Tweets.csv')
data[['airline_sentiment', 'text']].head()
```

Any command-line input or output is written as follows:

```
ip install pandas numpy scikit-learn matplotlib torch transformers
```

> **Tips or important notes**
> Appear like this.

Sections

In this book, you will find several headings that appear frequently (*Getting ready*, *How to do it...*, *How it works...*, *There's more...*, and *See also*).

To give clear instructions on how to complete a recipe, use these sections as follows:

Getting ready

This section tells you what to expect in the recipe and describes how to set up any software or any preliminary settings required for the recipe.

How to do it...

This section contains the steps required to follow the recipe.

How it works...

This section usually consists of a detailed explanation of what happened in the previous section.

There's more...

This section consists of additional information about the recipe in order to make you more knowledgeable about the recipe.

See also

This section provides helpful links to other useful information for the recipe.

Get in touch

Feedback from our readers is always welcome.

General feedback: If you have questions about any aspect of this book, mention the book title in the subject of your message and email us at customercare@packtpub.com.

Errata: Although we have taken every care to ensure the accuracy of our content, mistakes do happen. If you have found a mistake in this book, we would be grateful if you would report this to us. Please visit www.packtpub.com/support/errata, selecting your book, clicking on the Errata Submission Form link, and entering the details.

Piracy: If you come across any illegal copies of our works in any form on the internet, we would be grateful if you would provide us with the location address or website name. Please contact us at copyright@packt.com with a link to the material.

If you are interested in becoming an author: If there is a topic that you have expertise in and you are interested in either writing or contributing to a book, please visit authors.packtpub.com.

Share Your Thoughts

Once you've read *The Regularization Cookbook*, we'd love to hear your thoughts! Scan the QR code below to go straight to the Amazon review page for this book and share your feedback.

https://packt.link/r/1837634084

Your review is important to us and the tech community and will help us make sure we're delivering excellent quality content.

Download a free PDF copy of this book

Thanks for purchasing this book!

Do you like to read on the go but are unable to carry your print books everywhere?

Is your eBook purchase not compatible with the device of your choice?

Don't worry, now with every Packt book you get a DRM-free PDF version of that book at no cost.

Read anywhere, any place, on any device. Search, copy, and paste code from your favorite technical books directly into your application.

The perks don't stop there, you can get exclusive access to discounts, newsletters, and great free content in your inbox daily

Follow these simple steps to get the benefits:

1. Scan the QR code or visit the link below

https://packt.link/free-ebook/9781837634088

2. Submit your proof of purchase
3. That's it! We'll send your free PDF and other benefits to your email directly

An Overview of Regularization

Let's embark on a journey into the world of regularization in machine learning. I hope you will learn a lot and find as much joy in reading this book as I did in writing it.

Regularization is important for any individual willing to deploy robust **machine learning** (**ML**) models.

This chapter will introduce some context and key concepts about regularization before diving deeper into it in the next chapters. At this point, you may have many questions about this book and about regularization in general. What is regularization? Why do we need regularization for production-grade ML models? How do we diagnose the need for regularization? What are the limits of regularization? What are the approaches to regularization?

All the foundational knowledge about regularization will be provided in this chapter in the hope of answering all these questions. Not only will this give you a high-level understanding of what regularization is but it will also allow you to fully appreciate the methods and techniques proposed in the next chapters of this book.

In this chapter, we are going to cover the following main topics:

- Introducing regularization
- Developing intuition about regularization on a toy dataset
- Introducing the key concepts of underfitting, overfitting, bias, and variance

Technical requirements

In this chapter, you will have the opportunity to generate a toy dataset, display it, and train basic linear regression on that data. Therefore, the following Python libraries will be required:

- NumPy
- Matplotlib
- scikit-learn

Introducing regularization

> *"Regularization in ML is a technique used to improve the generalization performance of a model by adding additional constraints to the model's parameters. This forces the model to use simpler representations and helps reduce the risk of overfitting.*
>
> *Regularization can also help improve the performance of a model on unseen data by encouraging the model to learn more relevant, generalizable features."*

This definition of regularization, arguably good enough, was actually generated by the famous GPT-3 model when given the following prompt: *Detailed definition of regularization in machine learning*. Even more astonishing, this definition passed several plagiarism tests, meaning it's actually fully original text. Do not worry if you do not yet understand all the words in this definition from GPT-3; it is not meant for beginners. But you will fully understand it by the end of this chapter.

> **Note**
>
> **GPT-3**, short for **Generative Pre-trained Transformer 3**, is a 175 billion-parameter model proposed by OpenAI and is available for use at `platform.openai.com/playground`.

You can easily imagine that, to get such a result, not only has GPT-3 been trained on a large amount of data but it is really carefully regularized, so that it won't just reproduce a learned text but will instead generate a new one.

This is exactly what regularization is about: being able to generalize and produce acceptable results when faced with an unknown situation.

Why is regularization so crucial to ML? The key to successfully deploying ML in production lies in the model's ability to effectively adapt to and accommodate new data. Once a model is in production, it will not be fed with well-known, average input data. Most likely, the production model will face unseen data, exceptional scenarios, a drift in feature distribution, or evolving customer behavior. While a well-regularized ML model may not guarantee its robustness in handling various scenarios, a poorly regularized model is almost certain to encounter failure upon its initial deployment in production.

Let's now have a look at a few examples of models that failed during deployment in recent years, so that we can fully understand why regularization is so important.

Examples of models that did not pass the deployment test

The last few years have been full of examples of models that failed in the first days of deployment. According to a Gartner report from 2020 (`https://www.gartner.com/en/newsroom/press-releases/2020-10-19-gartner-identifies-the-top-strategic-technology-trends-for-2021#:~:text=Gartner%20research%20shows%20only%2053,a%20production%2Dgrade%20AI%20pipeline`), more than 50% of AI prototypes will *not* make it to production deployment. Not all failures were only due to regularization issues, but some certainly were.

Let's have a quick look at some failed attempts at deploying models in production over the last few years:

- Amazon had to stop using its AI recruitment model because it was reportedly discriminating against women (https://finance.yahoo.com/news/amazon-reportedly-killed-ai-recruitment-100042269.html?guccounter=1&guce_referrer=aHR0cHM6Ly9hbmFsZXRpY3NpbmRpYW1hZy5jb20v&guce_referrer_sig=AQAAACNWCozxgjh8_DkmyT59IZEGsn3qlmfu2pVu6IxMu5B0ExzHJVkat UuBmpO3zGcWp-0nvgWJ9yqR9eaQU-20-DvgJzJdR7xj9U8faNpVUTPo00gND-W5WWPh_wGNLNTASitfnb-MnStbjZaNN_O3EbWHDarh0_cAzXza31yeYcEe)
- Microsoft's chatbot Tay was shut down after only 16 hours of production after posting offensive tweets (https://en.wikipedia.org/wiki/Tay_(chatbot))
- IBM's Watson was providing unsafe cancer treatment recommendations to patients (https://www.theverge.com/2018/7/26/17619382/ibms-watson-cancer-ai-healthcare-science)

Those are just a few examples that made the headlines from tech giants. The number of projects that have experienced failure yet remain undisclosed to the public is staggering. These failures often involve smaller companies and confidential initiatives. But still, there are several lessons to learn from those examples:

- **Amazon's case**: The input data was biased against women, as was the model
- **Microsoft's case**: The model was probably too sensitive to new data since it was feeding on new tweets
- **IBM's case**: The model was perhaps trained on too much synthetic or unrealistic data, and not able to adapt to edge cases and unseen data

Regularization serves as a valuable approach to enhance the success rate of ML models in production. Effective regularization techniques can prevent AI recruitment models from exhibiting gender biases, either by eliminating certain features or incorporating synthetic data. Additionally, proper regularization enables chatbots to maintain an appropriate level of sensitivity toward new tweets. It also equips models to handle edge cases and previously unseen data proficiently, even when trained on synthetic data.

As a disclaimer, there may be many other ways to overcome or prevent these kinds of failures that are not mutually exclusive with regularization. For example, having good-quality data is key. Everyone in the AI field knows the adage *garbage in, garbage out*.

MLOps (a field that is getting more and more mature every day) and ML engineering best practices are also key to success for many projects. Subject matter knowledge can sometimes also make a difference.

Depending on the context of the project, many other parameters may impact the success of an ML project, but anything besides regularization is beyond the scope of this book.

Now that we understand the need for regularization for production-level ML models, let's take a step back and gain some intuition about what regularization is with a simple example.

Intuition about regularization

Regularization has been defined and mentioned already in this book, but let's try now to develop some intuition about what it really is.

Let us consider a typical, real-world use case: the real estate price per square meter, as a function of the surface of an apartment (or house) in the city of Paris. The goal, from a business perspective, is to be able to predict the price per square meter, given the apartment's surface.

We will first need some imports, as well as a helper function to plot the data more conveniently.

The plot_data() function simply plots the provided data and adds axis labels and a legend if needed:

```
import numpy as np
import matplotlib.pyplot as plt
from sklearn.linear_model import LinearRegression
np.random.seed(42)
def plot_data(surface: np.array, price: np.array,
    fit: np.array = None, legend: bool = False):
    plt.scatter(surface, price, label='data')
    if fit is not None:
        plt.plot(surface, fit, label='fit')
    if legend:
        plt.legend()
    plt.ylim(11300, 11550)
    plt.xlabel('Surface (m$^{2}$)')
    plt.ylabel('Price (€/m$^{2}$)')
    plt.grid(True)
    plt.show()
```

The following code will now allow us to generate and display our first toy dataset:

```
# Define the surfaces and prices
surface = np.array([15, 17, 20, 22, 25, 28]).reshape(-1, 1)
price = 12000 - surface*50 + np.square(
    surface) + np.random.normal(0, 30, surface.shape)
# Plot the data
plot_data(surface, price)
```

Here is the plot for it:

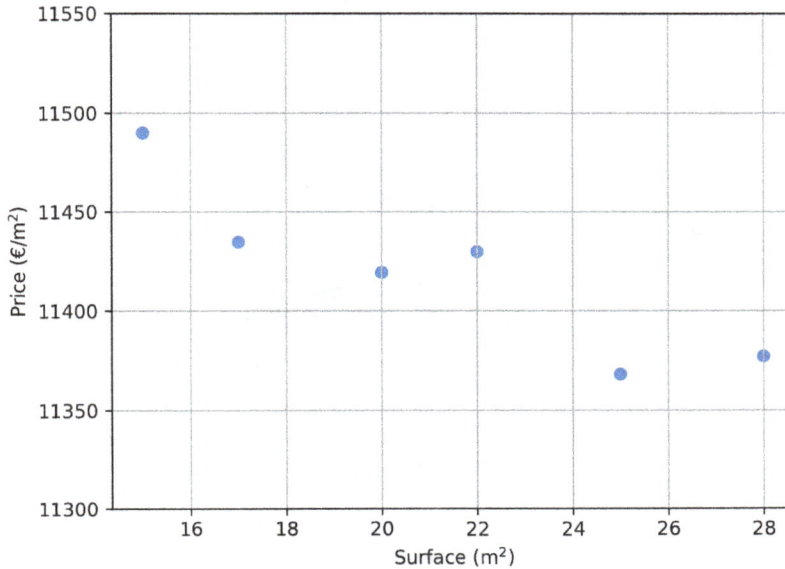

Figure 1.1 – Price per square meter as a function of the apartment surface

Even if this is a toy dataset, for the sake of pedagogy, we can assume this data would have been collected on the real estate market.

We can notice a downward trend in the price per square meter as the apartment surface increases. Indeed, in Paris, the small surfaces are much more in demand (perhaps because there are many students or because the price is more affordable). That could explain why the price per square meter is actually higher for smaller surfaces.

For simplicity, we will omit all the typical ML workflow. Here, we just perform a linear regression on this data and display the result with the following code:

```
# Perform a linear regression on the data
lr = LinearRegression()
lr.fit(surface, price)
# Compute prediction
y_pred = lr.predict(surface)
# Plot data
plot_data(surface, price, y_pred, True)
```

Here is the output:

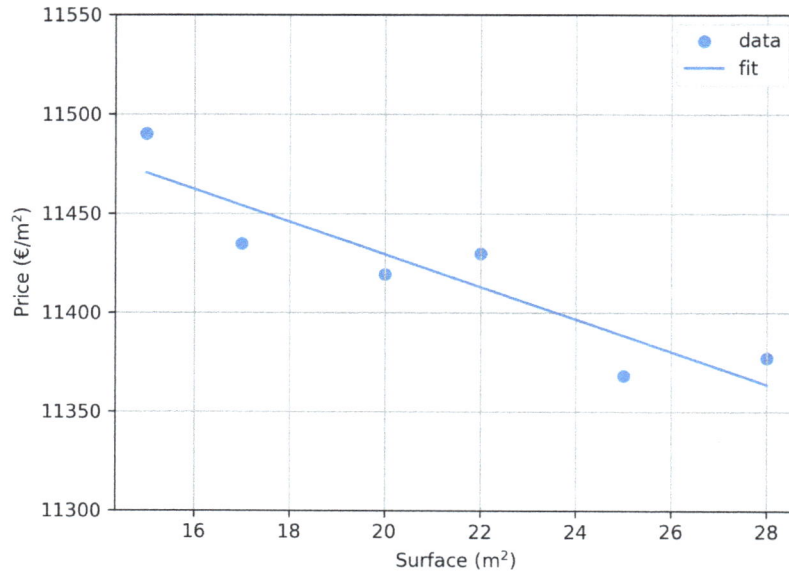

Figure 1.2 – Price per square meter as a function of the apartment surface and resulting fit curve

Good enough! The fit seems to capture the downward trend. While it's not so close to all the given data samples, the business seems happy about it for the moment, as the model performances are within their expectations.

Thanks to this new model, the company has now acquired a few more clients. From those clients, the company collected some new data from larger apartment sales, so our dataset now looks like the following:

```
# Generate data
updated_surface = np.array([15, 17, 20, 22, 25, 28, 30, 33,
    35, 37]).reshape(-1, 1)
updated_price = 12000 - updated_surface*50 + np.square(
    updated_surface) + np.random.normal(0, 30, updated_surface.shape)
# Plot data
plot_data(updated_surface, updated_price)
```

Here is the plot for it:

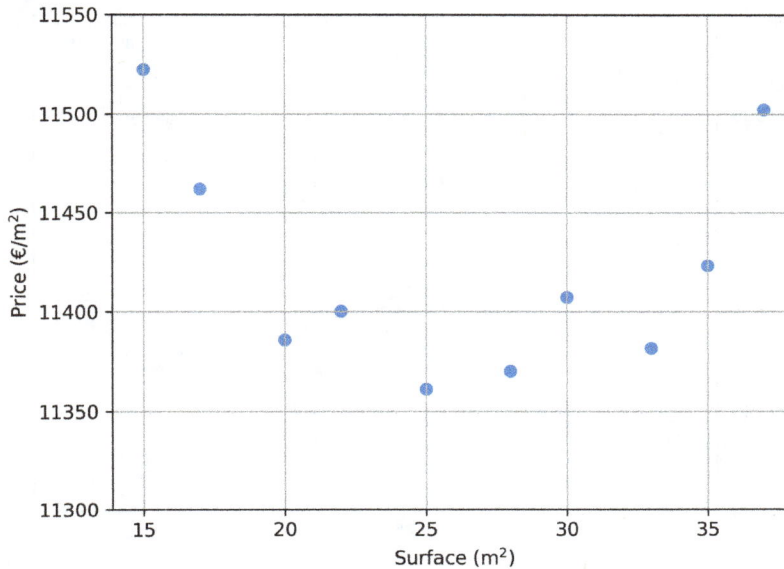

Figure 1.3 – Updated price per square meter as a function of the apartment surface

This actually changes everything; this is the typical failing deployment test. There is no global downward trend anymore: with larger apartment surfaces, the price per square meter actually seems to follow an upward trend. One simple business explanation could be the following: larger apartments may be less common and thus have a higher price.

Without confidence, for the sake of trying, we try to reuse the exact same method as we did previously: linear regression. The result would be the following:

```
# Perform linear regression and plot result
lr = LinearRegression()
lr.fit(updated_surface, updated_price)
y_pred_updated = lr.predict(updated_surface)
plot_data(updated_surface, updated_price, y_pred_updated, True)
```

Here is the plot:

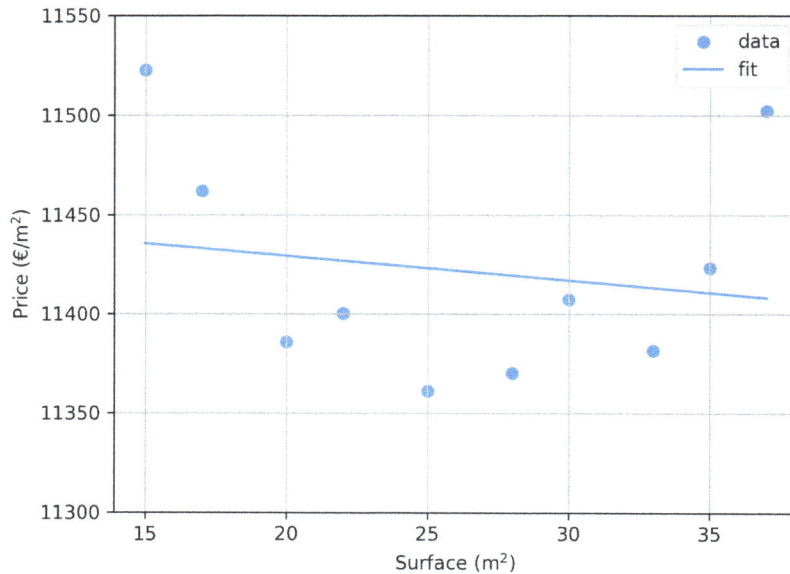

Figure 1.4 – Example of underfitting: the data complexity is not fully captured by the model

As expected, linear regression was not able to capture the complexity of the data anymore, leading to this situation. This is called **underfitting**; the model is not able to fully capture the complexity of the data. Indeed, with just the surface as a parameter, the best the model can do is a straight line, which is not enough for this data.

One way for linear regression to capture more complexity is to provide more features. Given that our current input data is limited to the surface, a potential straightforward approach could involve utilizing the raised-to-the-power surface. For the sake of this example, let's take a rather extreme approach and add all the features from `power1` to `power15`, and make them fit this dataset. This can be done pretty easily with the following code:

```
# Compute power up to 15
x_power15 = np.concatenate([np.power(
    updated_surface, i+1) for i in range(15)], 1)
# Perform linear regression and plot result
lr = LinearRegression()
lr.fit(x_power15, updated_price)
y_pred_power15 = lr.predict(x_power15)
plot_data(updated_surface, updated_price, y_pred_power15, True)
```

Here is the output for it:

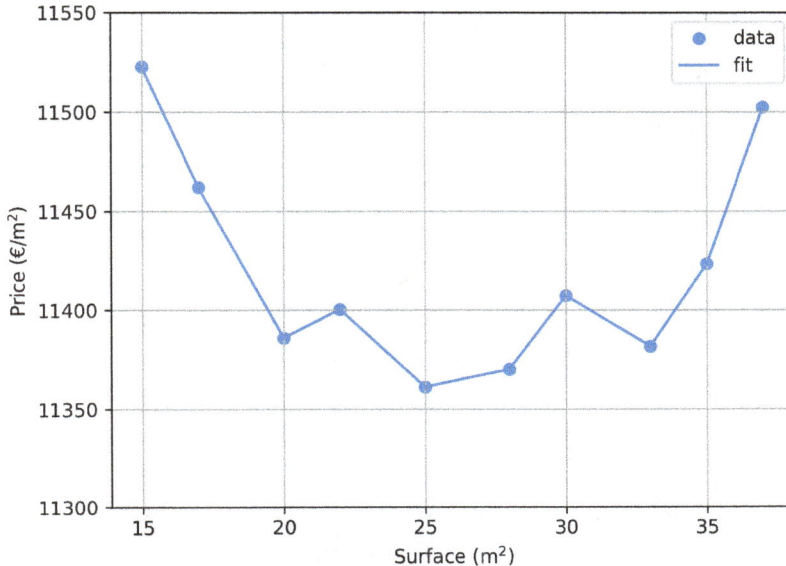

Figure 1.5 – Example of overfitting: the model is capturing the noise in the data

As we could expect by adding so many **degrees of freedom** in our model, the fit is now going exactly through all of the data points. Indeed, without going into the mathematical details, the model has more parameters than data points on which it trains, and is capable of going through all those points. Is it a good fit, though? We can imagine it does not only capture the global trend of the data but also the noise. This is called **overfitting**: the model is too close to the data, and may not be able to make correct predictions for new, unseen data. Any new data point would not be on the curve, and the behavior outside the training range is totally unpredictable.

Finally, a more reasonable approach to this situation would be only taking the surface up to the power of 2, for example:

```
# Compute power up to 2
x_power2 = np.concatenate([np.power(
    updated_surface, i+1) for i in range(2)], 1)
# Perform linear regression and plot result
lr = LinearRegression()
lr.fit(x_power2, updated_price)
y_pred_power2 = lr.predict(x_power2)
plot_data(updated_surface, updated_price, y_pred_power2, True)
```

Here is the output for it:

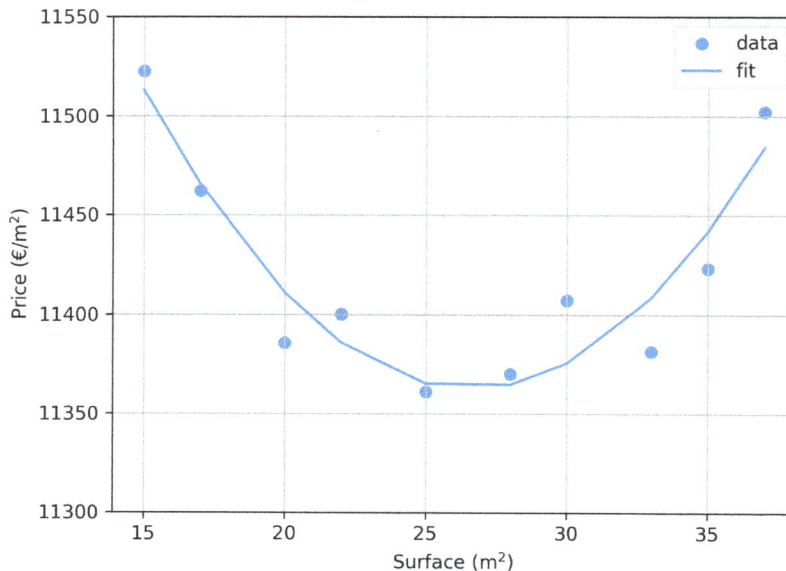

Figure 1.6 – Example of right fitting: the model is capturing the overall trend but not the noise

The fit seems much more acceptable now. It does capture the global trend of the data: at first downward for small surfaces, and then upward for larger surfaces. Moreover, it does not try to capture the noise coming from all the data points, making it more robust for predicting new, unseen data. Finally, the behavior is quite predictable beyond the training range (for example, surface $< 15m^2$ or surface $> 40m^2$).

This is typically the kind of desired behavior from a good model: neither underfitting nor overfitting. Removing some of the raised-to-the-power features allowed our model to generalize better; we effectively regularized our model.

To summarize this example, we explored several concepts here:

- We visualized examples of underfitting, overfitting, and well-regularized models
- By adding a raised-to-the-power surface to our features, we were able to go from underfitting to overfitting
- Finally, by removing most of the raised-to-the-power features (keeping only square features), we were able to go from overfitting to a well-regularized model, effectively adding regularization

Hopefully, you now have a good understanding of underfitting, overfitting, and regularization, as well as why this is so important in ML. We can now build upon this by providing a more formal definition of the key concepts of regularization.

Key concepts of regularization

Having gained some intuition regarding what constitutes a suitable fit, as well as understanding examples of underfitting and overfitting, let us now delve into a more precise definition and explore key concepts that enable us to better comprehend regularization.

Bias and variance

Bias and variance are two key concepts when talking about regularization. We can define two main kinds of errors a model can have:

- **Bias** is how bad a model is at capturing the general behavior of the data
- **Variance** is how bad a model is at being robust to small input data fluctuations

Those two concepts, in general, are not mutually exclusive. If we take a step back from ML, there is a very common figure to visualize bias and variance, assuming the model's goal is to hit the center of a target:

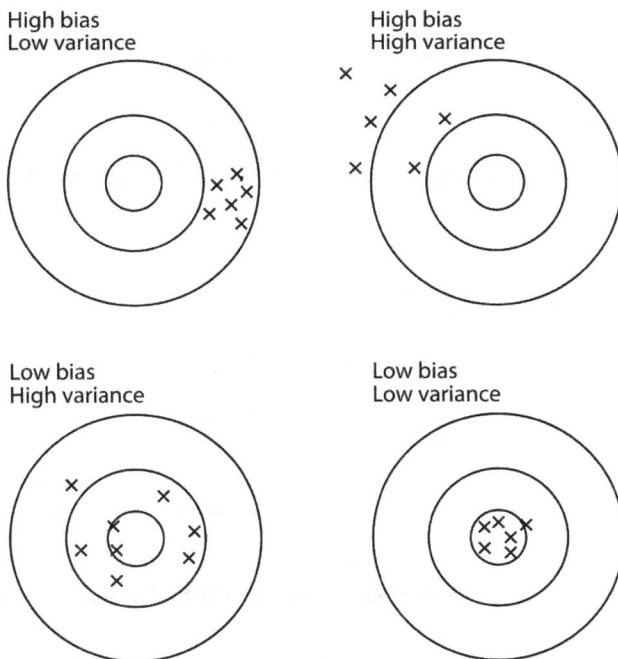

Figure 1.7 – Visualization of bias and variance

Let's describe those four cases:

- **High bias and low variance**: The model is hitting away from the center of the target, but in a very consistent manner

- **Low bias and high variance**: The model is, on average, hitting the center of the target, but is quite noisy and inconsistent in doing so

- **High bias and high variance**: The model is hitting away from the center in a noisy way

- **Low bias and low variance**: The best of both worlds – the model is hitting the center of the target consistently

Underfitting and overfitting

We saw a very classic approach to bias and variance definition.

But now, what does that mean in terms of ML? How does that relate to regularization? Well, before we get there, we will first revisit bias and variance in a more typical ML case: linear regression of real estate prices.

Let's have a look at how a model would behave in all those cases on our data.

High bias and low variance

The model is robust to data fluctuations but could not capture the high-level behavior of the data. Refer to the following graph:

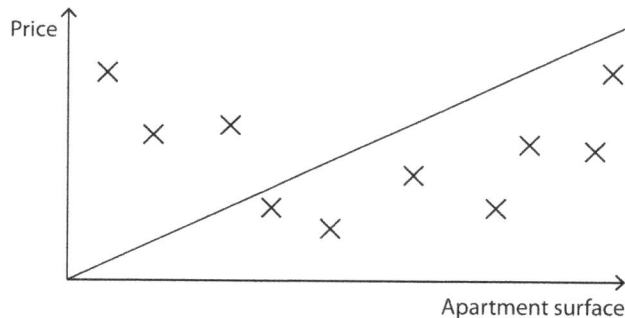

Figure 1.8 – High bias and low variance in practice for linear regression

This is *underfitting*, as we faced earlier, in *Figure 1.4*.

Low bias and high variance

The model did capture the global behavior of the data, but could not stay robust to input data fluctuations. Refer to the following graph:

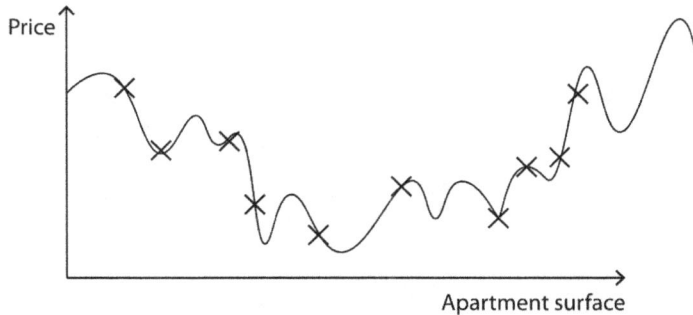

Figure 1.9 – Low bias and high variance in practice for linear regression

This is *overfitting*, the case we faced previously, in *Figure 1.5*.

High bias and high variance

The model could neither capture the global behavior nor be robust enough to input data fluctuations. This case never happens, or high variance is hidden behind high bias, but it could look something like the following:

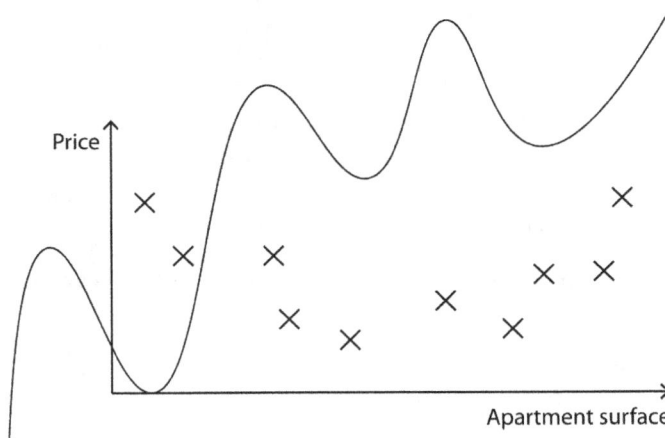

Figure 1.10 – High bias and high variance in practice for linear regression:
this most likely never actually happens on such data

Low bias and low variance

The model could both capture the global data behavior and be robust enough for data fluctuation. This is the end goal when training a model. This is the case we faced in *Figure 1.6*.

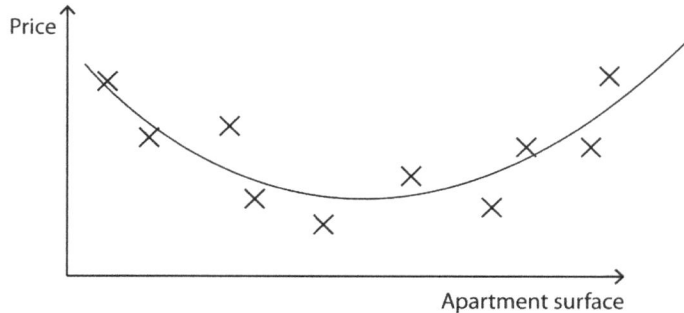

Figure 1.11 – Low bias and low variance in practice for linear regression: the ultimate goal

Of course, the goal is almost always to get both low bias and low variance, even if it's not always possible. Let's see how regularization is a means toward this goal.

Regularization – from overfitting to underfitting

In light of all those examples, we can now get a really clear understanding of what regularization is.

If we look again at the definition provided by GPT-3, regularization is what allows us to prevent a model from overfitting by adding constraints to the model. Indeed, *adding regularization allows us to reduce variance* in a model, and therefore, to have a less overfitting model.

We can go one step further: what if regularization is added to an already well-trained model (that is, low bias and low variance)? In other words, what happens if constraints are added to a model that works well? Intuitively, it would degrade the overall performance. It would not allow the model to fully grasp the data behavior, and thus add bias to the model.

Indeed, here comes a fundamental drawback of regularization: **adding regularization increases model bias**.

This can be summarized in one figure:

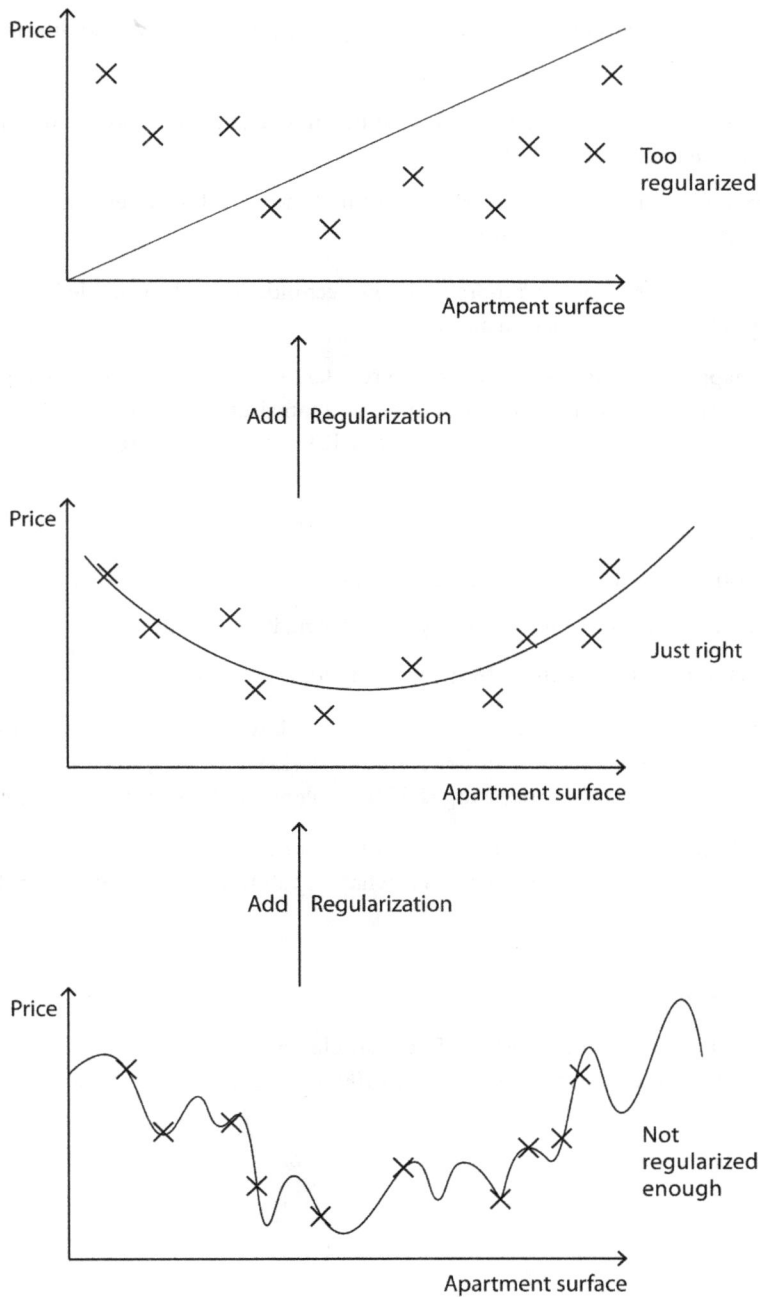

Figure 1.12 – A high variance model (bottom); the same model with more regularization and the right level of fitting (middle); and the same model with even more regularization, now underfitting (top)

This is what is called the **bias-variance trade-off**, a very important concept. Indeed, adding regularization is always a balance:

- We need to have enough regularization so that the model generalizes well and is not sensitive to small data fluctuations and noise
- We need to not have too much regularization so that the model is free enough to fully capture the complexity of the data in all cases

As we go further in the book, more and more tools and techniques will be provided to diagnose our model and find the right bias-variance balance.

Throughout the chapters, we will see many ways to regularize a model. We think of regularization as just adding direct constraints to a model, but there are many indirect regularization methods that may help a model better generalize. A non-exhaustive list of the existing regularization methods could be the following:

- Adding constraints to the model architecture
- Adding constraints to the model training, such as the loss
- Adding constraints from the input data by engineering it differently
- Adding constraints from the input by generating more samples

Other regularization methods may be proposed, but the book will mostly focus on those methods for various cases, such as structured and unstructured data, linear models, tree-based models, deep learning models, **natural language processing (NLP)** problems, and computer vision problems.

As good as a model can be with the right regularization method, most tasks have a hard limit on the possible performances a model can achieve: this is what we call **unavoidable bias**. Let's have a look at what it is.

Unavoidable bias

In almost any task, there is unavoidable bias. For example, in *Figure 1.13*, there are both Shetland Sheepdogs and Rough Collie dogs. Can you say with 100% accuracy which is which?

Figure 1.13 – Random pictures of both Shetland Sheepdogs and Rough Collie dogs

From the preceding figure, can you tell which is which? Most likely not. If you are a trained expert in dogs, you may have a lower error rate. But the odds are that given a large enough number of images, you might be wrong with some images. The lowest level of error an expert can achieve is what is called **human-level error**. In most cases, human-level error gives a very good idea of the lowest error a model can achieve. Anytime you are evaluating a model, it is a good idea to know (or to wonder) what a human could possibly do on such a task.

Indeed, some tasks are much easier than others:

- Humans perform well at classifying dogs and cats, as does AI
- Humans perform well at classifying songs, as does AI
- Humans perform quite poorly at hiring people (at least, not all recruiters will agree on a candidate), as does AI

Another possible source to compute the unavoidable bias is the Bayes error. Most commonly impossible to compute on complex AI tasks, the Bayes error is the lowest possible error rate a classifier can achieve. Most of the time, the Bayes error is lower than the human-level error, but much harder to estimate. This would be the actual theoretical limitation of any model performance.

The Bayes error and the human-level error are unavoidable biases. They indicate the irreducible error of any model and are a key concept when evaluating whether a model needs more or less regularization.

Diagnosing bias and variance

We usually define bias and variance using figures with a more or less accurate fit on some data, as we did earlier on the apartment price data. Although useful to explain the concepts, in real life, we mostly have highly dimensional datasets. By using a simple Titanic dataset, we provided a dozen features, so making this kind of visual inspection is simply impossible.

Let's assume we are training a dogs and cats classification model with balanced data. A good method is to compare the evaluation metric (whether it be accuracy, F-score, or whatever you deemed relevant) for both the training set and validation set.

> **Note**
>
> If the concepts of evaluation metrics or training and validation sets are not clear, they will be explained in more detail in *Chapter 2*. Put simply, the model is trained on the training set. The metric is the value used to evaluate the trained model and is computed on both the training and validation sets.

For example, let's assume we get the following results:

	Training set	Validation set
Accuracy	0.85	0.84

Figure 1.14 – Hypothetical accuracy on training and validation sets

If we think about the expected human-level error for such a task, we expect a much higher accuracy. Indeed, most humans can recognize a dog from a cat with very high confidence.

So, in this case, the performances of training and validation sets are far below the human-level error rate. This is typical of a **high-bias** scenario: the evaluation metric is bad on both training and validation sets. In such cases, the model needs to be less regularized.

Let's now assume that after adding lower regularization (perhaps adding raised-to-the-power features as we did in the *Intuition about regularization* section), we have the following results:

	Training set	Validation set
Accuracy	1	0.89

Figure 1.15 – Hypothetical accuracy after adding more features

Those results are better; the validation set now has an accuracy of 89%. Nevertheless, there are two issues here:

- The score on the train set is way too good: it is literally perfect
- The score on the validation is still far below the human-level error rate, which we would expect to be at least 95%

This is typical of a **high variance** scenario: results are really good (usually too good) on the training set, and far below on the validation set. In such cases, the model needs more regularization.

After adding regularization, let's assume we now have the following results:

	Training set	Validation set
Accuracy	0.98	0.97

Figure 1.16 – Hypothetical accuracy after adding regularization

This seems much better: both training and validation sets have an accuracy that seems close to human-level performance. Perhaps with more data, a better model, or some other improvements, results could get a little better, but overall, this seems like a solid result.

In most cases, diagnosing high bias and high variance is simple, and the method is always the same:

1. Evaluate your model on both the training and validation set.
2. Compare the results with each other, as well as with the human-level error rate.

From that point, there are mostly three cases:

- **High bias/underfitting**: Both training and validation sets exhibit poor performance
- **High variance/overfitting**: The training set performance is far better than the validation set performance; validation set performance is well below the human-level error rate
- **Good fit**: Both training and validation sets exhibit performance close to the human-level error rate

> **Note**
> Most of the time, it is proposed to use this technique with training and test sets, instead of training and validation sets. While the reasoning holds true, doing such optimization on the test set directly may lead to overfitting, and thus overestimate the actual performance of a model. In this book, though, we will use the test set for simplification in the next chapters.

Having all the key concepts of regularization in mind, you might now start to understand why regularization might indeed require a whole book. While diagnosing a need for regularization is usually fairly easy, choosing the right regularization method can be really challenging. Let's now categorize the regularization approaches that will be covered in this book.

Regularization – a multi-dimensional problem

Having the right diagnosis for a model is crucial, as it allows us to choose the strategy more carefully to improve the model. But from any diagnosis, many paths are possible to improve the model. Those paths can be separated into three main categories, as proposed in the following figure:

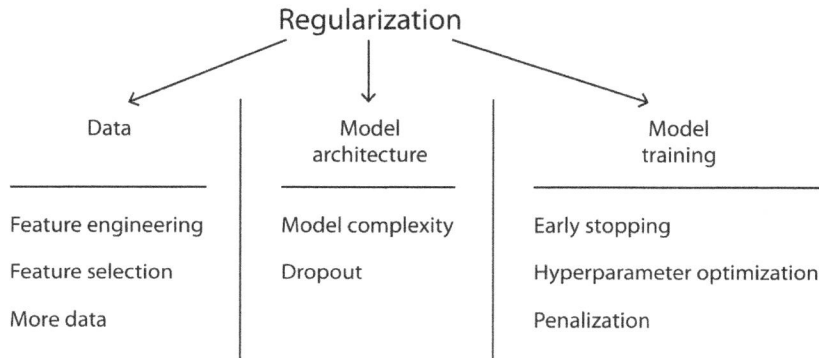

Figure 1.17 – A proposed categorization of regularization types: data, model architecture, and model training

At the data level, we may have the following tools for regularization:

- Adding more data, either synthetic or real
- Adding more features
- Feature engineering
- Data preprocessing

Indeed, the data is of extreme importance in ML in general, and regularization is no exception. We will see many examples throughout the book of regularizing data.

At the model level, the following methods may be used for regularization:

- Choosing a more or less simple architecture
- In deep learning, many architectural designs allow regularization (for example, dropout)

The model complexity may strongly impact regularization. An overly complicated architecture can easily lead to overfitting, while a too simplistic one may underfit, as depicted in the following figure:

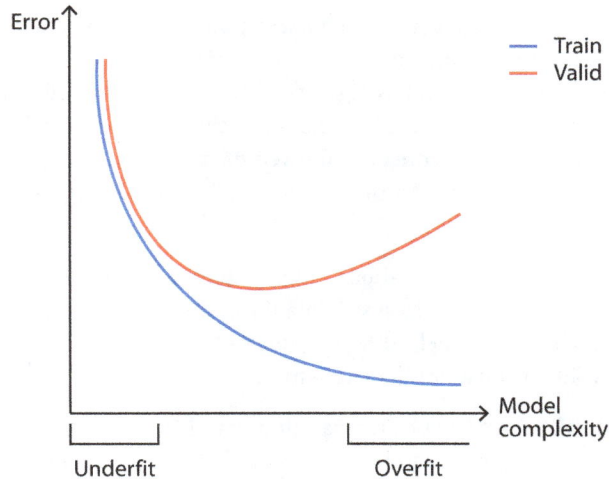

Figure 1.18 – Possible visualization of error as a function of model complexity, for both training and validation sets

Finally, at the training level, some of the methods to regularize are as follows:

- Adding penalization
- Weight initialization
- Transfer learning
- Early stopping

Early stopping is a very common way to avoid overfitting by preventing the model from getting too close to the training set.

There may be many ways to regularize, as it is a multi-dimensional problem: data, model architecture, and model training are just high-level categories. Even though those categories are just some examples and more may exist or be defined, most – if not all – of the techniques this book will cover will fall into one of those categories.

Summary

We started this chapter by demonstrating, with several real-world examples, that regularization is the key to success in ML in a production environment. Along with several other methods and best practices, a robustly regularized model is necessary for production. In production, unseen data and edge cases will appear on a regular basis, thus any deployed model must have an acceptable response to such cases.

We then walked through some key concepts of regularization. Overfitting and underfitting are two common problems in ML and relate somehow to bias and variance. Indeed, an overfitting model has high variance, while an underfitting model has high bias. Thus, to perform well, a model is required to have low bias and low variance. We explained how, no matter how good a model can get, unavoidable bias limits its performance. Those key concepts allowed us to propose a method to diagnose bias and variance using the performance of both the training and validation sets, as well as human-level error estimation.

This led us to what regularization is: regularization is adding constraints to a model so that it generalizes well to new data, and is not too sensitive to small data fluctuations. Regularization is a great tool to make an overfitting model a robust model. Although, due to the bias-variance trade-off, we must not regularize too much to avoid having an underfitting model.

Finally, we categorized the different ways of regularizing that this book will cover. They mainly fall into three categories: the data, the model architecture, and the model training.

This chapter did not include any recipes in order to build the foundational knowledge that will be required to fully understand the remainder of this book, but the next chapters will comprise recipes and will be more solution-oriented.

2

Machine Learning Refresher

Machine learning (ML) is much more than just models. It is about following a certain process and best practices. This chapter will provide a refresher on these: from loading data and model evaluation to model training and optimization, the main steps and methods will be explained here.

In this chapter, we are going to cover the following main topics:

- Loading data
- Splitting data
- Preparing quantitative data
- Preparing qualitative data
- Training a model
- Evaluating a model
- Performing hyperparameter optimization

Even though the recipes in this chapter are independent from a methodological standpoint, they build upon each other and are meant to be executed sequentially.

Technical requirements

In this chapter, you will need to be able to run code to load datasets, prepare data, and train, optimize, and evaluate ML models. To do so, you will need the following libraries:

- **numpy**
- **pandas**
- **scikit-learn**

They can be installed using `pip` with the following command line:

```
pip install numpy pandas scikit-learn
```

> **Note**
>
> In this book, some best practices such as using virtual environments won't be explicitly mentioned. However, it is highly recommended that you use a virtual environment before installing any library using `pip` or any other package manager.

Loading data

The primary focus of this recipe is to load data from a CSV file. However, this is not the only thing that this recipe covers. Since the data is usually the first step in any ML project, this recipe is also a good opportunity to give a quick recap of the ML workflow, as well as the different types of data.

Getting ready

Before loading the data, we should keep in mind that an ML model follows a two-step process:

1. Train a model on a given dataset to create a new model.
2. Reuse the previously trained model to infer predictions on new data.

These two steps are summarized in the following figure:

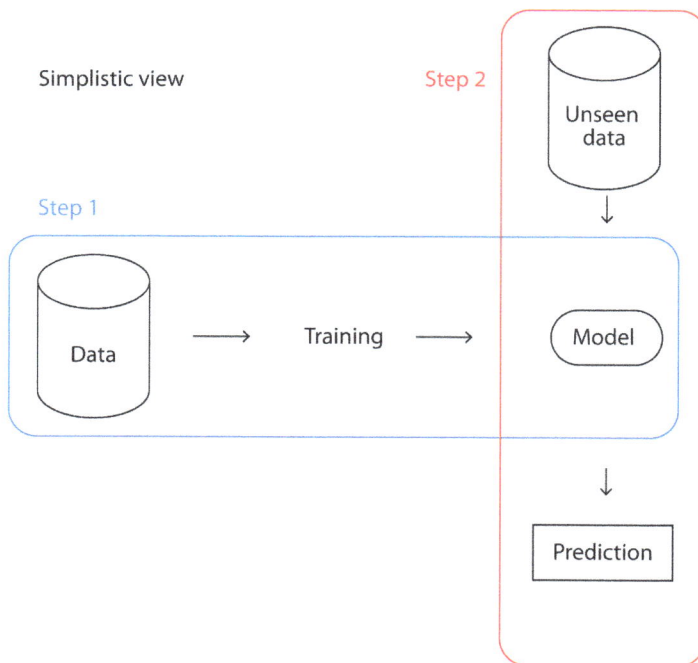

Figure 2.1 – A simple view of the two-step ML process

Of course, in most cases, this is a rather simplistic view. A more detailed view can be seen in Figure 2.2:

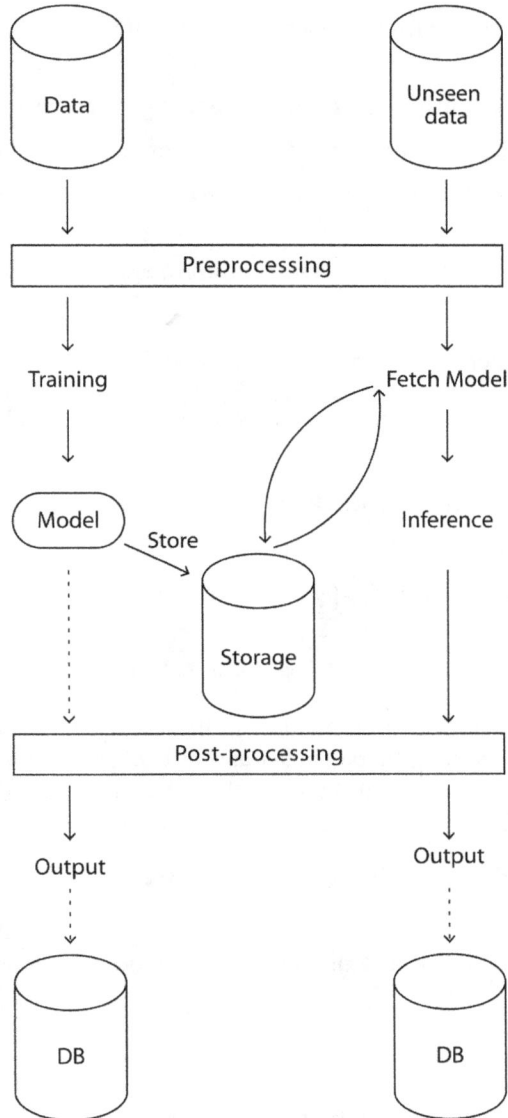

Figure 2.2 – A more complete view of the ML process

Let's take a closer look at the training part of the ML process shown in *Figure 2.2*:

1. First, training data is queried from a data source (this can be a database, a data lake, an open dataset, and so on).
2. The data is preprocessed, such as via feature engineering, rescaling, and so on.
3. A model is trained and stored (on a data lake, locally, on the edge, and so on).
4. Optionally, the output of this model is post-processed – for example, via formatting, heuristics, business rules, and more.
5. Optionally again, this model (with or without postprocessing) is stored in a database for later reference or evaluation if needed.

Now, let's take a look at the inference part of the ML process:

1. The data is queried from a data source (a database, an API query, and so on).
2. The data goes through the same preprocessing step as the training data.
3. The trained model is fetched if it doesn't already exist locally.
4. The model is used to infer output.
5. Optionally, the output of the model is post-processed via the same post-processing step as the training data.
6. Optionally, the output is stored in a database for monitoring and later reference.

Even in this schema, many steps were not mentioned: splitting data for training purposes, using evaluation metrics, cross-validation, hyperparameter optimization, and others. This chapter will dive into the more training-specific steps and apply them to the very common but practical Titanic dataset, a binary classification problem. But first, we need to load the data.

To do so, you must download the **Titanic dataset training set** locally. This can be performed with the following command line:

```
wget https://github.com/PacktPublishing/The-Regularization-Cookbook/
blob/main/chapter_02/train.csv
```

How to do it...

This recipe is about loading a CSV file and displaying a few lines of code so that we can have a first glance at what it is about:

1. The first step is to import the required libraries. Here, the only library we need is pandas:

```
import pandas as pd
```

2. Now, we can load the data using the `read_csv` function provided by pandas. The first argument is the path to the file. Assuming the file is named `train.csv` and located in the current folder, we only have to provide `train.csv` as an argument:

```
# Load the data as a DataFrame
df = pd.read_csv('train.csv')
```

The returned object is a `dataframe` object, which provides many useful methods for data processing.

3. Now, we can display the first five lines of the loaded file using the `.head()` method:

```
# Display the first 5 rows of the dataset
df.head()
```

This code will output the following:

```
   PassengerId  Survived  Pclass  \
0            1         0       3
1            2         1       1
2            3         1       3
3            4         1       1
4            5         0       3

                         Name                Sex   Age   SibSp  \
0   Braund, Mr. Owen Harris              male  22.0       1
1   Cumings, Mrs. John Bradley (Florence Briggs Th...
                                    female  38.0       1
2   Heikkinen, Miss. Laina  female  26.0       0
3   Futrelle, Mrs. Jacques Heath (Lily May Peel)
                                    female  35.0       1
4   Allen, Mr. William Henry       male  35.0       0

   Parch       Ticket     Fare   Cabin       Embarked
0  0              A/5  21171   7.2500  NaN              S
1  0         PC 17599   71.2833   C85          C
2  0     STON/O2. 3101282   7.9250  NaN       S
3  0          113803   53.1000  C123          S
4  0          373450   8.0500   NaN   S
```

Here is a description of the data types in each column:

*   `PassengerId` (qualitative): A unique, arbitrary ID for each passenger.

*   `Survived` (qualitative): 1 for yes, 0 for no. This is our label, so this is a binary classification problem.

*   `Pclass` (quantitative, discrete): The class, which is arguably quantitative. Is class 1 better than class 2? Most likely yes.

- `Name` (unstructured): The name and title of the passenger.

- `Sex` (qualitative): The registered sex of the passenger, either male or female.

- `Age` (quantitative, discrete): The age of the passenger.

- `SibSp` (quantitative, discrete): The number of siblings and spouses on board.

- `Parch` (quantitative, discrete): The number of parents and children on board.

- `Ticket` (unstructured): The ticket reference.

- `Fare` (quantitative, continuous): The ticket price.

- `Cabin` (unstructured): The cabin number, which is arguably unstructured. It can be seen as a qualitative feature with high cardinality.

- `Embarked` (qualitative): The embarked city, either Southampton (S), Cherbourg (C), or Queenstown (Q).

There's more...

Let's talk about the different types of data that are available. Data is a very generic word and can describe many things. We are surrounded by data all the time. One way to specify data is using opposites.

Data can be *structured* or *unstructured*:

- Structured data comes in the form of tables, databases, Excel files, CSV files, and JSON files.

- Unstructured data does not fit in a table: it can be text, sound, image, videos, and so on. Even if we tend to have tabular representation, this kind of data does not naturally fit in an Excel table.

Data can be *quantitative* or *qualitative*.

Quantitative data is ordered. Here are some examples:

- €100 is greater than €10

- 1.8 meters is taller than 1.6 meters

- 18 years old is younger than 80 years old

Qualitative data has no intrinsic order, as shown here:

- Blue is not intrinsically better than red

- A dog is not intrinsically greater than a cat

- A kitchen is not intrinsically more useful than a bathroom

These are not mutually exclusive. An object can have both quantitative and qualitative features, as can be seen in the case of the car in the following figure:

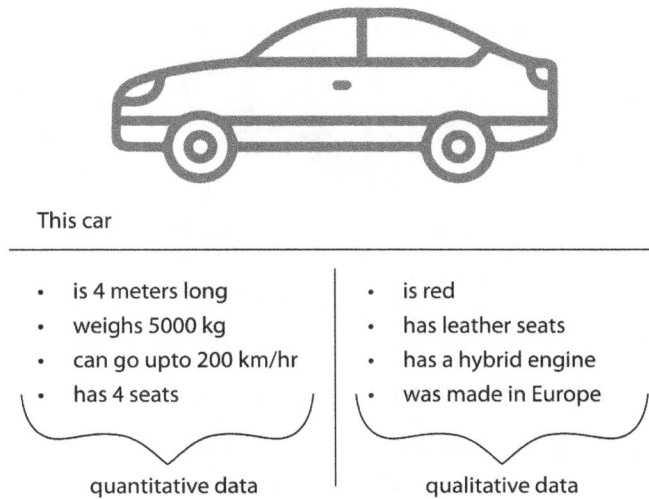

This car

quantitative data	qualitative data
• is 4 meters long	• is red
• weighs 5000 kg	• has leather seats
• can go upto 200 km/hr	• has a hybrid engine
• has 4 seats	• was made in Europe

Figure 2.3 – A single object depicted by both quantitative (left) and qualitative (right) features

Finally, data can be *continuous* or *discrete*.

Some data is continuous, as follows:

- A weight
- A volume
- A price

On the other hand, some data is discrete:

- A color
- A football score
- A nationality

> **Note**
>
> Discrete != qualitative.
>
> For example, a football score is discrete, but there is an intrinsic order: 3 points is more than 2.

See also

The pandas `read_csv` function has a lot of flexibility as it can use other separators, handle headers, and much more. This is described in the official documentation: `https://pandas.pydata.org/docs/reference/api/pandas.read_csv.html`.

The pandas library allows I/O operations that have different types of inputs. For more information, have a look at the official documentation: `https://pandas.pydata.org/docs/reference/io.html`.

Splitting data

After loading data, splitting it is a crucial step. This recipe will explain why we need to split data, as well as how to do it.

Getting ready

Why do we need to split data? An ML model is quite like a student.

You provide a student with many lectures and exercises, with or without the answers. But more often than not, students are evaluated on a completely new problem. To make sure they fully understand the concepts and methods, they not only learn the exercises and solutions – they also understand the underlying concepts.

An ML model is no different: you train the model on training data and then evaluate it on test data. This way, you make sure the model fully understands the task and generalizes well to new, unseen data.

So, the dataset is usually split into *train* and *test* sets:

- The train set must be as large as possible to give as many samples as possible to the model
- The test set must be large enough to be statistically significant in evaluating the model

Typical splits can be anywhere between 80% to 20% for rather small datasets (for example, hundreds of samples), and 99% to 1% for very large datasets (for example, millions of samples and more).

For this recipe and the others in this chapter, it is assumed that the code has been executed in the same notebook as the previous recipe since each recipe reuses the code from the previous ones.

How to do it...

Here are the steps to try out this recipe:

1. You can split the data rather easily with scikit-learn and the `train_test_split()` function:

    ```
    # Import the train_test_split function
    from sklearn.model_selection import train_test_split
    # Split the data
    X_train, X_test, y_train, y_test = train_test_split(
        df.drop(columns=['Survived']), df['Survived'],
        test_size=0.2, stratify=df['Survived'],
        random_state=0)
    ```

This function uses the following parameters as input:

- X: All columns but the 'Survived' label
- y: The 'Survived' label column
- test_size: This is 0.2, which means the training size will be 80%
- stratify: This specifies the 'Survived' column to ensure the same label balance is used in both splits
- random_state: 0 is any integer to ensure reproducibility

It returns the following outputs:

- X_train: The train split of X
- X_test: The test split of X
- y_train: The training split of y, associated with X_train
- y_test: The test split of y, associated with X_test

> **Note**
> The stratify option is not mandatory but can be critical to ensure a balanced split of any qualitative feature, not just the labels, as is the case with imbalanced data.

This split should be done as early as possible when performing data processing so that you avoid any potential data leakage. From now on, all the preprocessing will be computed on the train set, and only then applied to the test set, in agreement with *Figure 2.2*.

See also

See the official documentation for the train_test_split function: https://scikit-learn.org/stable/modules/generated/sklearn.model_selection.train_test_split.html.

Preparing quantitative data

Depending on the type of data, how the features must be prepared may differ. In this recipe, we'll cover how to prepare quantitative data, including missing data imputation and rescaling.

Getting ready

In the Titanic dataset, as well as any other dataset, there may be missing data. There are several ways to deal with missing data. For example, you can drop a column or a row, or impute a value. There are many imputation techniques, some of which are more or less sophisticated. scikit-learn supplies several implementations of imputers, such as `SimpleImputer` and `KNNImputer`.

As we will see in this recipe, using `SimpleImputer`, we can impute the missing quantitative data with the mean value.

Once the missing data has been handled, we can prepare the quantitative data by rescaling it so that all the data is at the same scale.

Several rescaling strategies exist, such as min-max scaling, robust scaling, standard scaling, and others.

In this recipe, we will use **standard scaling**. So, for each feature, we will subtract the mean value of this feature, and then divide it by the standard deviation of that feature:

$$X = \frac{X - X.mean()}{X.std()}$$

Fortunately, scikit-learn provides a fully working implementation via `StandardScaler`.

How to do it...

We will sequentially handle missing values and rescale the data in this recipe:

1. Import the required classes – `SimpleImputer` for missing data imputation and `StandardScaler` for rescaling:

    ```
    from sklearn.impute import SimpleImputer
    from sklearn.preprocessing import StandardScaler
    ```

2. Select the quantitative features we want to keep. Here, we will keep `'Pclass'`, `'Age'`, `'Fare'`, `'SibSp'`, and `'Parch'` and store these features in new variables for both the train and test sets:

    ```
    quanti_columns = ['Pclass', 'Age', 'Fare', 'SibSp', 'Parch']
    # Get the quantitative columns
    X_train_quanti = X_train[quanti_columns]
    X_test_quanti = X_test[quanti_columns]
    ```

3. Instantiate the simple imputer with a mean strategy. Here, the missing value of a feature will be replaced with the mean value of that feature:

```
# Impute missing quantitative values with mean feature value
quanti_imputer = SimpleImputer(strategy='mean')
```

4. Fit the imputer on the train set and apply it to the test set so that it avoids leakage in the imputation:

```
# Fit and impute the training set
X_train_quanti = quanti_imputer.fit_transform(X_train_quanti)
# Just impute the test set
X_test_quanti = quanti_imputer.transform(X_test_quanti)
```

5. Now that imputation has been performed, instantiate the `scaler` object:

```
# Instantiate the standard scaler
scaler = StandardScaler()
```

6. Finally, fit and apply the standard scaler to the train set, and then apply it to the test set:

```
# Fit and transform the training set
X_train_quanti = scaler.fit_transform(X_train_quanti)
# Just transform the test set
X_test_quanti = scaler.transform(X_test_quanti)
```

We now have quantitative data with no missing values, fully rescaled, with no data leakage.

There's more...

In this recipe, we used the simple imputer, assuming there was missing data. In practice, it is highly recommended that you look at the data first to check whether there are missing values, as well as how many. It is possible to look at the number of missing values per column with the following code snippet:

```
# Display the number of missing data for each column
X_train[quanti_columns].isna().sum()
```

This will output the following:

```
Pclass        0
Age         146
Fare          0
SibSp         0
Parch         0
```

Thanks to this, we know that the Age feature has 146 missing values, while the other features have no missing data.

See also

A few imputers are available in scikit-learn. The list is available here: https://scikit-learn. org/stable/modules/classes.html#module-sklearn.impute.

There are many ways to scale data, and you can find the methods that are available in scikit-learn here: https://scikit-learn.org/stable/modules/classes.html#module-sklearn.preprocessing.

You might be interested in looking at this comparison of several scalers on some given data: https:// scikit-learn.org/stable/auto_examples/preprocessing/plot_all_scaling. html#sphx-glr-auto-examples-preprocessing-plot-all-scaling-py.

Preparing qualitative data

In this recipe, we will prepare qualitative data, including missing value imputation and encoding.

Getting ready

Qualitative data requires different treatment from quantitative data. Imputing missing values with the mean value of a feature would make no sense (and would not work with non-numeric data): it makes more sense, for example, to use the most frequent value or the mode of a feature. The SimpleImputer class allows us to do such things.

The same goes for rescaling: it would make no sense to rescale qualitative data. Instead, it is more common to encode it. One of the most typical techniques is called **one-hot encoding**.

The idea is to transform each of the categories, over a total possible N categories, in a vector holding a 1 and N-1 zeros. In our example, the Embarked feature's one-hot encoding would be as follows:

- 'C' = *[1, 0, 0]*
- 'Q' = *[0, 1, 0]*
- 'S' = *[0, 0, 1]*

> **Note**
>
> Having N columns for N categories is not necessarily optimal. What happens if, in the preceding example, we remove the first column? If the value is not 'Q' = *[1, 0]* nor 'S' = *[0, 1]*, then it must be 'C' = *[0, 0]*. There is no need to add one more column to have all the necessary information. This can be generalized to N categories only requiring N-1 columns to have all the information, which is why one-hot encoding functions usually allow you to drop a column.

The `sklearn` class' `OneHotEncoder` allows us to do this. It also allows us to deal with unknown categories that may appear in the test set (or the production environment) with several strategies, such as an error, ignore, or infrequent class. Finally, it allows us to drop the first column after encoding.

How to do it...

Just like in the preceding recipe, we will handle any missing data and the features will be one-hot encoded:

1. Import the necessary classes – `SimpleImputer` for missing data imputation (already imported in the previous recipe) and `OneHotEncoder` for encoding. We also need to import numpy so that we can concatenate the qualitative and quantitative data that's been prepared at the end of this recipe:

    ```
    import numpy as np
    from sklearn.impute import SimpleImputer
    from sklearn.preprocessing import OneHotEncoder
    ```

2. Select the qualitative features we want to keep: `'Sex'` and `'Embarked'`. Then, store these features in new variables for both the train and test sets:

    ```
    quali_columns = ['Sex', 'Embarked']
    # Get the quantitative columns
    X_train_quali = X_train[quali_columns]
    X_test_quali = X_test[quali_columns]
    ```

3. Instantiate `SimpleImputer` with `most_frequent` strategy. Any missing values will be replaced by the most frequent ones:

    ```
    # Impute missing qualitative values with most frequent feature
    value
    quali_imputer =SimpleImputer(strategy='most_frequent')
    ```

4. Fit and transform the imputer on the train set, and then transform the test set:

    ```
    # Fit and impute the training set
    X_train_quali = quali_imputer.fit_transform(X_train_quali)
    # Just impute the test set
    X_test_quali = quali_imputer.transform(X_test_quali)
    ```

5. Instantiate the encoder. Here, we will specify the following parameters:

   - `drop='first'`: This will drop the first columns of the encoding

   - `handle_unknown='ignore'`: If a new value appears in the test set (or in production), it will be encoded as zeros:

     ```
     # Instantiate the encoder
     encoder=OneHotEncoder(drop='first', handle_unknown='ignore')
     ```

6. Fit and transform the encoder on the training set, and then transform the test set using this encoder:

   ```
   # Fit and transform the training set
   X_train_quali = encoder.fit_transform(X_train_quali).toarray()
   # Just encode the test set
   X_test_quali = encoder.transform(X_test_quali).toarray()
   ```

> **Note**
>
> We need to use `.toarray()` out of the encoder because the array is a sparse matrix object by default and cannot be concatenated in that form with the other features.

7. With that, all the data has been prepared – both quantitative and qualitative (considering this recipe and the previous one). It is now possible to concatenate this data before training a model:

   ```
   # Concatenate the data back together
   X_train = np.concatenate([X_train_quanti,
       X_train_quali], axis=1)
   X_test = np.concatenate([X_test_quanti, X_test_quali], axis=1)
   ```

There's more...

It is possible to save the data as a pickle file, either to share it or save it and avoid having to prepare it again. The following code will allow us to do this:

```
import pickle

pickle.dump((X_train, X_test, y_train, y_test),
    open('prepared_titanic.pkl', 'wb'))
```

We now have fully prepared data that can be used to train ML models.

> **Note**
>
> Several steps have been omitted or simplified here for more clarity. Data may need more preparation, such as more thorough missing value imputation, outlier and duplicate detection (and perhaps removal), feature engineering, and so on. It is assumed that you already have some sense of those aspects and are encouraged to read other materials about this topic if required.

See also

This more general documentation about missing data imputation is worth looking at: `https://scikit-learn.org/stable/modules/impute.html`.

Finally, this more general documentation about data preprocessing can be very useful: `https://scikit-learn.org/stable/modules/preprocessing.html`.

Training a model

Once data has been fully cleaned and prepared, it is fairly easy to train a model thanks to scikit-learn. In this recipe, before training a logistic regression model on the Titanic dataset, we will quickly recap the ML paradigm and the different types of ML we can use.

Getting ready

If you were asked how to differentiate a car from a truck, you may be tempted to provide a list of rules, such as the number of wheels, size, weight, and so on. By doing so, you would be able to provide a set of explicit rules that would allow anyone to identify a car and a truck as different types of vehicles.

Traditional programming is not so different. While developing algorithms, programmers often build explicit rules, which allow them to map from data input (for example, a vehicle) to answers (for example, a car). We can summarize this paradigm as *data + rules = answers*.

If we were to train an ML model to discriminate cars from trucks, we would use another strategy: we would feed an ML algorithm with many pieces of data and their associated answers, expecting the model to learn to correct rules by itself. This is a different approach that can be summarized as *data + answers = rules*. This paradigm difference is summarized in *Figure 2.4*. As little as it might look to ML practitioners, it changes everything in terms of regularization:

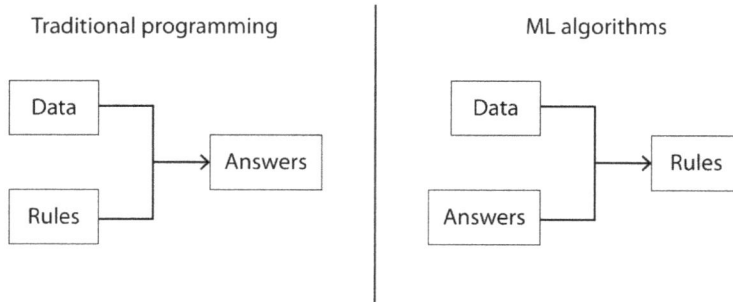

Figure 2.4 – Comparing traditional programming with ML algorithms

Regularizing traditional algorithms is conceptually straightforward. For example, what if the rules for defining a truck overlap with the bus definition? If so, we can add the fact that buses have lots of windows.

Regularization in ML is intrinsically implicit. What if the model in this case does not discriminate between buses and trucks?

- Should we add more data?
- Is the model complex enough to capture such a difference?
- Is it underfitting or overfitting?

This fundamental property of ML makes regularization complex.

ML can be applied to many tasks. Anyone who uses ML knows there is not just one type of ML model.

Arguably, most ML models fall into three main categories:

- **Supervised learning**
- **Unsupervised learning**
- **Reinforcement learning**

As is usually the case for categories, the landscape is more complex, with sub-categories and methods overlapping several categories. But this is beyond the scope of this book.

This book will focus on regularization for supervised learning. In supervised learning, the problem is usually quite easy to specify: we have input features, X (for example, apartment surface), and labels, y (for example, apartment price). The goal is to train a model so that it's robust enough to predict y, given X.

The two major types of ML are classification and regression:

- **Classification**: The labels are made of qualitative data. For example, the task is predicting between two or more classes such as car, bus, and truck.

- **Regression**: The labels are made of quantitative data. For example, the task is predicting an actual value, such as an apartment price.

Again, the line can be blurry; some tasks can be solved with classification while the labels are quantitative data, while others tasks can be both classification and regression ones. See *Figure 2.5*:

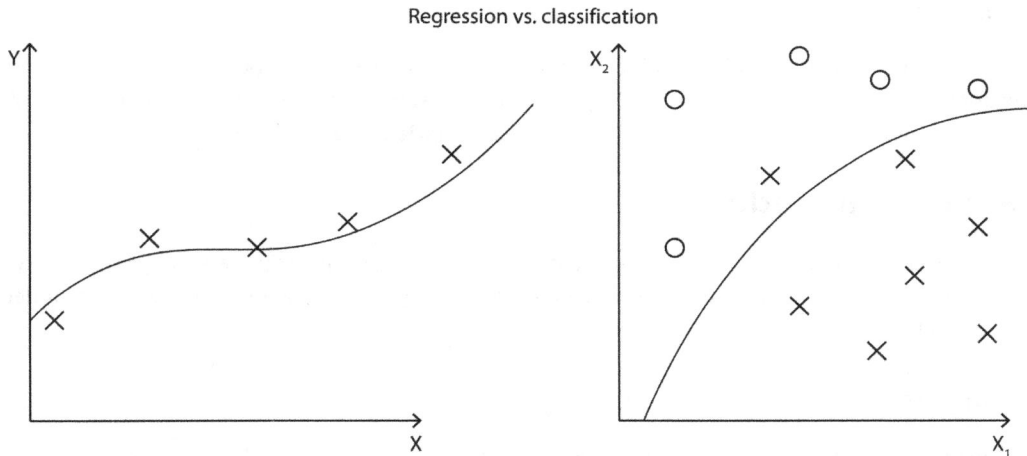

Figure 2.5 – Regularization versus classification

How to do it...

Assuming we want to train a logistic regression model (which will be explained properly in the next chapter), the scikit-learn library provides the LogisticRegression class, along with the fit() and predict() methods. Let's learn how to use it:

1. Import the LogisticRegression class:

```
from sklearn.linear_model import LogisticRegression
```

2. Instantiate a LogisticRegression object:

```
# Instantiate the model
lr = LogisticRegression()
```

3. Fit the model on the train set:

```
# Fit on the training data
lr.fit(X_train, y_train)
```

4. Optionally, compute predictions by using that model on the test set:

```
# Compute and store predictions on the test data
y_pred = lr.predict(X_test)
```

See also

Even though more details will be provided in the next chapter, you might be interested in looking at the documentation of the `LogisticRegression` class: `https://scikit-learn.org/stable/modules/generated/sklearn.linear_model.LogisticRegression.html`.

Evaluating a model

Once the model has been trained, it is important to evaluate it. In this recipe, we will provide a few insights about a few typical metrics for both classification and regression, before evaluating our model on the test set.

Getting ready

Many evaluation metrics exist. If we think about predicting a binary classification and take a step back, there are only four cases:

- **False positive** (FP): Positive prediction, negative ground truth
- **True positive** (TP): Positive prediction, positive ground truth
- **True negative** (TN): Negative prediction, negative ground truth
- **False negative** (FN): Negative prediction, positive ground truth:

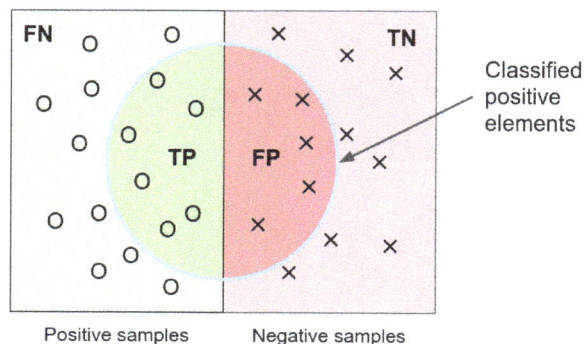

Figure 2.6 – Representation of false positive, true positive, true negative, and false negative

Based on this, we can define a wide range of evaluation metrics.

One of the most common metrics is accuracy, which is the ratio of good predictions. The definition of accuracy is as follows:

$$Accuracy \ = \ \frac{TP + TN}{FP + TP + TN + FN}$$

> **Note**
>
> Although very common, the accuracy may be misleading, especially for imbalanced labels. For example, let's assume an extreme case where 99% of Titanic passengers survived, and we have a model that predicts that every passenger survived. Our model would have a 99% accuracy but would be wrong for 100% of passengers who did not survive.

There are several other very common metrics, such as precision, recall, and the F1 score.

Precision is most suited when you're trying to maximize the true positives and minimize the false positives – for example, making sure you detect only surviving passengers:

$$Precision = \frac{TP}{TP + FP}$$

Recall is most suited when you're trying to maximize the true positives and minimize the false negatives – for example, making sure you don't miss any surviving passengers:

$$Recall = \frac{TP}{TP + FN}$$

The F1 score is just a combination of the precision and recall metrics as a harmonic mean:

$$F1\text{-}score = 2 * \frac{Precision * Recall}{Precision + Recall}$$

Another useful classification evaluation metric is the **Receiver Operating Characteristic Area Under Curve (ROC AUC)** score.

All these metrics behave similarly: when there are values between 0 and 1, the higher the value, the better the model. Some are also more robust to imbalanced labels, especially the F1 score and ROC AUC.

For regression tasks, the most used metrics are the **mean squared error** (MSE) and the R2 score.

The MSE is the averaged square difference between the predictions and the ground truth:

$$MSE = \frac{1}{m} \sum_{i=1}^{m} (y_i - \hat{y}_i)^2$$

Here, m is the number of samples, $\hat{y}$ is the predictions, and y is the ground truth:

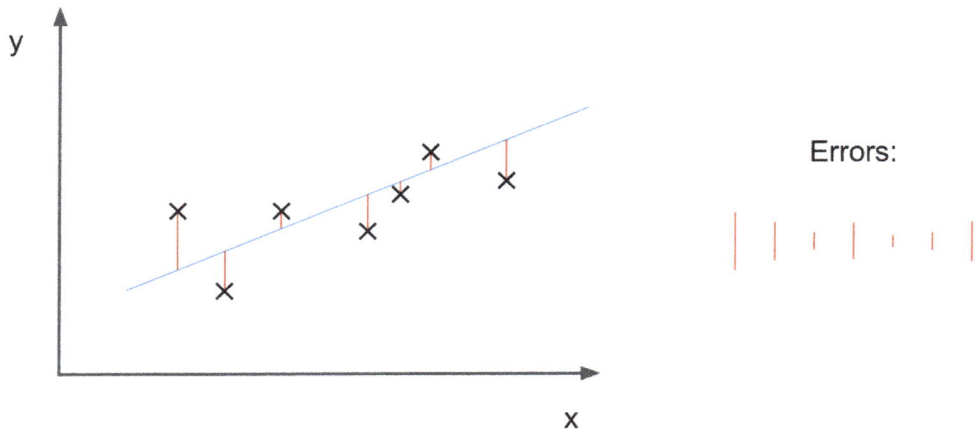

Figure 2.7 – Visualization of the errors for a regression task

In terms of the R2 score, it is a metric that can be negative and is defined as follows:

$$R^2 = 1 - \frac{\sum_{i=1}^{m}(y_i - \hat{y}_i)^2}{\sum_{i=1}^{m}(y_i - \bar{y})^2}$$

> **Note**
> While the R2 score is a typical evaluation metric (the closer to 1, the better), the MSE is more typical of a loss function (the closer to 0, the better).

How to do it...

Assuming our chosen evaluation metric here is accuracy, a very simple way to evaluate our model is to use the `accuracy_score()` function:

```
from sklearn.metrics import accuracy_score
# Compute the accuracy on test of our model
print('accuracy on test set:', accuracy_score(y_pred,
    y_test))
```

This outputs the following:

```
accuracy on test set: 0.7877094972067039
```

Here, the `accuracy_score()` function provides an accuracy of 78.77%, meaning about 79% of our model's predictions are right.

See also

Here is a list of the available metrics in scikit-learn: `https://scikit-learn.org/stable/modules/classes.html#module-sklearn.metrics`.

Performing hyperparameter optimization

In this recipe, we will explain what hyperparameter optimization is and some related concepts: the definition of a **hyperparameter**, **cross-validation**, and various **hyperparameter optimization methods**. We will then perform a grid search to optimize the hyperparameters of the logistic regression task on the Titanic dataset.

Getting ready

Most of the time, in ML, we do not simply train a model on the training set and evaluate it against the test set.

This is because, like most other algorithms, ML algorithms can be fine-tuned. This fine-tuning process allows us to optimize hyperparameters to achieve the best possible results. This sometimes acts as leverage so that we can regularize a model.

> **Note**
> In ML, hyperparameters can be tuned by humans, unlike parameters, which are learned through the model training process, and thus can't be tuned.

To properly optimize hyperparameters, a third split has to be introduced: the validation set.

This means there are now three splits:

- **The training set**: Where the model is trained
- **The validation set**: Where the hyperparameters are optimized
- **The test set**: Where the model is evaluated

You could create such a set by splitting `X_train` into `X_train` and `X_valid` with the `train_test_split()` function from scikit-learn.

But in practice, most people just use cross-validation and do not bother creating this validation set. The k-fold cross-validation method allows us to make *k* splits out of the training set and divide it, as presented in *Figure 2.8*:

Figure 2.8 – Typical split between training, validation, and test sets, without cross-validation (top) and with cross-validation (bottom)

In doing so, not just one model is trained, but k, for a given set of hyperparameters. The performances are averaged over those k models, based on a chosen metric (for example, accuracy, MSE, and so on).

Several sets of hyperparameters can then be tested, and the one that shows the best performance is selected. After selecting the best hyperparameter set, the model is trained one more time on the entire train set to maximize the data for training purposes.

Finally, you can implement several strategies to optimize the hyperparameters, as follows:

- **Grid search**: Test all combinations of the provided values of hyperparameters
- **Random search**: Randomly search combinations of hyperparameters
- **Bayesian search**: Perform Bayesian optimization on the hyperparameters

How to do it...

While being rather complicated to explain conceptually, hyperparameter optimization with cross-validation is super easy to implement. In this recipe, we'll assume that we want to optimize a logistic regression model to predict whether a passenger would have survived:

1. First, we need to import the `GridSearchCV` class from `sklearn.model_selection`.
2. We would like to test the following hyperparameter values for C: `[0.01, 0.03, 0.1]`. We must define a parameter grid with the hyperparameter as the key and the list of values to test as the value.

The C hyperparameter is the inverse of the penalization strength: the higher C is, the lower the regularization. See the next chapter for more details:

```
# Define the hyperparameters we want to test
param_grid = { 'C': [0.01, 0.03, 0.1] }
```

3. Finally, let's assume we want to optimize our model on accuracy, with five cross-validation folds. To do this, we will instantiate the GridSearchCV object and provide the following arguments:

    *   The model to optimize, which is a LogisticRegression instance

    *   The parameter grid, param_grid, which we defined previously

    *   The scoring on which to optimize – that is, accuracy

    *   The number of cross-validation folds, which has been set to 5 here

4. We must also set return_train_score to True to get some useful information we can use later:

```
# Instantiate the grid search object
grid = GridSearchCV(
    LogisticRegression(),
    param_grid,
    scoring='accuracy',
    cv=5,
    return_train_score=True
)
```

5. Finally, all we have to do is train this object on the train set. This will automatically make all the computations and store the results:

```
# Fit and wait
grid.fit(X_train, y_train)
GridSearchCV(cv=5, estimator=LogisticRegression(),
    param_grid={'C': [0.01, 0.03, 0.1]},
    return_train_score=True, scoring='accuracy')
```

> **Note**
>
> Depending on the input dataset and the number of tested hyperparameters, the fit may take some time.

Once the fit has been completed, you can get a lot of useful information, such as the following:

*   The hyperparameter set via the .best_params attribute

*   The best accuracy score via the .best_score attribute

*   The cross-validation results via the .cv_results attribute

6. Finally, you can infer the model that was trained with optimized hyperparameters using the `.predict()` method:

```
y_pred = grid.predict(X_test)
```

7. Optionally, you can evaluate the chosen model with the accuracy score:

```
print('Hyperparameter optimized accuracy:',
    accuracy_score(y_pred, y_test))
```

This provides the following output:

Hyperparameter optimized accuracy: 0.781229050279329

Thanks to the tools provided by scikit-learn, it is fairly easy to have a well-optimized model and evaluate it against several metrics. In the next recipe, we'll learn how to diagnose bias and variance based on such an evaluation.

See also

The documentation for GridSearchCV can be found at `https://scikit-learn.org/stable/modules/generated/sklearn.model_selection.GridSearchCV.html`.

3

Regularization with Linear Models

A huge part of **machine learning** (ML) is made up of linear models. Although sometimes considered less powerful than their nonlinear counterparts (such as tree-based models or deep learning models), linear models do address many concrete, valuable problems. Customer churn and advertising optimization are just a couple of problems where linear models may be the right solution.

In this chapter, we will cover the following recipes:

- Training a linear regression with scikit-learn
- Regularizing with ridge regression
- Regularizing with lasso regression
- Regularizing with elastic net regression
- Training a logistic regression model
- Regularizing a logistic regression model
- Choosing the right regularization

By the end of this chapter, we will have learned how to use and regularize some of the most commonly used linear models.

Technical requirements

In this chapter, besides loading data, you will learn how to fit and compute inferences with several linear models. In order to do so, the following libraries are required:

- NumPy
- Matplotlib
- Scikit-learn

Training a linear regression model with scikit-learn

Linear regression is one the most basic ML models we can use, but it is very useful. Most people used linear regression in high school without talking about ML, and still use it on a regular basis within spreadsheets. In this recipe, we will explain the basics of linear regression, and then train and evaluate a linear regression model using scikit-learn on the California housing dataset.

Getting ready

Linear regression is not a complicated model, but it is still useful to understand what is under the hood to get the best out of it.

The way linear regression works is pretty straightforward. Heading back to the real estate price example, if we consider a feature x such as the apartment surface and a label y such as the apartment price, a common solution would be to find a and b such that $y = ax + b$.

Unfortunately, this is not so simple in real life. There is usually no a and b that makes this equality always respected. It is more likely that we can define a function $h(x)$ that aims to give a value as close as possible to y.

Also, we may have not just one feature x, but several features $x1, x2,..., x_n$, representing apartment surface, location, floor, number of rooms, exponent features, and so on.

By this logic, we would end up with a prediction $h(x)$ that may look like the following:

$$h(x) = \sum_j w_j x_j + b$$

w_j is the weight associated to the feature x_j xj, and b is a bias term. This is just a generalization of the

previous $y = ax + b$ to n features. This formula allows a linear regression to predict virtually any real number.

The goal of our ML model is to find the set of w and b values that minimizes prediction errors on the training set. By this, we mean finding the parameters w and b so that $h(x)$ and y are as close as possible.

One way to achieve that is to minimize the loss L, that can be defined here as a slightly modified **mean squared error** (**MSE**):

$$L = \frac{1}{2m} \sum_i (y^{(i)} - h(x)^{(i)})^2$$

$y^{(i)}$ is the ground truth of the sample i in the training set, and m is the number of samples in the training set.

> **Note**
>
> The loss is usually a representation of the difference between the ground truth and the predictions. Hence, minimizing the loss allows the model to predict values that are as close as possible to the ground truth.

Minimizing this mean squared error would allow us to find the set of w and b values so that the prediction $h(x)$ is as close as possible to the ground truth y. Schematically, this can be represented as finding the w that minimizes the loss, as shown in the following figure:

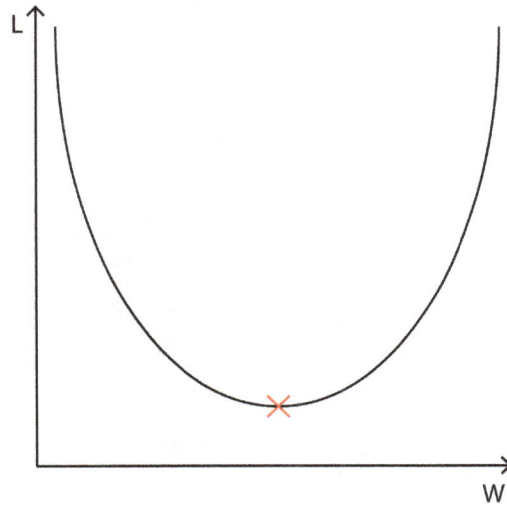

Figure 3.1 – Loss function as a function of a parameter theta, having a global minimum at the cross

The next question is: how do we find the set of values to minimize the loss? There are several ways of solving this problem. One commonly used technique in ML is gradient descent.

What is gradient descent? In a few words, it is going down the curve to the minimum value in the preceding figure.

How does this work? It's a multi-step process:

1. Start with random values of the parameters w and b. Random values are usually defined using normal distribution centered on zero. This is why having scaled features may help significantly for convergence.

2. Compute the loss for the given data and current values of w and b. As defined earlier, we may use the mean squared error to compute the loss L.

    The following figure is a good representation of the situation at this point:

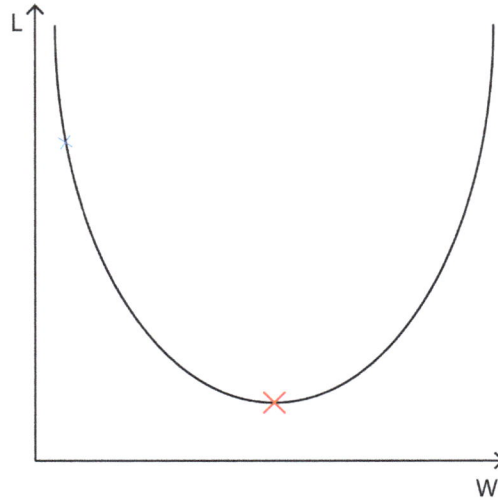

Figure 3.2 – The loss function with the global minimum at the red cross, and a possible random initial state at the blue cross

3. Compute the loss gradient with respect to each parameter $\frac{\partial \mathcal{L}}{\partial w_j}, \frac{\partial \mathcal{L}}{\partial b}$. This is nothing more than

the slope of the loss for a given parameter, which can be computed with the following equations:

$$\frac{\partial \mathcal{L}}{\partial w_j} = -\frac{1}{m}\sum_i (y^{(i)} - h(x)^{(i)})x_j^{(i)}$$

$$\frac{\partial \mathcal{L}}{\partial b} = -\frac{1}{m}\sum_i (y^{(i)} - h(x)^{(i)})$$

> **Note**
>
> One may notice that the slope is expected to decrease as we get closer to the minimum. Indeed, as we get close to the minimum, the error tends to zero and so does the slope, based on these equations.

4. Apply gradient descent to parameters. Apply the gradient descent to parameters, with a user-defined learning rate α. This is computed using the following formulas:

$$w_j = w_j - \alpha \frac{\partial \mathcal{L}}{\partial w_j}$$

$$b = b - \alpha \frac{\partial \mathcal{L}}{\partial b}$$

This allows us to take a step toward the minimum, as represented in the following figure:

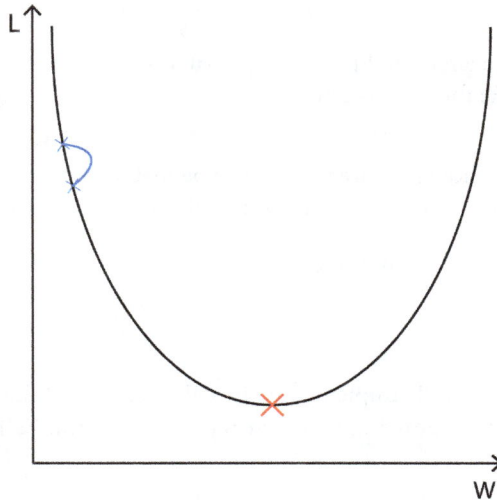

Figure 3.3 – The gradient descent allows us to take one step down the loss
function, allowing us to get closer to the global minimum

5. Iterate through *steps 2 to 4* until convergence or max iteration. This would allow us to reach
    the optimal parameters, as represented in the following figure:

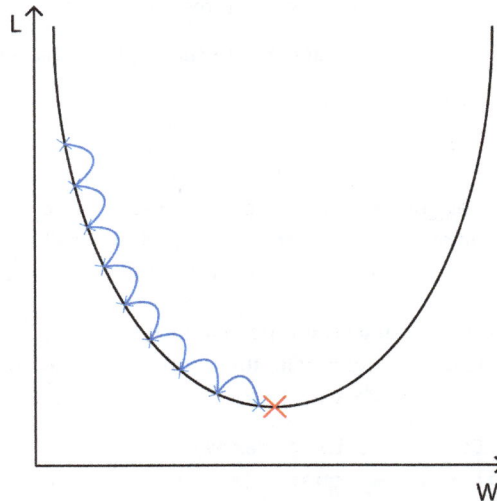

Figure 3.4 – With enough iterations and a convex loss function, the
parameters will converge to the global minimum

> **Note**
>
> A learning rate α that is too large would miss the global minimum, or even diverge, while one that is too small would take forever to converge.

To complete this recipe, the following libraries have to be installed: numpy, matplotlib, and sklearn. They can be installed with pip in the terminal, with the following command line:

```
pip install -U numpy sklearn matplotlib
```

How to do it...

Fortunately, all this procedure is fully implemented in scikit-learn, and the only thing you need to do is fully reuse this library. Let's now train a linear regression on the California housing dataset provided by scikit-learn.

1. **Make the required imports**: Here, we import NumPy for data manipulation. From scikit-learn, we do several imports:

    - fetch_california_housing: A function that allows us to load the dataset
    - train_test_split: A function that allows us to split the data
    - StandardScaler: A class that allows us to rescale the data
    - LinearRegression: The class that contains the implementation of the linear regression

    Here is what the code looks like:

    ```
    import numpy as np
    from sklearn.datasets import fetch_california_housing
    from sklearn.model_selection import train_test_split
    from sklearn.preprocessing import StandardScaler
    from sklearn.linear_model import LinearRegression
    ```

2. **Load the California housing dataset**: For pedagogical purposes, we will do a bit of feature engineering here and calculate the sum of all the features raised to the power of two: to do so, we simply concatenate the features X with X*X:

    ```
    # Load the California housing dataset
    X, y = fetch_california_housing(return_X_y=True)
    X = np.concatenate([X, X*X], axis=1)
    ```

3. **Split the data**: We split the data using the `train_test_split` function, with `test_size=0.2`, meaning we end up having 80% of the data in the training set, and 20% in the test set, split at random. This is shown here:

```
# Split the data
X_train, X_test, y_train, y_test = train_test_split(
    X, y, test_size=0.2, random_state=0)
```

4. **Prepare the data**: Since we have only quantitative features here, the only preparation we need is rescaling. We can use the standard scaler of scikit-learn. We need to instantiate it, then fit it on the training set and transform the training set, and finally we transform the test set. Feel free to use any other rescaler:

```
# Rescale the data
scaler = StandardScaler()
X_train = scaler.fit_transform(X_train)
X_test = scaler.transform(X_test)
```

5. Fit the model on the training set. The model must be instantiated, and here we use the default parameters, so nothing is specified. Once the model is instantiated, we can use the `.fit()` method on the training set:

```
# Fit the linear regression model
lr = LinearRegression()
lr.fit(X_train, y_train)
```

6. Evaluate the model on both the training and test set. Here, we use the `.score()` method of the `LinearRegression` class, which provides $R2$-score, but you can use any other metric provided in `sklearn.metrics` that suits regression:

```
# Print the R2-score on train and test
print('R2-score on train set:', lr.score(X_train, y_train))
print('R2-score on test set:', lr.score(X_test, y_test))
```

Here is the output:

```
R2-score on train set: 0.6323843381852894 R2-score on test set:
-1.2472000127402643
```

As we can see, there is a significant difference between the train and test set's scores, indicating model overfitting on the train set. To address this problem, regularization techniques will be proposed in the following recipes.

There's more...

Once the model has been trained, we can access all the parameters *w* (here, 16 values for 16 input features) as well as the intercept *b*, with the attributes, `.coef_`, and the `.intercept_` value of the `LinearRegression` object respectively:

```
print('w values:', lr.coef_)
print('b value:', lr.intercept_)
```

Here is the output:

```
w values: [ 1.12882772e+00 -6.48931138e-02 -4.04087026e-
01  4.87937619e-01  -1.69895164e-03 -4.09553062e-01 -3.72826365e+00
-8.38728583e+00  -2.67065542e-01  2.04856554e-01  2.46387700e-01
-3.19674747e-01   2.58750270e-03  3.91054062e-01  2.82040287e+00
-7.50771410e+00] b value: 2.072498958939411
```

If we plot these values, we will notice that their values range between approximately -8 and 2 on this dataset:

```
import matplotlib.pyplot as plt
plt.bar(np.arange(len(lr.coef_)), lr.coef_)
plt.xlabel('feature index')
plt.ylabel('weight value')
plt.show()
```

Here is a visual representation of this:

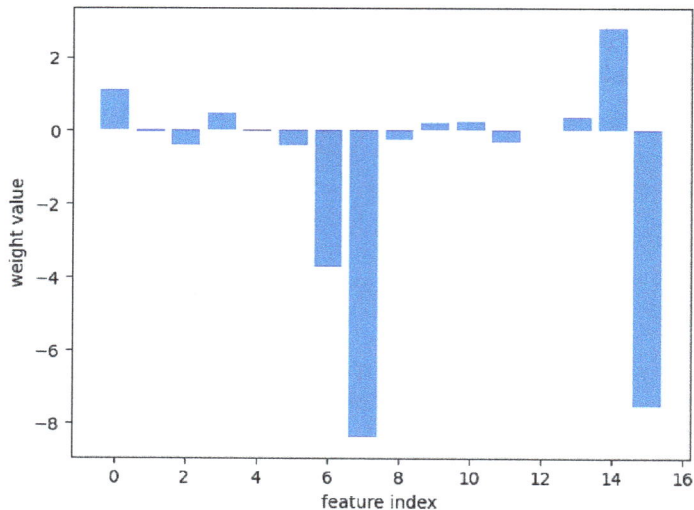

Figure 3.5 – Learned values of each weight of the linear regression model. The range of the values is quite large, from -8 to 2

See also

To have a full understanding of how to use linear regression using scikit-learn, it is good practice to check the official documentation of the class: `https://scikit-learn.org/stable/modules/generated/sklearn.linear_model.LinearRegression.html`.

We now have a good understanding of how linear regression works, and we will see in the next section how to regularize it with penalization.

Regularizing with ridge regression

A very common and useful way to regularize a linear regression is through penalization of the loss function. In this recipe, after reviewing what it means to add penalization to the loss function in the case of ridge regression, we will train a ridge model on the same California housing dataset as in the previous recipe, and see how it can improve the score thanks to regularization.

Getting ready

One way to make sure that a model's parameters are not going to overfit is to keep them close to zero: if the parameters do not have the possibility to evolve freely, they are less likely to overfit.

To that end, ridge regression adds a new term (regularization term) to the loss L_{Ridge}:

$$L_{Ridge} = \frac{1}{2m} \sum_{i} (y^{(i)} - h(x)^{(i)})^2 + \lambda |w|^2$$

Where $|w|^2$ is the $L2$ norm of w:

$$|w|^2 = \sum_{j} w_j^2$$

With this loss, we intuitively understand that high values of weights w are not possible, and thus overfitting is less likely. Also, λ is a hyperparameter (it can be fine-tuned) allowing us to control the regularization level:

- A high value of λ means high regularization
- A value of $\lambda=0$ means no regularization, for example, regular linear regression

The gradient descent formulas are slightly updated to the following:

$$\frac{\partial \mathcal{L}}{\partial w_j} = -\frac{1}{m}\sum_i (y^{(i)} - h(x)^{(i)})x_j^{(i)} + 2\lambda w_j$$

$$\frac{\partial \mathcal{L}}{\partial b} = -\frac{1}{m}\sum_i (y^{(i)} - h(x)^{(i)}) + 2\lambda b$$

Let's now see how to use ridge regression with scikit-learn: the same libraries required in the previous recipe must be installed: `numpy`, `sklearn`, and `matplotlib`.

Also, we assume the data is already downloaded and prepared from the previous recipe. To download, split and prepare the data, refer to the previous recipe.

How to do it...

Let's assume we are reusing the same data from the previous recipe. We will just train and evaluate another model on this exact same data, including the feature engineering with the squared features. The related implementation in scikit-learn for ridge regression is the `Ridge` class, where the `alpha` class attribute is equivalent to the λ in the preceding equation. Let's use it:

1. Import the `Ridge` class from scikit-learn:

    ```
    from sklearn.linear_model import Ridge
    ```

2. We then instantiate a ridge model. A regularization parameter of `alpha=5000` has been selected here, but every dataset may need a very specific hyperparameter value to perform best. Next, train the model on the training set (previously prepared) with the `.fit()` method, which is shown as follows:

    ```
    # Fit the Ridge model ridge = Ridge(alpha=5000)
    ridge.fit(X_train, y_train)
    Ridge(alpha=5000)
    ```

3. We then evaluate the model. Here, we compute and display the R2-score provided by the `.score()` of the ridge class, but any other regression metric could be used:

    ```
    # Print the R2-score on train and test
    print('R2-score on train set:', ridge.score(X_train, y_train))
    print('R2-score on test set:', ridge.score(X_test, y_test))
    ```

Here is the output:

```
R2-score on train set: 0.5398290317808138 R2-score on test set:
0.5034148460338739
```

We will notice that we are getting better results on the test set compared to the linear regression model (with no regularization) by allowing the R2-score on the test set to be slightly above 0.5.

There's more...

We can also print the weights and plot them, and compare those values to the ones of regular linear regression, as follows:

```
print('theta values:', ridge.coef_)
print('b value:', ridge.intercept_)
```

Here is the output:

```
theta values: [
0.43456599  0.06311698  0.00463607  0.00963748  0.00896739
-0.05894055  -0.17177956
-0.15109744  0.22933247  0.08516982  0.01842825
-0.01049763  -0.00358684  0.03935491 -0.17562536  0.1507696 ] b value:
2.07249895893891
```

For visualization, we can also plot these values with the following code:

```
plt.bar(np.arange(len(ridge.coef_)), ridge.coef_)
plt.xlabel('feature index') plt.ylabel('weight value')
plt.show()
```

This code outputs the following:

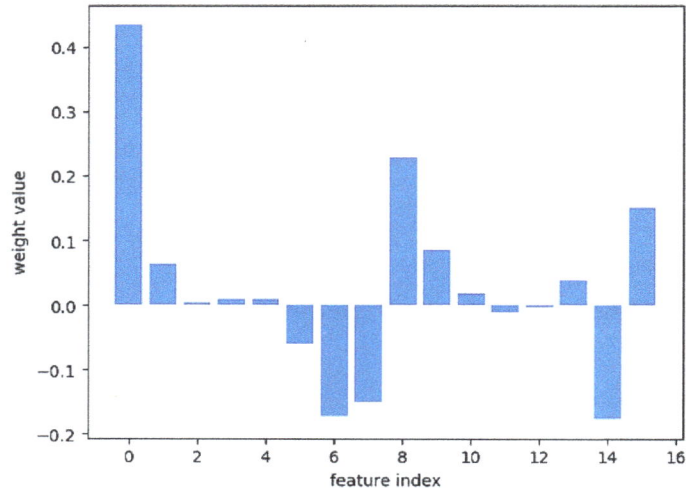

Figure 3.6 – Learned values of each weight of the ridge regression model.
Note that some are positive, some negative, and none are purely equal to
zero. Also, the range is much smaller than without regularization

The weight values are now ranging from -0.2 to .0.5, which is indeed much smaller than with no penalization.

As expected, this adding regularization results in a test set R2-score that is much better and closer to the train set than without regularization.

See also

For more information about all the possible parameters and hyperparameters of the Ridge regression, you can take a look at the official scikit-learn documentation: https://scikit-learn.org/stable/modules/generated/sklearn.linear_model.LinearRegression.html.

Regularizing with lasso regression

Lasso regression stands for **Least Absolute Shrinkage and Selection Operator**. This is a regularization method that is conceptually very close to ridge regression. In some cases, lasso regression outperforms ridge regression, which is why it's useful to know what it does and how to use it. In this recipe, we will briefly explain what lasso regression is and then train a model using scikit-learn on the same California housing dataset.

Getting ready

Instead of using the L2-norm, lasso uses the L1-norm, so that the loss L_{Lasso} is the following:

$$L_{Lasso} = \frac{1}{2m} \sum_{i} (y^{(i)} - h(x)^{(i)})^2 + \lambda|w|$$

While ridge regression tends to decrease weights close to zero quite smoothly, lasso is more drastic. Lasso, having a much steeper loss, tends to set weights to zero quite quickly.

Just like the ridge regression recipe, we'll use the same libraries and assume they are installed: numpy, sklearn, and matplotlib. Also, we'll assume the data is already downloaded and prepared.

How to do it...

The scikit-learn implementation of lasso is available with the Lasso class. Like in the Ridge class, alpha is the term that allows control of this regularization:

- The value of alpha is 0 means no regularization
- A large value of alpha means high regularization

Again, we will reuse the same, already prepared dataset that we used for linear regression and ridge regression:

1. Import the Lasso class from scikit-learn:

   ```
   from sklearn.linear_model import Lasso
   ```

2. We instantiate a lasso model with a value of alpha=0.2, which provides pretty good results and low overfitting, as we will see right away. However, feel free to test other values, as each dataset may have its very unique optimal value. Next, train the model on the training set using the .fit() method of the Lasso class:

   ```
   # Fit the Lasso model lasso = Lasso(alpha=0.02)
   lasso.fit(X_train, y_train)
   ```

3. Evaluate the lasso model on the training and test datasets, using the R2-score, implemented in the .score() method of the Lasso class:

   ```
   # Print the R2-score on train and test
   print('R2-score on train set:', lasso.score(X_train, y_train))
   print('R2-score on test set:', lasso.score(X_test, y_test))
   ```

This code outputs the following:

```
R2-score on train set: 0.5949103710772492
R2-score on test set: 0.57350350155955
```

Here we see an improvement when compared to the linear regression with no penalization, and we have improved against the ridge regression too, having an R2-score of about 0.57 on the test set.

There's more...

If we plot again for the weights instances, we now have the following values:

```
plt.bar(np.arange(len(lasso.coef_)), lasso.coef_)
plt.xlabel('feature index')
plt.ylabel('weight value')
plt.show()
```

This outputs the plot in *Figure 3.14*:

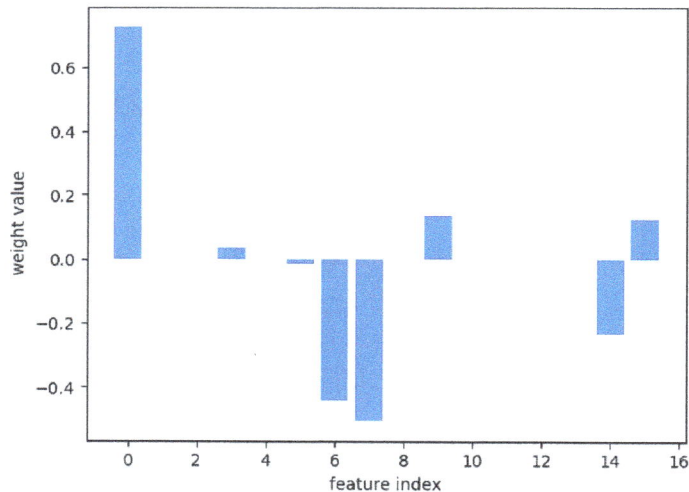

Figure 3.7 – Learned values of each weight of the lasso model. Note that, unlike the ridge model, several weights are set to zero

As expected, some values are set to 0, with an overall range of -0.5 to 0.7.

Lasso regularization allowed us to significantly improve performance on the test set in this case, while also reducing overfitting.

See also

Take a look at the official documentation for more information on the lasso class: `https://scikit-learn.org/stable/modules/generated/sklearn.linear_model.Lasso.html`.

Another technique called group lasso can be useful for regularization; find out more about it here: `https://group-lasso.readthedocs.io/en/latest/`.

Regularizing with elastic net regression

Elastic net regression, besides having a very fancy name, is nothing more than a combination of ridge and lasso penalization. It's a regularization method that can be of help in some specific cases. Let's have a look at what it means in terms of loss, and then train a model on the California housing dataset.

Getting ready

The idea with elastic net is to have both L1 and L2 regularization.

This means that the loss $L_{ElasticNet}$ is the following:

$$L_{ElasticNet} = \frac{1}{2m}\sum_i (y^{(i)} - h(x)^{(i)})^2 + \lambda_1|w| + 0.5\lambda_2|w|^2$$

The two hyperparameters, λ_1 and λ_2, can be fine-tuned.

We won't go into detail on the equations for the gradient descent, since deriving them is straightforward as soon as ridge and lasso are clear.

To train a model, we again need the `sklearn` library, which we already installed in previous recipes. Also, we again assume that the California housing dataset is already downloaded and prepared.

How to do it...

In scikit-learn, elastic net is implemented in the `ElasticNet` class. However, instead of having two hyperparameters, λ_1 and λ_2, they are using two hyperparameters, `alpha` and `l1_ratio`:

- $alpha = \lambda_1 + \lambda_2$
- $l1_{ratio} = \lambda_1 / (\lambda_1 + \lambda_2)$

Let's now apply this to our already prepared California housing dataset:

1. Import the `ElasticNet` class from scikit-learn:

    ```
    from sklearn.linear_model import ElasticNet
    ```

2. Instantiate an elastic net model. Values of `alpha=0.1` and `l1_ratio=0.5` have been chosen, but other values can be tested. Then train the model on the training set, using the `.fit()` method of the `ElasticNet` class:

```
# Fit the LASSO model
Elastic = ElasticNet(alpha=0.1, l1_ratio=0.5)
elastic.fit(X_train, y_train)
ElasticNet(alpha=0.1)
```

3. Evaluate the elastic net model on the training and test dataset, using the R2-score computed with the `.score()` method:

```
# Print the R2-score on train and test
print('R2-score on train set:', elastic.score(
    X_train, y_train))
print('R2-score on test set:', elastic.score(
    X_test, y_test))
```

Here is the output for it:

```
R2-score on train set: 0.539957010948829
R2-score on test set: 0.5134203748307193
```

In this case, the results are not an improvement upon lasso regularization: perhaps a better fine-tuning of the hyperparameters is required to achieve equivalent performance.

While more complicated to fine-tune because of having two hyperparameters, elastic net regression may offer more flexibility in regularization than the ridge or lasso regularizations. The use of hyperparameter optimization is the recommended method to find the right set of hyperparameters for any specific task.

> **Note**
>
> In practice, elastic net regression is probably less widely used than the ridge and lasso regressions.

See also

The official documentation of scikit-learn for elastic net regression can be found here: `https://scikit-learn.org/stable/modules/generated/sklearn.linear_model.ElasticNet.html`.

Training a logistic regression model

Logistic regression is really close to linear regression conceptually. Once linear regression is fully understood, logistic regression is just a couple of tricks away. But unlike linear regression, logistic regression is most commonly used for classification tasks.

Let's first explain what logistic regression is, and then train a model on the breast cancer dataset using scikit-learn.

Getting ready

Unlike linear regression, logistic regression's output is limited to a range of 0 to 1. The first idea is exactly the same as linear regression, having for each feature x_j a parameter θ_j:

$$z = \sum_j \theta_j x_j + b$$

There is one more step to limit the range to 0 to 1, which is to apply the logistic function to this output z. As a reminder, the logistic function (also called the sigmoid function, although it's a more generic function) is the following:

$$\sigma(x) = \frac{1}{1 + e^{-x}}$$

The logistic function has an S-shape, with values ranging from 0 to 1, with a value of 0.5 on $x = 0$, as shown here in *Figure 3.8*:

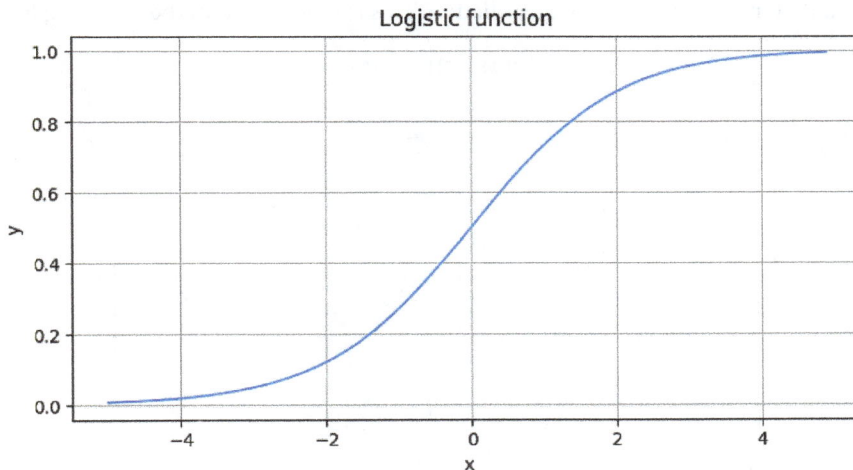

Figure 3.8 – Logistic function, ranging from 0 to 1 through 0.5 for x = 0

Finally, by applying the sigmoid function, we have the logistic regression prediction $h(x)$ as the following:

$$h(x) = \frac{1}{1 + e^{-z}} = \frac{1}{1 + e^{-\sum_j \theta_j x_j + b}}$$

This ensures an output value $h(x)$ in the range of 0 to 1. But it does not yet allow us to have a classification. The final step is to apply a threshold (for example, 0.5) to have a classification prediction:

$$\hat{y} = \begin{cases} 1 & \text{if } h(x) > 0.5 \text{ otherwise} \\ 0 \end{cases}$$

As we did for linear regression, we now need to define a loss L that is to be minimized, in order to optimize the parameters θ_j and b. The commonly used loss is the so-called **binary cross entropy**:

$$L = -\frac{1}{m}\sum_i y^{(i)} \log(h(x^{(i)}) + (1 - y^{(i)}) \log(1 - h(x^{(i)}))$$

We can see four extreme cases here:

- *if $y = 1$ and $h(x) \simeq 1$: $L \simeq 0$*
- *if $y = 1$ and $h(x) \simeq 0$: $L \simeq +\infty$*
- *if $y = 0$ and $h(x) \simeq 0$: $L \simeq 0$*
- *if $y = 0$ and $h(x) \simeq 1$: $L \simeq +\infty$*

So now for us to have the expected behavior, which is, a loss that tends to 0 indicating a highly accurate prediction, and that increases for a wrong prediction. This is represented in the following figure:

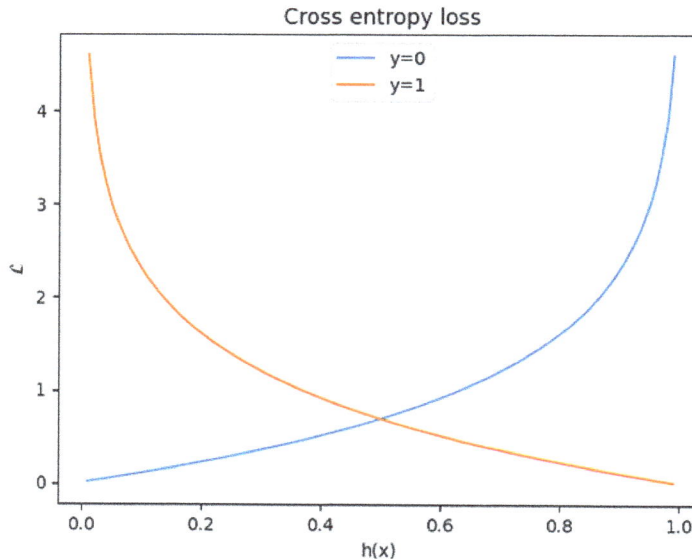

Figure 3.9 – Binary cross entropy, minimizing the error for both y = 0 and y = 1

Again, one way to optimize logistic regression is through gradient descent, the exact same way as for linear regression. As a matter of fact, the equations are exactly the same, even if we can't prove it here.

To get ready for this recipe, all we need is to have the `scikit-learn` library installed.

How to do it...

Logistic regression is fully implemented in scikit-learn as the `LogisticRegression` class. Unlike linear regression in scikit-learn, logistic regression has regularization implemented directly into one single class. The following parameters are the ones to tweak in order to fine-tune regularization:

- `penalty`: This can be either `'l1'`, `'l2'`, `'elasticnet'`, or `'none'`
- `C`: This is a float, inverse to regularization strength; the smaller the value, the greater the regularization

In this recipe, we will apply logistic regression with no regularization to the breast cancer dataset provided by scikit-learn. We will first load and prepare the data, and then train and evaluate the logistic regression model, as follows:

1. Import the `load_breast_cancer` function, which will allow us to load the dataset, and the `LogisticRegression` class:

   ```
   from sklearn.datasets import load_breast_cancer
   from sklearn.linear_model import LogisticRegression
   ```

2. Load the dataset using the `load_breast_cancer` function:

   ```
   # Load the dataset
   X, y = load_breast_cancer(return_X_y=True)
   ```

3. Split the dataset using the `train_test_split` function. We assume it has been imported in a previous recipe. Otherwise, you will need to import it with `from sklearn.model_selection import train_test_split`. We choose `test_size=0.2` so that we have a training size of 80% and a test size of 20%:

   ```
   # Split the data
   X_train, X_test, y_train, y_test = train_test_split(
       X, y, test_size=0.2, random_state=42)
   ```

4. We then prepare the data. Since the dataset is only composed of quantitative data, we just apply rescaling using the standard scaler: we instantiate it, fit it on the training set, and transform this same training set with `fit_transform()`. Finally, we rescale the test set with `.transform()`.

Again, we assume the standard scaler has been imported from a previous recipe, otherwise, we need to import it with `from sklearn.preprocessing import StandardScaler`:

```
# Rescale the data
scaler = StandardScaler()
X_train = scaler.fit_transform(X_train)
X_test = scaler.transform(X_test)
```

5. Then we instantiate the logistic regression, and we specify `penalty='none'` here so that we don't use any penalization of the loss for pedagogical reasons. Check out the next recipe to see how penalization works. Then we train the logistic regression model on the training set with the `.fit()` method:

```
# Fit the logistic regression model with no regularization
lr = LogisticRegression(penalty='none')
lr.fit(X_train, y_train)
LogisticRegression(penalty='none')
```

6. Evaluate the model on both the training and the test set. The `.score()` method of the `LogisticRegression` class uses the accuracy metric, but any other metric can be used:

```
# Print the accuracy score on train and test
print('Accuracy on train set:', lr.score(X_train, y_train))
print('Accuracy on test set:', lr.score(X_test, y_test))
```

This code outputs the following:

```
Accuracy on train set: 1.0 Accuracy on test set:
0.9385964912280702
```

We are facing rather strong overfitting here, with classification accuracy of 100% on the training set but only about 94% on the test set. This is a good start, but in the next recipe, we will use regularization to help improve the test accuracy.

Regularizing a logistic regression model

Logistic regression uses the same trick as linear regression to add regularization: it adds penalization to the loss. In this recipe, we will first briefly explain how penalization affects the loss, and how to add regularization using scikit-learn on the breast cancer dataset that we prepared in the previous recipe.

Getting ready

Just like linear regression, it is very easy to add a regularization term to the loss L, either an L1- or L2-norm of the parameters w. For example, the loss with an L2-norm would be the following:

$$L = -\frac{1}{m}\sum_i y^{(i)} \log(h(x^{(i)}) + (1 - y^{(i)}) \log\left(1 - h(x^{(i)})\right) + \frac{1}{C}|w|^2$$

As we did for ridge regression, we've added a squared sum of the weights, with a hyperparameter in front of it. To keep as close as possible to the scikit-learn implementation, we will use 1/C instead of λ for the regularization hyperparameter, but the idea remains the same.

In this recipe, we assume the following libraries are already installed from previous recipes: `sklearn` and `matplotlib`. Also, we assume the data from the breast cancer dataset is already loaded and prepared from the previous recipe, so that we can directly reuse it.

How to do it...

Let's now try to improve the test accuracy we had on the previous recipe by adding L2 regularization:

1. Instantiate the logistic regression model. Here, we choose an L2 penalization and a regularization value of C=0.1. A lower value of C means greater regularization:

    ```
    lr = LogisticRegression(penalty='l2', C=0.1)
    ```

2. Fit the logistic regression model on the training set, with the `.fit()` method:

    ```
    lr.fit(X_train, y_train)
    LogisticRegression(C=0.1)
    ```

3. Evaluate the model on both training and test set. We use the `.score()` method here, providing the accuracy score:

    ```
    # Print the accuracy score on train and test
    print('Accuracy on train set:', lr.score(
        X_train, y_train))
    print('Accuracy on test set:', lr.score(
        X_test, y_test))
    Accuracy on train set: 0.9802197802197802
    Accuracy on test set: 0.9824561403508771
    ```

As we can see here, adding L2 regularization allowed us to climb up to 98% accuracy on the test set, which is quite an improvement from about 94% without regularization.

> **Another reminder**
> The best way to find the right regularization value for C is with hyperparameter optimization.

Out of curiosity, we can plot the train and test accuracy here for several values of regularization strength:

```
accuracy_train = []
accuracy_test = []
c_values = [0.001,0.003,0.01,0.03,0.1,0.3,1,3,10,30]
for c in c_values:
    lr = LogisticRegression(penalty='l2', C=c)
    lr.fit(X_train, y_train)
    accuracy_train.append(lr.score(X_train, y_train))
    accuracy_test.append(lr.score(X_test, y_test))
plt.plot(c_values, accuracy_train, label='train')
plt.plot(c_values, accuracy_test, label='test')
plt.legend()
plt.xlabel('C: inverse of regularization strength')
plt.ylabel('Accuracy')
plt.xscale('log')
plt.show()
```

Here is the graph for it:

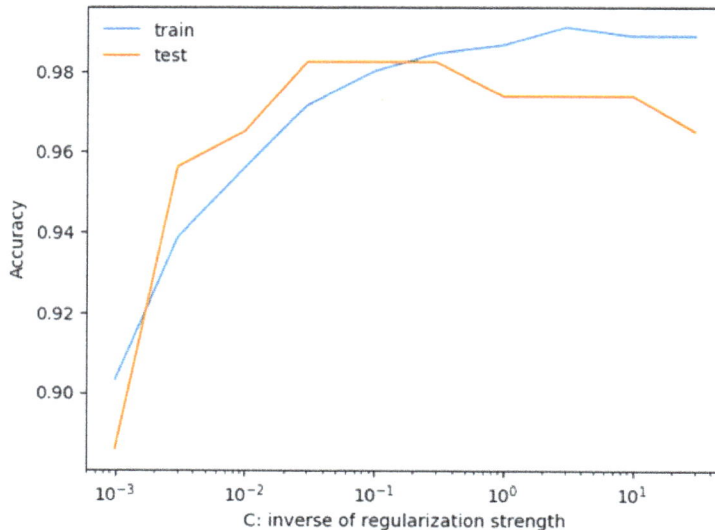

Figure 3.10 – Accuracy as a function of the C parameter, for both the training and test sets

This plot can actually be read from right to left. We can see that as the value of C decreases (thus increasing regularization), the train accuracy keeps on decreasing, as the regularization gets higher and higher. On the other hand, decreasing C (thus increasing regularization) first allows us to improve the test results: the model is generalizing more and more. But at some point, adding regularization (decreasing C more) does not help any further, and even hurts the test accuracy. Indeed, adding too much regularization creates a high-bias model.

There's more...

The `LogisticRegression` implementation of scikit-learn allows us to not only use L2 penalization but also L1 and elastic net, just like linear regression, which allows us to have some flexibility on the best regularization for any given dataset and task.

The official documentation can be checked for more details: `https://scikit-learn.org/stable/modules/generated/sklearn.linear_model.LogisticRegression.html`.

Choosing the right regularization

Linear models share this regularization method with L1 and L2 penalization. The only difference in the implementation is the fact that linear regression has its own class for each regularization type, as mentioned here:

- `LinearRegression` for no regularization
- `RidgeRegression` for L2 regularization
- `Lasso` for L1 regularization
- `ElasticNet` for both L1 and L2

Logistic regression has an integrated implementation, passing L1 or L2 as the class parameter.

> **Note**
> With **Support Vector Machines** (**SVMs**), the scikit-learn's implementation provides a C parameter for L2 regularization for both the `SVC` classification class and the `SVR` regression class.

But for linear regression as well as logistic regression, one question remains: should we use L1 or L2 regularization?

In this recipe, we will provide some practical tips about whether to use L1 or L2 penalization in some cases and then we will perform a grid search on the breast cancer dataset using logistic regression to find the best regularization.

Getting ready

There is no absolute answer to what penalization is best. Most of the time, the only way to find out is by doing hyperparameter optimization. But in some cases, data itself or external constraints may give hints about which regularization to use. Let's have a quick look. First, let's compare L1 and L2 regularization, then let's explore hyperparameter optimization.

L1 versus L2 regularization

L1 and L2 regularization have intrinsic differences that can sometimes help us make an educated guess upfront, and then save computational time. Let's have a look at these differences.

As discussed, L1 regularization tends to set some weights to zero, and thus may allow feature selection. Thus, if we have a dataset with many features, it can be helpful to have this information, and then just remove those features in the future.

Also, L1 regularization uses the absolute value of weights in the loss, making it more robust to outliers compared to L2 regularization. If our dataset could contain outliers, it is to be considered.

Finally, in terms of computing speed, L2 regularization is adding a quadratic term and is as a consequence less computationally expensive than L1 regularization. If training speed is of concern, L2 regularization should be considered before L1.

In a nutshell, we can consider the following cases to be ones where the choice can be made up front based on data or other constraints such as computation resources:

- The data contains numerous features, many of lesser importance: L1 regularization
- The data contains many outliers: L1 regularization
- Training computation resources are of concern: L2 regularization

Hyperparameter optimization

A more practical and perhaps pragmatic approach is just to do hyperparameter optimization, with L1 or L2 being a hyperparameter (elastic net regression could be added too).

We will use hyperparameter optimization with grid search as implemented by scikit-learn, optimizing a logistic regression model with both L1 and L2 regularization on the breast cancer dataset.

For this recipe, we expect the `scikit-learn` library to be installed from a previous recipe. We also assume that the breast cancer data is already loaded and prepared from the *Training a logistic regression model* recipe.

How to do it...

We will perform a grid search on a given set of hyperparameters, more specifically, L1 and L2 penalization with several values of penalization, C:

1. First, we need to import the `GridSearchCV` class from `sklearn`:

   ```
   from sklearn.model_selection import GridSearchCV
   ```

2. Then we define the parameter grid. This is the space of hyperparameters that we want to test. Here, we will try both L1 and L2 penalization, and for each penalization, we will try the C values in `[0.01, 0.03, 0.06, 0.1, 0.3, 0.6]` as follows:

   ```
   # Define the hyperparameters we want to test param_grid = {
       'penalty': ['l1', 'l2'],
       'C': [0.01, 0.03, 0.06, 0.1, 0.3, 0.6] }
   ```

3. Next, we instantiate the grid search object. Several parameters are passed here:

   - **The model to optimize**: `LogisticRegression`, for which we specify the solver to be `liblinear` (check the *See also* section for more information) so that it can handle both L1 and L2 penalization. We assume the class is already imported from a previous recipe; otherwise, you can import it with `from sklearn.linear_model import LogisticRegression`.

   - **The parameter grid**: A dictionary with hyperparameters to explore as keys and the associated list of values to explore as dictionary values. We defined it in the previous step, containing all the values we want to test.

   - **The scoring to optimize**: We choose here `scoring='accuracy'`, but it can be any other relevant metric.

   - **The number of cross-validation folds**: We choose 5 cross-validation folds here with `cv=5`, as it is pretty standard, but depending on the size of the dataset, other values may be just fine too:

   ```
   # Instantiate the grid search
   object grid = GridSearchCV(
       LogisticRegression(solver='liblinear'),
       param_grid,
       scoring='accuracy',
       cv=5 )
   ```

4. Train the grid on the training data with the `.fit()` method. Then we can display the best hyperparameters found out of curiosity, using the `best_params_` attribute:

```
# Fit and wait
grid.fit(X_train, y_train)
# Print the best set of hyperparameters
print('best hyperparameters:', grid.best_params_)
best hyperparameters: {'C': 0.06, 'penalty': 'l2'}
```

In this case, the hyperparameters appear to be `C=0.06` with L2 penalization. We can now evaluate the model.

5. Evaluate the model on both the training and the test sets, using the `.score()` method that computes accuracy. Using `.score()` or `.predict()` directly on the grid object will automatically compute the best model predictions:

```
# Print the accuracy score on train and test
print('Accuracy on train set:', grid.score(
    X_train, y_train))
print('Accuracy on test set:', grid.score(
    X_test, y_test))
Accuracy on train set: 0.9824175824175824
Accuracy on test set: 0.9912280701754386
```

In this case, this improved the performance on the test set, although it's not always that easy. But the method remains the same and can be applied to any dataset.

See also

The grid search official documentation can be found here: `https://scikit-learn.org/stable/modules/generated/sklearn.model_selection.GridSearchCV.html`.

More information about the solvers and penalizations available in scikit-learn can be found here: `https://scikit-learn.org/stable/modules/linear_model.html#logistic-regression`.

4

Regularization with Tree-Based Models

Tree-based models using ensemble learning such as Random Forest or Gradient Boosting are often seen as easy-to-use, state-of-the-art models for regular machine learning tasks.

Many Kaggle competitions have been won with such models, as they can be quite robust and efficient at finding complex patterns in data. Knowing how to regularize and fine-tune them is key to having the very best performance.

In this chapter, we'll look at the following recipes:

- Building a classification tree
- Building regression trees
- Regularizing a decision tree
- Training a Random Forest algorithm
- Regularization of Random Forest
- Training a boosting model with XGBoost
- Regularization with XGBoost

Technical requirements

In this chapter, you will train and fine-tune several decision tree-based models, as well as visualize a tree. The following libraries will be required for this chapter:

- NumPy
- Matplotlib
- Scikit-learn

- Graphviz

- XGBoost

- pickle

Building a classification tree

Decision trees are a separate class of models in machine learning. Although a decision tree alone can be considered a weak learner, combined with the power of ensemble learning such as bagging or boosting, decision trees get great performances. Before digging into ensemble learning models and how to regularize them, in this recipe, we will review how decision trees work and how to use them on a classification task on the iris dataset.

To give an intuition of the power of decision trees, let's consider a use case. We would like to know whether to sell ice creams on the beach based on two input features: sun and temperature.

We have the data in *Figure 4.1* and would like to train a model on it.

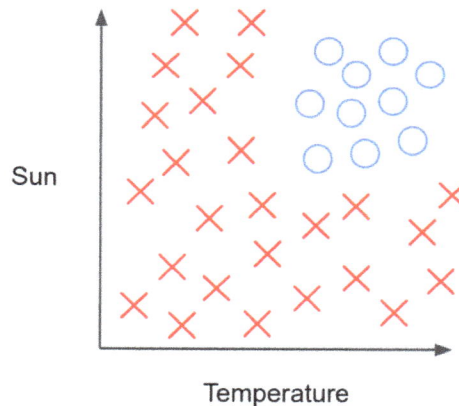

Figure 4.1 – A circle if we should sell ice creams as a function of
sun and temperature and a cross if we shouldn't

For a human, this seems quite easy. For a linear model though, not so much. If we try to use logistic regression on this data, it will end up drawing a decision line such as the left in *Figure 4.2*. Even with features that are raised to a higher power level, the logistic regression would struggle and propose something such as the decision line on the right in *Figure 4.2*.

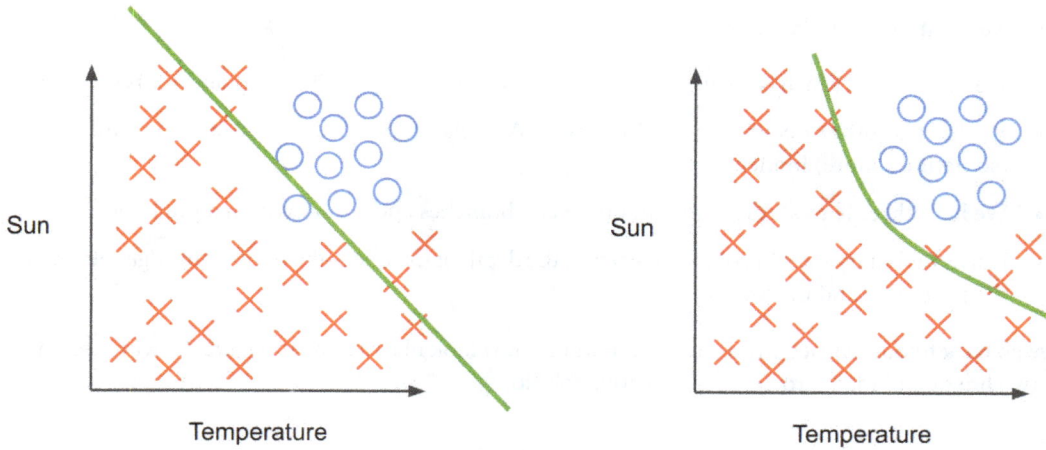

Figure 4.2 – Potential result of a linear model at classifying this dataset: on the left with raw features, on the right with higher power features

In a word, this data is not linearly separable. But it can be divided into two separate linearly separable problems:

- Is the weather sunny?

- Is the temperature warm?

If we fulfill those two conditions, then we should sell ice cream. This can be summarized as the tree in *Figure 4.3*:

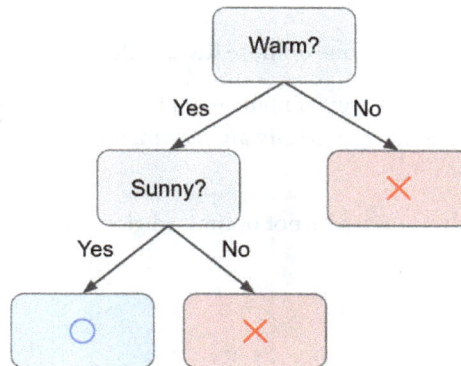

Figure 4.3 – A decision tree correctly classifying all data points

Let's cover a bit of vocabulary here:

- We have **Warm**, which is the first decision node and the root node with two branches: **Yes** and **No**.

- We have another decision node in **Sunny**. A decision node is any node containing two (sometimes more) branches.

- We have three leaves. A leaf does not have any branches and contains a final prediction.

- Just as for binary trees in computer science, the depth of the tree is the number of edges between the root node and the lowest leaf

If we go back to our dataset, the decision line would now look like the one in *Figure 4.4* with not one but two lines combined, providing an effective solution:

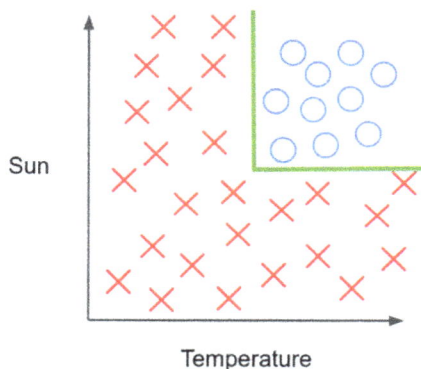

Figure 4.4 – Result of a decision tree at classifying this dataset

From now on, any new data will fall into one of those leaves, allowing it to be classified correctly.

This is the power of decision trees: they can compute complex, nonlinear rules, allowing them great flexibility. A decision tree is trained using a **greedy algorithm**, meaning it only tries to optimize one step at a time.

More specifically, it means the decision tree is not optimized globally, but one node at a time.

This can be seen as a recursive algorithm:

1. Take all samples in a node.

2. Find a threshold in a feature that minimizes the disorder of the splits. In other words, find the feature and threshold giving the best class separation.

3. Split this into two new nodes.

4. Go back to *step 1* until your node is pure (meaning that only one class remains) or any other condition, and thus a leaf.

But how do we actually choose the splits so that they are optimal? Of course, we use a loss function, which uses disorder measurement. Let's dig into those two topics before wrapping it up.

Disorder measurement

For a classification tree to be effective, it must have as little disorder as possible in its leaves. Indeed, in the previous example, we assumed all leaves are pure. They contain samples from only one class. In reality, leaves may be impure and contain samples from several classes.

> **Note**
> If after training a tree a leaf remains impure, we would use the majority class of that leaf for classification.

So, the idea is to minimize impurity, but how do we measure it? There are two ways: entropy and Gini impurity. Let's have a look at both.

Entropy

Entropy is a general word that is used in many contexts, such as physics and computer science. The entropy E we use here can be defined with the following equation, where p_i is the proportion of subclass p_i in a sample:

$$E = \sum_i p_i \log p_i$$

Let's consider a concrete example as depicted in *Figure 4.5*:

Figure 4.5 – A node with 10 samples of two classes: red and blue

In this example, the entropy would be the following:

$$E = -\frac{3}{10}\log 2\frac{3}{10} - \frac{7}{10}\log 2\frac{7}{10} \simeq 0.88$$

Indeed, we have the following:

- p_{blue}=3/10, since we have three blue samples out of 10

- p_{red}=7/10, since we have seven blue samples out of 10

If we look at extreme cases, we understand entropy is well suited to compute disorder:

- if p_{blue}=0, then p_{red}=1 and E = 0

- if p_{blue}pblue = p_{red} = 0.5, then E = 1

So, we understand that the entropy reaches a maximum value of one when the node contains perfectly mixed samples, and the entropy goes to zero when a node contains only one class. This is summarized by the curve in *Figure 4.6*:

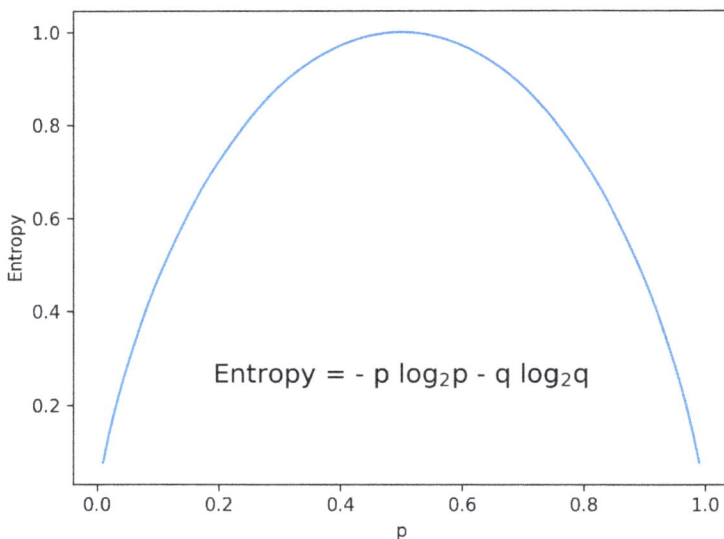

Figure 4.6 – Entropy as a function of p for two classes

Gini impurity

Gini impurity is another way to measure disorder. The formula of Gini impurity G is quite simple:

$$G = 1 - \sum p_i^2$$

Again, p_i is the proportion of class in the node.

Applied to the example node in *Figure 4.5*, the computation of the Gini impurity would lead to the following result:

$$G = 1 - \left(\frac{3}{10}\right)^2 - \left(\frac{7}{10}\right)^2 = 0.42$$

The result is quite different from entropy, but let's check that the properties remain the same with extreme values:

- if $p_{blue} = 0$, then $p_{red} = 1$ and G = 0
- if $p_{blue} = p_{red} = 0.5$, then G = 0.5

Indeed, the Gini impurity reaches a maximum value of 0.5 when the disorder is maximum and is equal to 0 when the node is pure.

Entropy or Gini?

That said, what should we use? Well, this can be seen as a hyperparameter, and scikit-learn's implementation allows one to choose between entropy and Gini.

In practice, the results are often the same for both. But Gini impurity is faster to compute (entropy involves more expensive log computations), so it is usually the first choice.

Loss function

We have a disorder measurement, but what is the loss we should minimize? The ultimate goal is to make splits that minimize the disorder, one node at a time.

Considering a decision node always has two children, we can define them as left and right nodes. Then, the loss for this node can be written like this:

$$L = \frac{m_{left}}{m} G_{left} + \frac{m_{right}}{m} G_{right}$$

Let's break down the formula:

- m, m_{left}, and m_{right} are the number of samples in each node respectively
- G_{left} and G_{right} are the Gini impurities of the left and right nodes

---

Note

Of course, this can be computed with entropy instead of Gini impurity.

Assuming we choose a split decision, we then have a parent node and two children nodes defined by the split in *Figure 4.7*:

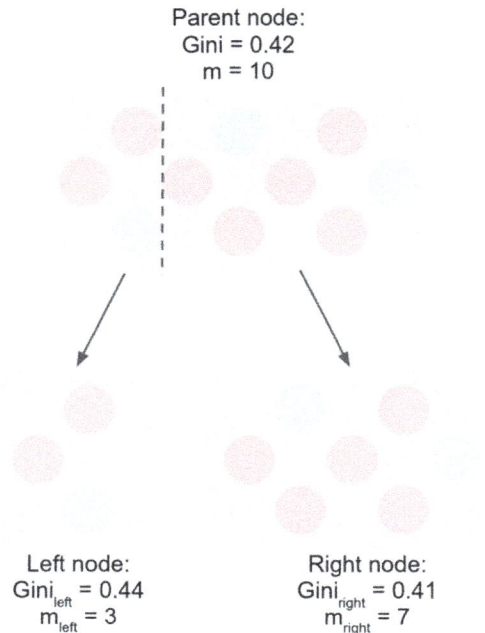

Figure 4.7 – One parent node and two children nodes, with their respective Gini impurities

In this case, the loss **L** would be the following:

$$L = \frac{3}{10}0.44 + \frac{7}{10}0.41 \simeq 0.419$$

Using this loss computation, we are now able to minimize the impurity (thus maximizing the purity of a node).

> **Note**
>
> What we defined here as the loss is only the loss at a node level. Indeed, as stated earlier, the decision tree is trained using a greedy approach, not a gradient descent.

Getting ready

Finally, before going into the practical details of this recipe, we need to have the following libraries installed: scikit-learn, `graphviz`, and `matplotlib`.

They can be installed with the following command line:

```
pip install scikit-learn graphviz matplotlib
```

How to do it...

Before actually training a decision tree, let's quickly go through all the steps to train a decision tree:

1. We have a node containing samples of N classes.

2. We iterate through all the features and all the possible values of a feature.

3. For each feature value, we compute the Gini impurity and the loss.

4. We keep the feature value with the lowest loss and split the node into two children nodes.

5. Go back to *step 1* with both nodes until a node is pure (or the stop condition is fulfilled).

Using this approach, the decision tree will eventually find the right set of decisions to successfully separate any classes. Then, two cases are possible for each leaf:

- If the leaf is pure, predict this class

- If the leaf is impure, predict the most represented class

> **Note**
>
> One way to test all the possible feature values is to use all the existing values in the dataset. Another is to use a linear split over the range of existing values in the dataset.

Let's now train a decision tree on the iris dataset with scikit-learn:

1. First, we need the required imports: `matplotlib` for data visualization (not necessary otherwise), `load_iris` for loading the dataset, `train_test_split` for splitting the data into training and test sets, and the `DecisionTreeClassifier` decision tree implementation from scikit-learn:

    ```python
    from matplotlib import pyplot as plt
    from sklearn.datasets import load_iris
    from sklearn.model_selection import train_test_split
    from sklearn.tree import DecisionTreeClassifier
    ```

2. We can now load the data using `load_iris`:

    ```python
    # Load the dataset
    X, y = load_iris(return_X_y=True)
    ```

3. We split the dataset into training and test sets with `train_test_split`, keeping the default parameters and only specifying the random state for reproducibility:

    ```python
    # Split the dataset
    X_train, X_test, y_train, y_test = train_test_split(
        X, y, random_state=0)
    ```

4. In this step, we display a two-dimensional projection of the data. This is just for pedagogical purposes but is not mandatory:

```
# Plot the training points
plt.scatter(X[:, 0], X[:, 1], c=y)
plt.xlabel('Sepal length')
plt.ylabel('Sepal width')
plt.show()
```

Here is the plot for it:

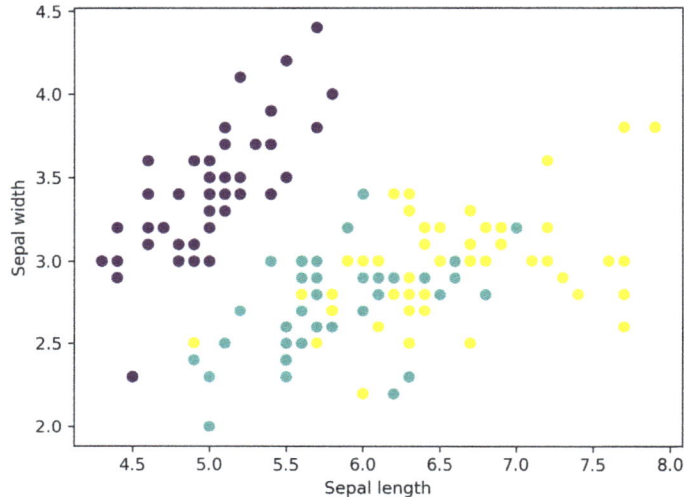

Figure 4.8 – The three iris classes as a function of the sepal width
and sepal length (plot produced by the code)

5. Instantiate the `DecisionTreeClassifier` model. We use the default parameters here:

```
# Instantiate the model
dt = DecisionTreeClassifier()
```

6. Train the model on the training set. We did not prepare the data with any preprocessing here because we have only quantitative features, and decision trees are not sensitive to scale, unlike linear models. But it would not hurt either to rescale the quantitative features:

```
# Fit the model on the training data
dt.fit(X_train, y_train)
```

7. Finally, we evaluate the accuracy of the model on both the training and test sets, using the
    `score()` method of the classification tree:

```
# Compute the accuracy on training and test sets
print('Accuracy on training set:', dt.score(
    X_train, y_train))
print('Accuracy on test set:', dt.score(
    X_test, y_test))
```

This prints the following output:

```
Accuracy on training set: 1.0
Accuracy on test set: 0.9736842105263158
```

We are achieving satisfactory results, even if we are clearly facing overfitting with 100% accuracy on the train set.

There's more...

Unlike linear models, there are no weights associated with each feature since a tree is made up of splits.

For visualization purposes, we can display the tree thanks to the `graphviz` library. This is mostly for pedagogical use or interest but is not necessarily useful otherwise:

```
from sklearn.tree import export_graphviz
import graphviz
# We load iris data again to retrieve features and classes names
iris = load_iris()
# We export the tree in graphviz format
graph_data = export_graphviz(
    dt,

    out_file=None,
    feature_names=iris.feature_names,
    class_names=iris.target_names,
    filled=True, rounded=True
)
# We load the tree again with graphviz library in order to display it
graphviz.Source(graph_data)
```

Here is the tree for it:

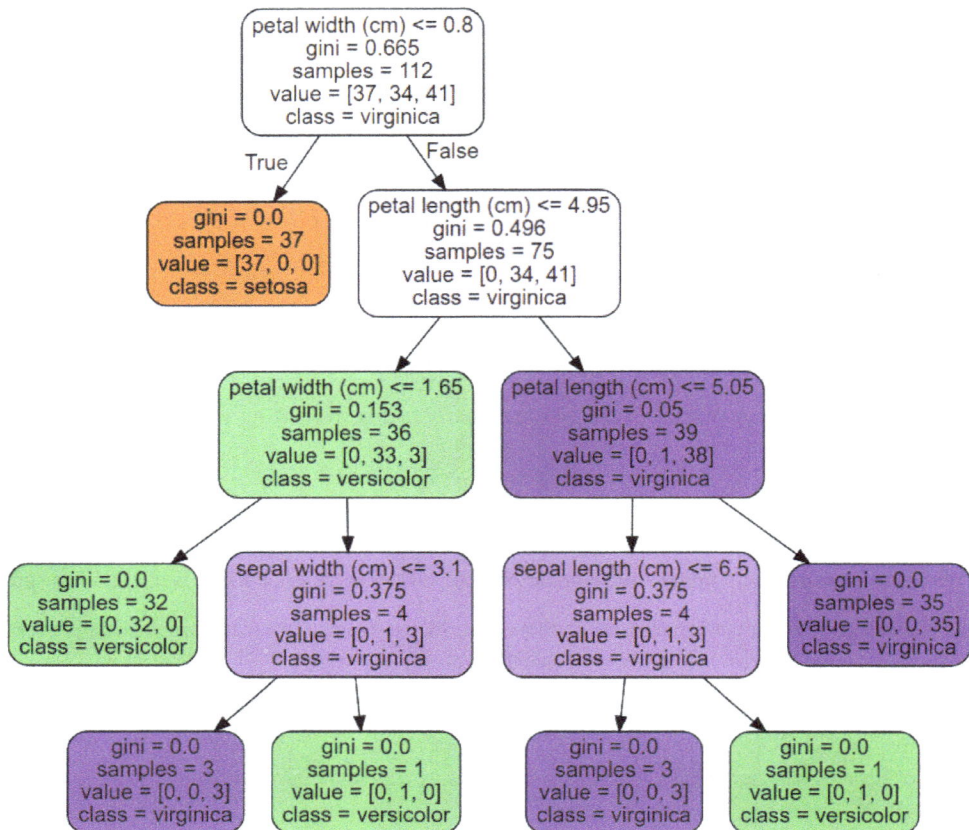

Figure 4.9 – Tree visualization produced by the graphviz library

From this tree visualization, we can see that the 37 samples of the setosa class are fully classified right away at the first decision node (considering the data visualization, this should not be a surprise). The samples of classes virginica and versicolor seem to be much more intertwined in the provided features, thus the tree requires many more decision nodes to fully discriminate them.

Unlike linear models, we do not have weights associated with each feature. But we can have a piece of somehow equivalent information, called feature importance, available with the `.feature_ importances` attribute:

```
import numpy as np
plt.bar(iris.feature_names, dt.feature_importances_)
plt.xticks(rotation=45))
plt.ylabel('Feature importance')
plt.title('Feature importance for the decision tree')
plt.show()
```

Here is the plot for it:

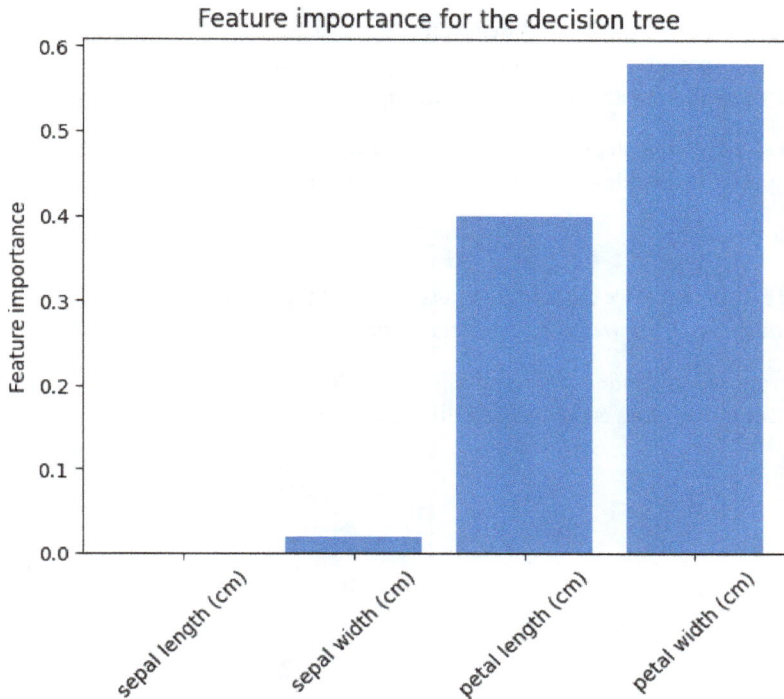

Figure 4.10 – Feature importance as a function of the feature name (histogram produced by the code)

This feature importance is relative (meaning the sum of all feature importance is equal to 1) and is computed based on the number of samples classified thanks to this feature.

> **Note**
>
> Feature importance is computed based on the amount of reduction of the metric used for splitting (for example, Gini impurity or entropy). If one single feature allows to make all the splits, then this feature will have an importance of 1.

See also

The sci-kit learning documentation on classification trees as available at the following URL: https://scikit-learn.org/stable/modules/generated/sklearn.tree.DecisionTreeClassifier.html.

Building regression trees

Before digging into the regularization of decision trees in general, let's have a recipe for regression trees. Indeed, all the explanations in the previous recipe were assuming we have a classification task. Let's explain how to apply it to a regression task and apply it to the California housing dataset.

For regression trees, only a few steps need to be modified compared to classification trees: the inference and the loss computation. Besides that, the overall principle is the same.

The inference

In order to make an inference, we can no longer use the most represented class in a leaf (or in the case of pure leaf, the only class). So, we use the average of the labels in each node.

In the example proposed in *Figure 4.11*, assuming this is a leaf, we would have an inference value that is the average of those 10 values equal to 14 in this case.

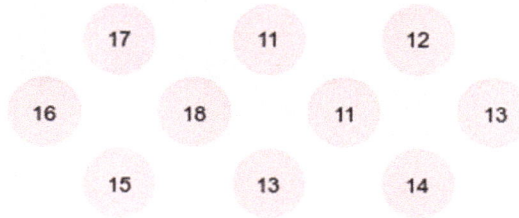

Figure 4.11 – The example of 10 samples with associated values

The loss

Instead of using a disorder measurement to compute the loss, in regression trees, the mean squared error is used. So, the loss is as follows:

$$L = \frac{m_{left}}{m} MSE_{left} + \frac{m_{right}}{m} MSE_{right}$$

Assume again a given split leading to the m_{left} samples in the left node and the m_{right} samples in the right node. The **MSE** for each split is computed using the average of the labels into that node:

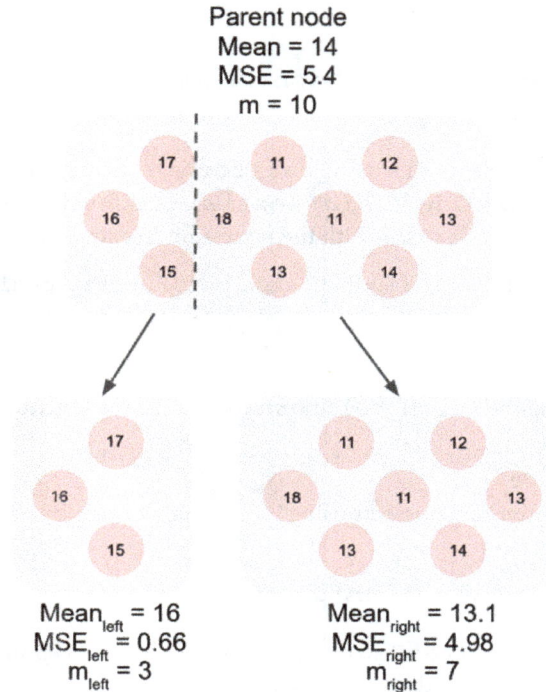

Figure 4.12 – Example of a node split on a regression task

If we take the example of the proposed split in *Figure 4.12*, we have all we need to compute the L loss:

$$L = \frac{3}{10}0.66 + \frac{7}{10}4.98 \simeq 3.67$$

Based on those two slight changes, we can train a regression tree with the same recursive, greedy algorithm as with classification trees.

Getting ready

Before getting practical, all we need for this recipe is to have the scikit-learn library installed. If not yet installed, just type the following command line in your terminal:

```
pip install scikit-learn
```

How to do it...

We will train a regression tree using the `DecisionTreeRegressor` class from scikit-learn on the California housing dataset:

1. First, the required imports: the `fetch_california_housing` function to load the California housing dataset, the `train_test_split` function to split data, and the `DecisionTreeRegressor` class with the regression tree implementation:

    ```
    from sklearn.datasets import fetch_california_housing
    from sklearn.model_selection import train_test_split
    from sklearn.tree import DecisionTreeRegressor
    ```

2. Load the dataset using the `fetch_california_housing` function:

    ```
    X, y = fetch_california_housing(return_X_y=True)
    ```

3. Split the data into training and test sets using the `train_test_split` function:

    ```
    X_train, X_test, y_train, y_test = train_test_split(
        X, y, test_size=0.2, random_state=0)
    ```

4. Instantiate the `DecisionTreeRegressor` object. We just keep the default parameters here, but they can be customized at this point:

    ```
    dt = DecisionTreeRegressor()
    ```

5. Train the regression tree on the training set using the `.fit()` method of the `DecisionTreeRegressor` class. Note that we do not apply any specification preprocessing to the data because we have only quantitative features, and decision trees are not sensitive to feature scale issues:

    ```
    dt.fit(X_train, y_train)
    DecisionTreeRegressor()
    ```

6. Evaluate the R2-score of the regression tree on both the training and test sets using the built-in `.score()` method of the model class:

    ```
    print('R2-score on training set:', dt.score(
        X_train, y_train))
    print('R2-score on test set:', dt.score(
        X_test, y_test))
    ```

    This would show something like this:

    ```
    R2-score on training set: 1.0
    R2-score on test set: 0.5923572475948657
    ```

As we can see, we face here a strong overfitting, having a perfect R2-score on the training set, while having a much worse (but still decent overall) R2-score on the test set.

See also

Look at the official `DecisionTreeRegressor` documentation more information: `https://scikit-learn.org/stable/modules/generated/sklearn.tree.DecisionTreeRegressor.html#sklearn-tree-decisiontreeregressor`.

Regularizing a decision tree

In this recipe, we will look at the means to regularize decision trees. We will review and comment on a couple of methods for reference and provide a few more to be explored.

Getting ready

Obviously, we cannot use L1 or L2 regularization as we did with linear models. Since we have no weights for the features and no overall loss such as the mean squared error or the binary cross entropy, it is not possible to apply this method here.

But we do have other ways to regularize, such as the max depth of the tree, the minimum number of samples per leaf, the minimum number of samples per split, the max number of features, or the minimum impurity decrease. In this recipe, we will look at those.

To do that, we only need the following libraries: scikit-learn, `matplotlib` and `NumPy`. Also, since we will provide some visualization to give some idea of regularization, we will use the following `plot_decision_function` function:

```
def plot_decision_function(dt, X, y):
    # Create figure to draw chart
    plt.figure(2, figsize=(8, 6))
    # We create a grid of points contained within [x_min,
       #x_max]x[y_min, y_max] with step h=0.02
    x0_min, x0_max = X[:, 0].min() - .5, X[:, 0].max() + .5
    x1_min, x1_max = X[:, 1].min() - .5, X[:, 1].max() + .5
    h = .02  # step size of the grid
    xx0, xx1 = np.meshgrid(np.arange(x0_min, x0_max, h),
        np.arange(x1_min, x1_max, h))

    # Retrieve predictions for each point of the grid
    Z_dt = dt.predict(np.c_[xx0.ravel(), xx1.ravel()])
    Z_dt = Z_dt.reshape(xx0.shape)
```

```
# Plot the decision boundary (label predicted assigned to a color)
plt.pcolormesh(xx0, xx1, Z_dt, cmap=plt.cm.Paired)

# Plot also the training points
plt.scatter(X[:, 0], X[:, 1], c=y, edgecolors='k',
    cmap=plt.cm.Paired)

# Format chart
plt.xlabel('Sepal length')
plt.ylabel('Sepal width')
plt.xticks(())
plt.yticks(())
plt.show()
```

This function will allow us to visualize the decision function of our decision tree and get a better understanding of what is overfitting and regularization when it comes to a classification tree.

How to do it...

We will give a recipe to regularize a decision tree based on the maximum depth, and then explore a few others in the *There's more* section.

The maximum depth is quite often one of the first hyperparameters to fine-tune when trying to regularize. Indeed, as we have seen earlier, decision trees can learn complex data patterns using more decision nodes. If not stopped, the decision trees may tend to overfit the data with too many consecutive decision nodes.

We will now train a classification tree with a limited maximum depth on the iris dataset:

1. Make the required imports:

    -   The load_iris function to load the dataset

    -   The train_test_split function to split the data into training and test sets

    -   The DecisionTreeClassifier class:

        ```
        from sklearn.datasets import load_iris
        from sklearn.model_selection import train_test_split
        from sklearn.tree import DecisionTreeClassifier
        ```

2. Load the dataset using the `load_iris` function. In order to be able to fully visualize the effects of regularization, we also keep only two features out of four, so that we can display them on a plot:

```
X, y = load_iris(return_X_y=True)
# Keep only 2 features
X = X[:, :2]
```

3. Split the data into training and test sets using the `train_test_split` function. We only specify the random state for reproducibility and let the other parameters be by default:

```
X_train, X_test, y_train, y_test = train_test_split(
    X, y, random_state=0)
```

4. Instantiate a decision tree object, limiting the maximum depth to five with the `max_depth=5` parameter. We also set the random state to 0 for reproducibility:

```
dt = DecisionTreeClassifier(max_depth=5,
    random_state=0)
```

5. Fit the classification tree on the training set using the `.fit()` method. As mentioned earlier, since the features are all quantitative and decision trees are not sensitive to the features scale, there is no need to apply rescaling:

```
dt.fit(X_train, y_train)
DecisionTreeClassifier(max_depth=5, random_state=0)
```

6. Evaluate the model accuracy, using the `.score()` method of the `DecisionTreeClassifier` class:

```
print('Accuracy on training set:', dt.score(
    X_train, y_train))
print('Accuracy on test set:', dt.score(
    X_test, y_test))
```

This would print the following output:

```
Accuracy on training set: 0.8660714285714286
Accuracy on test set: 0.6578947368421053
```

How it works...

In order to have a better understanding of how it works, let's look at the two dimensions of the iris dataset we retained. We will use the `plot_decision_function()` function defined in *Getting ready* to plot the decision function of a decision tree with no regularization (that is, default hyperparameters):

```
import numpy as np
from matplotlib import pyplot as plt
```

```
# Fit a decision tree over only 2 features
dt = DecisionTreeClassifier()
dt.fit(X_train[:, :2], y_train)
# Plot the decision tree decision function
plot_decision_function(dt, X_train[:, :2], y_train)
```

Here is the plot:

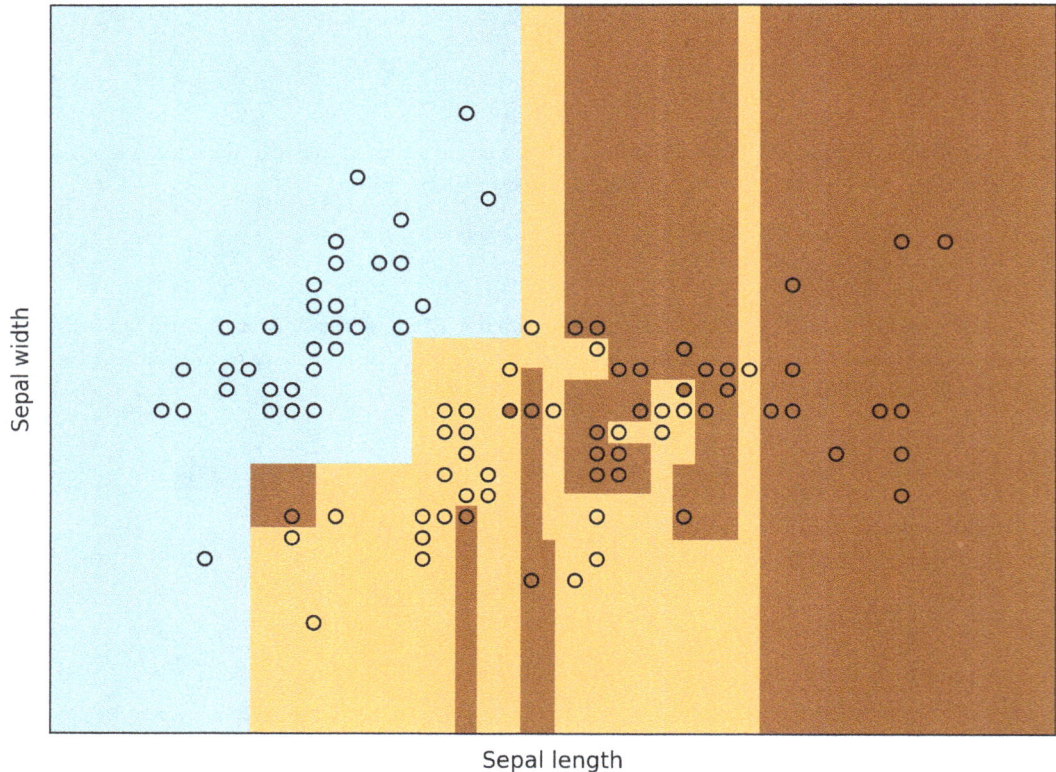

Figure 4.13 – Decision function of the model as a function of the sepal width and sepal length with a very complex and questionable decision function (plot produced by the code)

From this plot, we can deduce we are typically facing overfitting. Indeed, the boundaries are really specific, sometimes trying to make a complex pattern for only one sample, instead of focusing on the higher-level pattern.

Indeed, if we look at the accuracy score for both the training and test set, we have the following results:

```
# Compute the accuracy on training and test sets for only 2 features
print('Accuracy on training set:', dt.score(X_train[:, :2], y_train))
print('Accuracy on test set:', dt.score(X_test[:, :2], y_test))
```

We'll get the following output:

```
Accuracy on training set: 0.9375
Accuracy on test set: 0.631578947368421
```

While the accuracy is about 94% on the training set, it is only about 63% on the test set. There is overfitting, and regularization may be helpful.

> **Note**
>
> The accuracy is far lower than in the first recipe because we use only two features for visualization and pedagogical purposes. The reasoning remains true if we keep the four features, though.

Let's now add regularization by limiting the maximum depth of the decision tree as we did in this recipe:

```
max_depth: int, default=None
```

If the maximum depth of the tree is **None**, then nodes are expanded until all leaves are pure or until all leaves contain less than the `min_samples_split` samples.

It means that by default, the trees are expanded with no limit on the depth. The limit is then perhaps set by other factors and may go very deep. If we fix that by limiting the depth to 5, let's see the impact on the decision function:

```
# Fit a decision tree with max depth of 5 over only 2 features
dt = DecisionTreeClassifier(max_depth=5, random_state=0)
dt.fit(X_train[:, :2], y_train)
# Plot the decision tree decision function
plot_decision_function(dt, X_train[:, :2], y_train)
```

Here is the output:

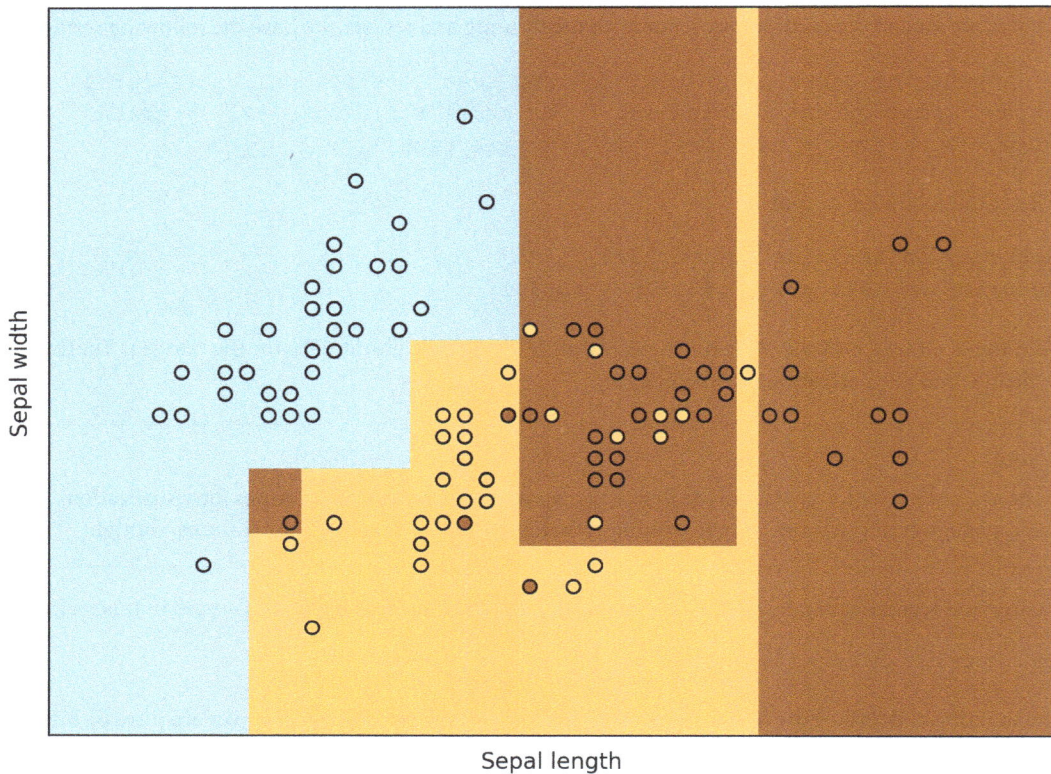

Figure 4.14 – Decision function with maximum depth regularization (plot produced by the code)

By limiting the max depth to 5, we get a less specific decision function, even if there seems to be some overfitting remaining. If we have a look again at the accuracy score, we can see that it actually helped slightly:

```
# Compute the accuracy on training and test sets for only 2 features
print('Accuracy on training set:', dt.score(X_train[:, :2], y_train))
print('Accuracy on test set:', dt.score(X_test[:, :2], y_test))
```

This would provide the following output:

```
Accuracy on training set: 0.8660714285714286
Accuracy on test set: 0.6578947368421053
```

Indeed, the accuracy score on the test set climbed from 63% to 66%, while the accuracy on the training set decreased from 95% to 87%. This is typically what we can expect from regularization: this added bias (and thus decreased training set performances) and decreased the variance (and thus allowed us to generalize better).

There's more...

The maximum depth hyperparameter is really convenient because it's easy to understand and fine-tune. But, there are many other hyperparameters that can help regularize decision trees. Let's review some of them here. We will focus on the minimum sample hyperparameters and then propose a few other hyperparameters.

Minimum samples

Other hyperparameters allowing us to regularize are the ones controlling the minimum number of samples per leaf and the minimum number of samples per split.

The idea is rather straightforward and intuitive but more subtle than the maximum depth. In the decision tree we visualized earlier in this chapter, we could see that the first splits classify substantial amounts of samples. The first split successfully classified 37 samples as setosa while keeping 75 in the other split. On the other end of the decision tree, the lowest nodes are sometimes splitting over only three or four samples.

Is splitting over only three samples significant? What if out of these three samples, there is an outlier? Generally, it does not sound like a promising idea to create a rule for only three samples if the end goal is to have a robust, well-generalizing model.

We have two different but somewhat related hyperparameters that allow us to deal with that:

- `min_samples_split`: The minimum samples required to split an internal node. If a float is provided, then it uses a fraction of the total number of samples.

- `min_samples_leaf`: The minimum samples required to be considered a leaf. If a float is provided, then it uses a fraction of the total number of samples.

While `min_samples_split` is acting at the decision node level, `min_samples_leaf` is acting only at the leaf level.

Let's see if that allows us to avoid overfitting in specific regions in our case. We set the minimum number of samples per split to 15 (while keeping all other parameters to default values). This is expected to regularize, since we know from the decision tree visualization some splits were for less than five samples:

```
# Fit a decision tree with min samples per split of 15 over only 2
features
dt = DecisionTreeClassifier(min_samples_split=15, random_state=0)
dt.fit(X_train[:, :2], y_train)
```

```
# Plot the decision tree decision function
plot_decision_function(dt, X_train[:, :2], y_train)
```

Here is the output:

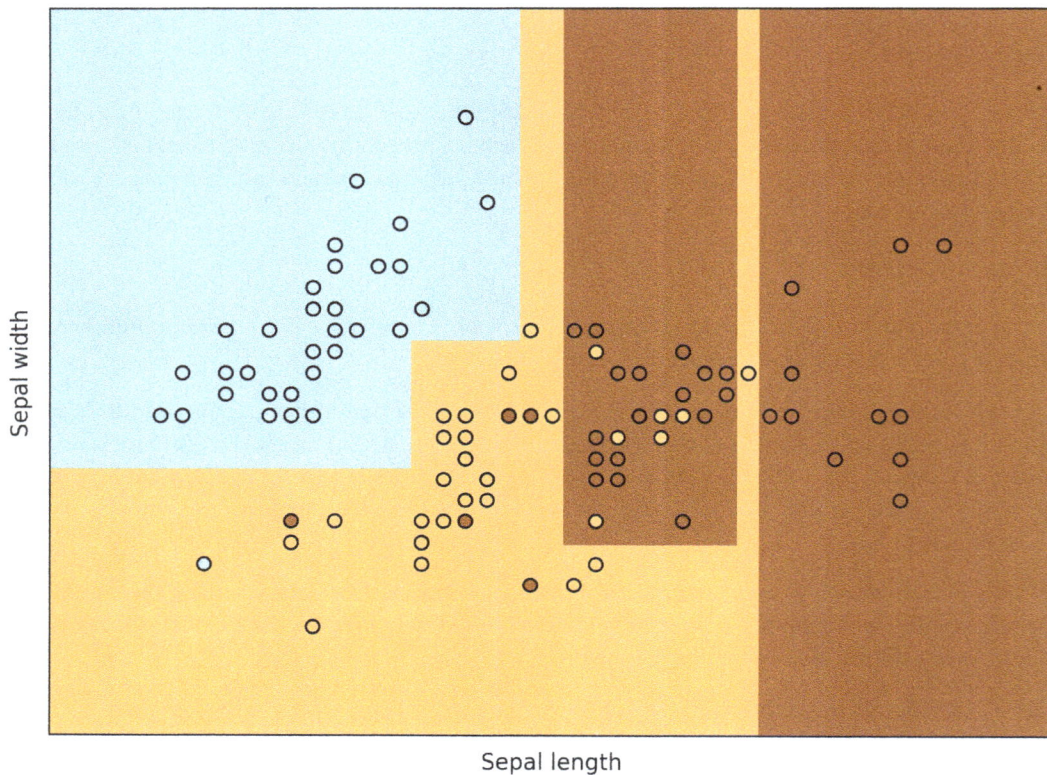

Figure 4.15 – Decision function with minimum samples per split regularization (plot produced by the code)

The resulting decision function is a bit different than when regularizing with the maximum depth and seems to be indeed more regularized than without constraint on this hyperparameter.

We can also look at the accuracy score to confirm the regularization was successful:

```
# Compute the accuracy on training and test sets for only 2 features
print('Accuracy on training set:', dt.score(X_train[:, :2],
    y_train))
print('Accuracy on test set:', dt.score(X_test[:, :2],
    y_test))
```

We'll get the following output:

```
Accuracy on training set: 0.85714285714285717
Accuracy on test set: 0.7368421052631579
```

Compared to default hyperparameters, the accuracy score on the test set climbed from 63% to 74%, while the accuracy on the training set decreased from 95% to 86%. Compared to the maximum depth hyperparameter, we added slightly more regularization, and got slightly better results on the test set.

In general, the hyperparameters on the number of samples (either per leaf or split) may allow a finer regularization than the maximum depth. Indeed, the max depth hyperparameter is setting a common hard limit to the whole decision tree. But it may happen that two nodes at the same depth level do not carry the same number of samples. One node may have hundreds of samples (and then a splitting is probably relevant), while another node may have just a few samples.

The criterion for a minimum number of samples on its side is more subtle: no matter the depth in the tree, if a node does not have enough samples, then we decide it is not worth splitting.

Other hyperparameters

Other hyperparameters can be used to regularize. We will not go through all the details for each of them, but rather list them and explain them briefly:

- `max_features`: By default, the decision tree is finding the best split among all features. You can choose to add randomness by setting another maximum number of features to use at each split. May add regularization by adding noise.

- `max_leaf_nodes`: Set a straight limit on the number of leaves in the tree. Somewhat like the max depth hyperparameter, it will regularize by limiting the number of splits, giving priority to nodes having the highest impurity reduction.

- `min_impurity_decrease`: This will split a node only if the impurity decrease is above the given threshold. This allows us to regularize by selecting highly impacting node splits only.

> **Note**
>
> Although we did not mention regression trees, the behavior and principles are analogous, and the same hyperparameters can be fine-tuned with the same behavior.

See also

The scikit-learn documentation is pretty explicit about all the hyperparameters and their potential impact: `https://scikit-learn.org/stable/modules/generated/sklearn.tree.DecisionTreeClassifier.html`.

Training the Random Forest algorithm

The Random Forest algorithm is an ensemble learning model, meaning it uses an ensemble of decision trees, hence *forest* in its name.

In this recipe, we will explain how it works and then train a Random Forest model on the California housing dataset.

Getting ready

Ensemble learning is based somehow on the idea of collective intelligence. Let's do a thought experiment to understand the power of collective intelligence.

Let's assume we have a bot that randomly answers correctly to any binary question 51% of the time. This would be considered inefficient and unreliable.

But now, let's also assume we are using not only one but an army of those randomly answering bots and use the majority vote as the final answer. If we have 1,000 of those bots, the majority vote will provide the right answer 75% of the time. If we have 10,000 bots, the majority vote will provide the right answer 97% of the time. This would turn a low-performing system into a remarkably high-performing system.

> **Note**
>
> A strong assumption was made for this example: each bot must be independent of the others. Otherwise, this example does not hold true. Indeed, the extreme counter-example would be that all bots are answering the same answer to any question, in which case, no matter how many bots you use, the accuracy remains at 51%.

This is the idea of collective intelligence, which relates somehow to human society. Most of the time, collective knowledge outperforms individual knowledge.

This is also the idea behind ensemble models: an ensemble of weak learners can become a powerful model. To do that with Random Forest, we need to define two key aspects:

- How to compute the majority vote
- How to ensure the independence of each decision tree in our model with randomness

Majority vote

To properly explain majority vote, let's imagine we have an ensemble of three decision trees trained on a binary classification task. On a given sample, the predictions are the following:

Tree	Predicted probability of class 1	Class predictions
Tree 1	0.05	0
Tree 2	0.6	1
Tree 3	0.55	1

Table 4.1 – Predictions

We have two pieces of information for each decision tree:

- The predicted probability of class 1
- The predicted class (usually computed as class 1 if probability > 0.5, class 0 otherwise)

> **Note**
> The `DecisionTreeClassifier.predict_proba()` method allows us to get the prediction probability. It is computed by using the proportion of the given class in the prediction leaf.

We could come up with many ways to compute the majority vote on such data, but let's explore two, hard vote and soft vote:

- A hard vote is the most intuitive one. This is the simple majority vote of the predicted classes. In our case, class 1 is predicted two times out of three. In this case, the hard majority vote is class 1.
- A soft vote uses the average probability and then applies a threshold. In our case, the average probability is 0.4, which is below the threshold of 0.5. In that case, the soft majority vote is class 0.

It is particularly interesting to note that, even if two out of three trees predicted class 1, the only tree that was really confident (having a high probability) was the tree predicting class 0.

A real-life example would be, when facing a question:

- Two friends give answer A, but are unsure
- One friend gives answer B but is highly confident

What would you do in such a case? The odds are you would listen to that highly confident friend. This is exactly what a soft majority vote is about: giving more power to highly confident trees. Most of the time, the soft vote outperforms the hard vote. Fortunately, Random Forest implementation in scikit-learn is based on the soft vote.

Bagging

Bagging is a key concept in Random Forest to ensure the independence of decision trees and is made of bootstrapping and aggregating. Let's see how those two steps are working together to get the best out of ensembling decision trees.

Bootstrapping is random sampling with replacement. In simple terms, if we apply bootstrapping to samples with replacement, it means we will randomly pick samples in the dataset with replacement. What *with replacement* means is that, once a sample has been picked, it is not removed from the dataset and may be picked again.

Let's assume we have an initial dataset of 10 samples, either blue or red. If we use bootstrapping to select 10 samples in this initial dataset, we may have some samples missing, and some samples appearing several times. If we do that three independent times, we may have three slightly different datasets, such as in *Figure 4.16*. We call those newly created datasets subsamples:

Boostrap

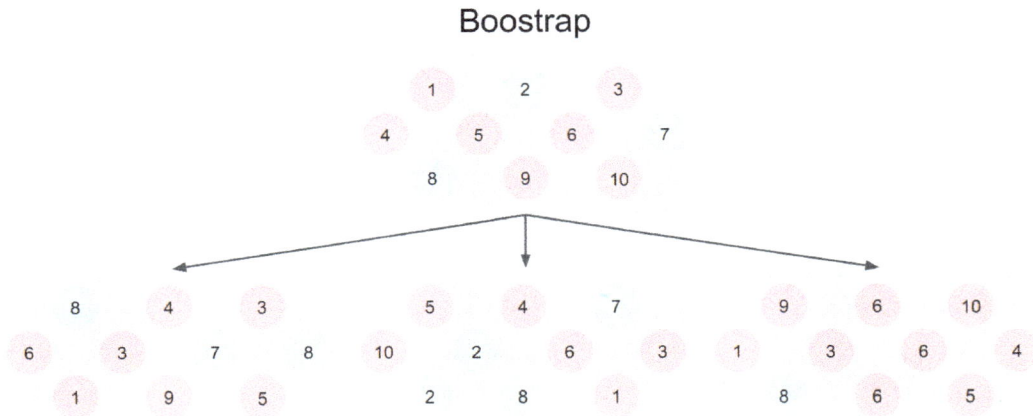

Figure 4.16 – An example of bootstrapping an initial dataset three times and selecting 10 samples for replacement (the three created subsamples are slightly different)

Since those subsamples are slightly different, in Random Forest, a decision tree is trained on each of those and ends up with hopefully independent models. The next step is the aggregating of those models' results, through a soft majority vote. Once those two steps (bootstrapping and aggregating) are combined, this is what we call bagging. *Figure 4.17* summarizes those two steps:

Bagging

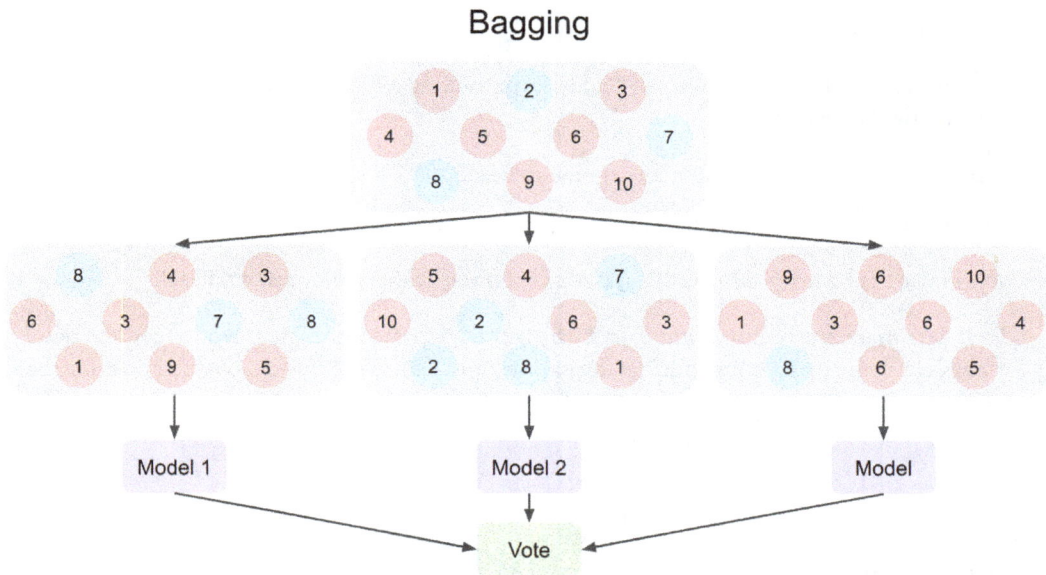

Figure 4.17 – Bootstrapping on the samples and then aggregating
the results to end with an ensemble model

As we have seen, the *random* in Random Forest comes from the bootstrapping of the samples, meaning we randomly select a subsample of the original dataset for each trained decision tree. But in reality, other levels of randomness have been omitted for pedagogical reasons. Without going into all the details, there are three levels of randomness in a Random Forest algorithm:

- **Bootstrapping of the samples**: Samples are selected with replacement
- **Bootstrapping of the features**: Features are selected with replacement
- **Feature selection of the best split of a node**: By default, in scikit-learn, all features are used, thus there is no randomness at this level

Now that we have a solid enough understanding of how Random Forest works, let's train a Random Forest algorithm on a regression task. To do so, we only need scikit-learn to be installed. If this has not already been done, just install it with the following command line:

```
pip install scikit-learn
```

How to do it...

As with other machine learning models in scikit-learn, training a Random Forest algorithm is quite easy. There are two main classes:

- RandomForestRegressor for regression tasks
- RandomForestClassifier for classification tasks

Here, we will use the RandomForestRegressor on the California housing dataset:

1. First, let's make the required imports: fetch_california_housing to load the data, train_test_split for splitting the dataset, and RandomForestRegressor for the model itself:

```
from sklearn.datasets import fetch_california_housing
from sklearn.model_selection import train_test_split
from sklearn.ensemble import RandomForestRegressor
```

2. Load the dataset using fetch_california_housing:

```
X, y = fetch_california_housing(return_X_y=True)
```

3. Split the data with train_test_split. Here, we just use the default parameters and set the random state to 0 for reproducibility:

```
X_train, X_test, y_train, y_test = train_test_split(
    X, y, random_state=0)
```

4. Instantiate the RandomForestRegressor model. We just keep the default parameters of the class here for simplicity; we only specify the random state:

```
rf = RandomForestRegressor(random_state=0)
```

5. Train the model on the training set with the .fit() method. This can take a few seconds to compute:

```
rf.fit(X_train, y_train)
RandomForestRegressor(random_state=0)
```

6. Evaluate the R2-score on both the training and test set using the .score() method:

```
# Display the accuracy on both training and test set
print('R2-score on training set:', rf.score(X_train, y_train))
print('R2-score on test set:', rf.score(X_test, y_test))
```

Our output would be as follows:

```
R2-score on training set: 0.9727159677969947
R2-score on test set: 0.7941678302821006
```

We have an R2-score on the training set of 97%, while on the test set, it is only 79%. This means we are facing overfitting, and we will see in the next recipe how to add regularization to such models.

See also

- The documentation of this class in scikit-learn: `https://scikit-learn.org/stable/modules/generated/sklearn.ensemble.RandomForestRegressor.html#sklearn-ensemble-randomforestregressor`

- Likewise, there is the documentation of the Random Forest classifier for classification tasks: `https://scikit-learn.org/stable/modules/generated/sklearn.ensemble.RandomForestClassifier.html`

Regularization of Random Forest

A Random Forest algorithm shares many hyperparameters with decision trees since a Random Forest is made up of trees. But a few more hyperparameters do exist, so in this recipe, we will present them and show how to use them to improve results on the California housing dataset regression.

Getting started

Random Forests are known to be quite prone to overfitting. Even if it's not a formal proof, in the previous recipe, we were indeed facing quite strong overfitting. But Random Forests, like decision trees, have many hyperparameters allowing us to try to reduce overfitting. As for a decision tree, we can use the following hyperparameters:

- Maximum depth
- Minimum samples per leaf
- Minimum samples per split
- `max_features`
- `max_leaf_nodes`
- `min_impurity_decrease`

But some other hyperparameters can be fine-tuned too:

- `n_estimators`: This is the number of decision trees trained in the random forest.
- `max_samples`: The number of samples to draw from the given dataset to train each decision tree. A lower value would add regularization.

Technically speaking, for this recipe, it is assumed that scikit-learn is installed.

How to do it...

In this recipe, we will try to add regularization by limiting the max number of features to the log of the total number of features. If you are reusing the same environment as for the previous recipe, you can jump directly to *step 4*:

1. As usual, let's make the required imports: `fetch_california_housing` to load the data, `train_test_split` for splitting the dataset, and `RandomForestRegressor` for the model itself:

    ```
    from sklearn.datasets import fetch_california_housing
    from sklearn.model_selection import train_test_split
    from sklearn.ensemble import RandomForestRegressor
    ```

2. Load the dataset using `fetch_california_housing`:

    ```
    X, y = fetch_california_housing(return_X_y=True)
    ```

3. Split the data with `train_test_split`. Here, we just use the default parameters, meaning we have a 75%25% split and set the random state to 0 for reproducibility:

    ```
    X_train, X_test, y_train, y_test = train_test_split(
        X, y, random_state=0)
    ```

4. Instantiate the `RandomForestRegressor` model. This time, we specify `max_features='log2'` so that for each split, only a random subset (of size `log2(n)`, assuming *n* features) of all the features is used:

    ```
    rf = RandomForestRegressor(max_features='log2', random_state=0)
    ```

5. Train the model on the training set with the `.fit()` method. This may take a few seconds to compute:

    ```
    rf.fit(X_train, y_train)
    RandomForestRegressor(max_features='log2', random_state=0)
    ```

6. Evaluate the R2-score on both the training and test set using the `.score()` method:

```
print('R2-score on training set:', rf.score(X_train,
    y_train))
print('R2-score on test set:', rf.score(X_test,
    y_test))
```

This would return the following:

```
R2-score on training set: 0.9748218476882353
R2-score on test set: 0.8137208340736402
```

Compared to the previous recipe with default hyperparameters, it improved the R2-score on the test set from 79% to 81%, while not significantly changing the score on the training set.

> **Note**
>
> In this case, as in many others in machine learning, it might be tricky (or sometimes impossible) to have the performances on the train and test set meeting halfway, meaning that even if the R2-score is 97% on training and 79% on the test set, there is absolutely no guarantee you can improve the R2-score on the test set. Sometimes, even the best hyperparameters are not the right key to improve performance.

In a word, all the regularization rules for decision trees can be applied to Random Forests, and a few more are available. As usual, an effective way to find the right set of hyperparameters is through hyperparameter optimization. Random Forest takes somewhat longer to train than a simple decision tree, so it may take some time.

Training a boosting model with XGBoost

Let's now see another application of decision trees: boosting. While bagging (used in Random Forest models) is training several trees in parallel, boosting is about training trees sequentially. In this recipe, we will have a quick review of what is boosting, and then train a boosting model with XGBoost, a widely used boosting library.

Getting ready

Let's have a look at introducing limits of bagging, then see how boosting may address some of those limits and how. Finally, let's train a model on the already prepared Titanic dataset with XGBoost.

Limits of bagging

Let's assume we have a binary classification task, and we trained Random Forest in three decision trees on two features. Bagging is expected to perform well if anywhere in the feature space, at least two out of three decision trees are right, as in *Figure 4.18*.

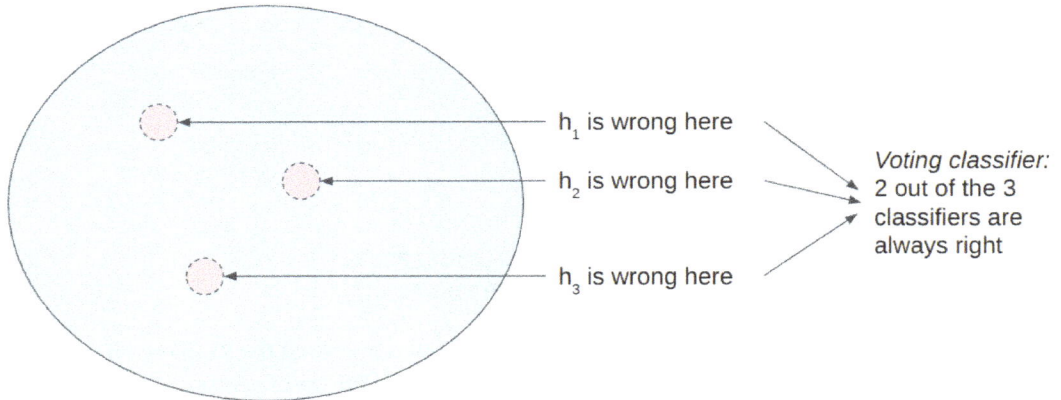

Figure 4.18 – The absence of overlap in dashed circle areas highlights decision tree errors, demonstrating Random Forest's strong performance

In *Figure 4.18*, we observe that the areas inside the dashed circles are where a decision tree is wrong. Since they don't overlap, at least two out of three decision trees are right everywhere. Thus, Random Forest is performing well.

Unfortunately, always having two out of three decision trees right is a strong assumption. What happens if only one or fewer decision tree is right in the feature space? As represented in *Figure 4.19*, the Random Forest algorithm starts performing poorly.

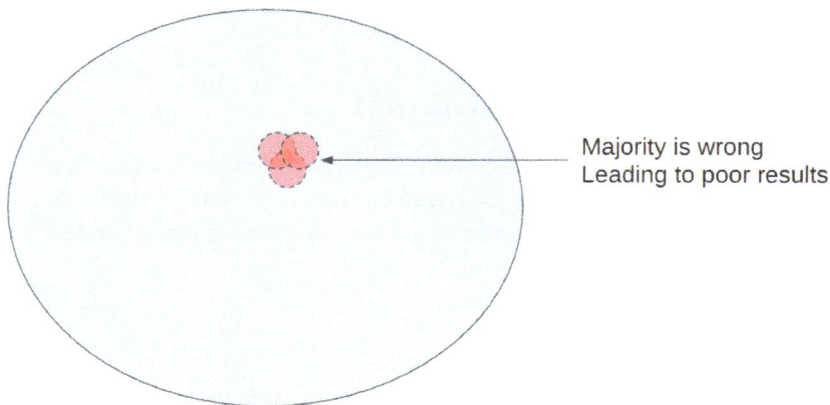

Figure 4.19 – When one or fewer out of three decision trees is right, Random Forest is performing poorly

Let's see how boosting can fix this issue by sequentially training decision trees to each try to fix the errors of the previous one.

> **Note**
>
> Those examples are simplified since Random Forest can be using soft vote, and thus predict a class that only a minority of trees predicted. But the principle remains true overall.

Gradient boosting principles

Gradient boosting has several implementations with a some differences: XGBoost, CatBoost, and LightGBM all have pros and cons, and the details of each are beyond the scope of this book. Rather, we will explain some general principles of the gradient boosting algorithm, enough to give a high-level understanding of the model.

The algorithm training can be summarized with the following steps:

1. Compute an average guess on the $\widehat{y_0}$ training set.
2. Compute the pseudo-residuals of each sample toward the last guessed prediction, $r_i = y_i - \hat{y}_i$.
3. Train a decision tree on the pseudo-residuals as labels, allowing to have predictions $\hat{r}_i$.
4. Compute the weight γ_i of that decision tree based on its performance.
5. Update the guess with the learning rate η, γ_i and these predicted pseudo-residuals:
   $\widehat{y_{i+1}} = \hat{y}_i + \eta \gamma_i \hat{r}_i$ o.
6. Go back to *step 2.* with the updated $\hat{y}_{i+1}$, iterate until reaching the maximum number of decision trees or another criterion.

In the end, the final prediction will be the following:

$$\hat{y} = \widehat{y_0} + \eta \sum_i \gamma_i \hat{r}_i$$

The η learning rate is a hyperparameter, typically 0.001.

> **Note**
>
> All boosting implementations are a bit different. For example, not all boosting implementations have weights γ associated with their trees. But since this is the case for XGBoost that we will use here, it is worth mentioning it for a better understanding.

In the end, having enough decision trees allows a model to perform well enough in most cases, hopefully. Unlike Random Forest, boosting models tend to avoid pitfalls such as having most number of wrong decision trees at the same place, since each decision tree is trying to fix remaining errors.

Also, boosting models tend to be more robust and generalized than Random Forest models, making them really powerful in many applications.

Finally, for this recipe, we will need the following libraries to be installed: `pickle` and `xgboost`. They can be installed using `pip` with the following command line:

```
pip install pickle xgboost
```

We will also reuse a prepared Titanic dataset from a previous recipe to avoid spending too much time on the data preparation. This data can be downloaded at `https://github.com/PacktPublishing/The-Regularization-Cookbook/blob/main/chapter_02/prepared_titanic.pkl` and should be added locally before doing the recipe with the following command line:

```
wget https://github.com/PacktPublishing/The-Regularization-Cookbook/blob/main/chapter_02/prepared_titanic.pkl
```

How to do it...

XGBoost is a very popular implementation of gradient boosting. It can be used with the same pattern as models in scikit-learn, using the following methods:

- `fit(X, y)` to train the model
- `predict(X)` to compute predictions
- `score(X, y)` to evaluate the model

Let's use it on the Titanic dataset with the default parameters downloaded locally:

1. The first step is the required imports. Here, we need pickle to read the data from the binary format and the XGBoost classification model class `XGBClassifier`:

    ```
    import pickle
    from xgboost import XGBClassifier
    ```

2. We load the already prepared data using pickle. Note that we already get a split dataset because it was actually saved this way:

    ```
    X_train, X_test, y_train, y_test = pickle.load(open(
        'prepared_titanic.pkl', 'rb'))
    ```

3. Instantiate the boosting model. We specify `use_label_encoder=False` because our qualitative features are actually already encoded with one hot encoding and because this feature is about to be deprecated:

    ```
    bst = XGBClassifier(use_label_encoder=False)
    ```

4. Train the model on the training set using the `.fit()` method, exactly like we would do for a scikit-learn model:

```
# Train the model on training set
bst.fit(X_train, y_train)
```

5. Compute the accuracy of the model on both training and test sets, using the `.score()` method. Again, this is the same as in scikit-learn:

```
print('Accuracy on training set:', bst.score(X_train,
    y_train))
print('Accuracy on test set:', bst.score(X_test,
    y_test))
```

We'll get the following now:

```
Accuracy on training set: 0.9747191011235955
Accuracy on test set: 0.8156424581005587
```

We notice overfitting: a 97% accuracy rate on the training set but only 81% on the test set. But in the end, the results of the test set are quite good, since it is quite hard to have much higher than 85% accuracy on the Titanic dataset.

See also

- The XGBoost documentation quality is arguably not as good as scikit-learn's, but it is still useful: `https://xgboost.readthedocs.io/en/stable/python/python_api.html#xgboost.XGBClassifier`.

- There is also the regression counterpart class XGBRegressor and its documentation: `https://xgboost.readthedocs.io/en/stable/python/python_api.html#xgboost.XGBRegressor`.

Regularization with XGBoost

After a recipe introducing boosting and the use of XGBoost for classification, let's now have a look at how to regularize such models. We will be using the same Titanic dataset and try to improve test accuracy.

Getting ready

Just like Random Forest, an XGBoost model is made of decision trees. Consequently, it has some hyperparameters such as the maximum depth of trees (`max_depth`) or the number of trees (`n_estimators`) that can allow to regularize in the same way. It also has several other hyperparameters related to the decision trees that can be fine-tuned:

- `subsample`: The number of samples to randomly draw for training, equivalent to `max_sample` for scikit-learn's decision trees. A smaller value may add regularization.

- `colsample_bytree`: The number of features to randomly draw (equivalent to scikit-learn's `max_features`) for each tree. A smaller value may add regularization.

- `colsample_bylevel`: The number of features to randomly draw at the tree level. A smaller value may add regularization.

- `colsample_bynode`: The number of features to randomly draw at the node level. A smaller value may add regularization.

Finally, more hyperparameters that are not shared with decision trees or Random Forest can allow fine-tuning the XGBoost models:

- `learning_rate`: The learning rate. A smaller learning rate may train even closer to the training set. Thus, a larger learning rate may regularize, although it may also degrade the performances

- `reg_alpha`: The strength of the L1 regularization. A higher value implies more regularization.

- `reg_lambda`: The strength of the L2 regularizations. A higher value implies more regularization.

Unlike other tree-based models seen so far, XGBoost allows L1 and L2 regularization too. Indeed, since each tree has an associated weight of γi, it is possible to add L1 or L2 regularization on those parameters, just the way it is done for linear models.

Those are the main hyperparameters to fine-tune to optimize and properly regularize an XGBoost whenever needed. Although it's a powerful, robust, and efficient model, it can be hard to fine-tune optimally, since there is quite a large number of hyperparameters.

More practically, in this recipe, we will only add L1 regularization. To do so, all we need is to have XGBoost installed, as well as the Titanic-prepared data, as for the previous recipe.

How to do it...

In this recipe, let's add L1 regularization with the parameter `reg_alpha` in order to add bias and hopefully reduce the variance of the model. We will reuse the `XGBClassifier` model on the prepared Titanic data, as we did for the previous recipe. If your environment is still the same as the previous recipe, you can jump to *step 3*.

1. As usual, we start with the required imports: `pickle` to read the data from the binary format and the XGBoost classification model class `XGBClassifier`:

    ```
    import pickle
    from xgboost import XGBClassifier
    ```

2. Then, we load the already prepared data using `pickle`. It assumes the `prepared_titanic.pkl` file is locally downloaded:

    ```
    X_train, X_test, y_train, y_test = pickle.load(open(
        'prepared_titanic.pkl', 'rb'))
    ```

3. Instantiate the boosting model. Besides specifying `use_label_encoder=False`, we now specify `reg_alpha=1` to add L1 regularization:

    ```
    bst = XGBClassifier(use_label_encoder=False, reg_alpha=1)
    ```

4. Train the model on the training set using the `.fit()` method:

    ```
    bst.fit(X_train, y_train)
    ```

5. Finally, compute the accuracy of the model on both training and test sets, using the `.score()` method:

    ```
    print('Accuracy on training set:', bst.score(X_train,
        y_train))
    print('Accuracy on test set:', bst.score(X_test,
        y_test))
    ```

    This would print the following:

    ```
    Accuracy on training set: 0.9410112359550562
    Accuracy on test set: 0.8435754189944135
    ```

Compared to the previous recipe with default hyperparameters, adding L1 penalization allowed us to improve the results. The accuracy scores are now about 84% on the test set, and they lowered to 94% on the training set, effectively adding regularization.

There's more...

Finding the best hyperparameter set with XGBoost can be tricky as there are many hyperparameters. Of course, using hyperparameter optimization techniques is required to gain some previous time.

For regression tasks, as mentioned earlier, the XGBoost library has an XGBRegressor class that makes them possible, with the same hyperparameters having the same effects on regularization.

5
Regularization with Data

Even though there are plenty of regularization methods for models (with each model having a unique set of hyperparameters), sometimes, the most effective regularization comes from the data itself. Indeed, sometimes, even the most powerful model can't have good performance if the data is not transformed properly beforehand.

In this chapter, we'll look at some methods that help regularize models from data:

- Hashing high cardinality features
- Aggregating features
- Undersampling an imbalanced dataset
- Oversampling an imbalanced dataset
- Resampling imbalanced data with SMOTE

Technical requirements

In this chapter, you will apply several tricks to data, as well as resample datasets or download new data via the command line. To do so, you will need the following libraries:

- NumPy
- pandas
- scikit-learn
- imbalanced-learn
- category_encoders
- Kaggle API

Hashing high cardinality features

High cardinality features are qualitative features with many possible values. High cardinality features may appear in many applications, such as a country in a customer database, a phone model in advertising, or vocabulary in NLP applications. High cardinality issues can be manifold: not only may they lead to a very highly dimensional dataset, but they can also evolve as more and more values become available. Indeed, even if the data for the number of countries or vocabulary is arguably quite stable, there are new phone models every week, if not every day.

Hashing is a very popular and useful way to deal with such problems. In this recipe, we'll see what it is and how to use it in practice on a dataset to predict whether employees will leave a company.

Getting started

Hashing is a very useful trick in computer science in general, and it is widely used in cryptography or blockchain, for example. It is also useful in machine learning when dealing with high cardinality features. It does not necessarily help with regularization per se, but it can sometimes be a side effect.

What is hashing?

Hashing is often used in machine learning at the production level for dealing with high cardinality features. High cardinality features tend to have a growing number of possible outcomes. This can include things such as a mobile phone model, a software version, an item ID, and so on. In such cases, using one-hot encoding on high cardinality features may lead to several problems:

- The required space is not fixed and can't be controlled
- We need to figure out how to encode a new value

Using hashing instead of one-hot encoding can address these limitations.

To do so, we must use a hash function that converts an input into a controlled output. One well-known hash function is md5. If we apply md5 to some strings, we'll get the following results:

```
from hashlib import md5
print('hashing of "regularization" ->',
    md5(b'regularization').hexdigest())
print('hashing of "regularized" ->',
    md5(b'regularized').hexdigest())
print('hashing of "machine learning" ->',
    md5(b'machine learning').hexdigest())
```

The output will look like this:

```
hashing of "regularization" -> 04ef847b5e35b165c190ced9d91f65da
hashing of "regularized" -> bb02c45d3c38892065ff71198e8d2f89
hashing of "machine learning" -> e04d1bcee667afb8622501b9a4b4654d
```

As we can see, hashing has several interesting properties:

- No matter the input size, the output size is fixed
- Two similar inputs may lead to very different outputs

These properties allow hash functions to be very effective when used with high cardinality features. All we have to do is this:

1. Choose a hash function.
2. Define the expected space dimension of the output.
3. Encode our feature with that function.

Of course, there are some drawbacks to hashing:

- There may be collisions – two different inputs may have the same output (even if this does not necessarily hurt performance if it's not that severe)
- We may want similar inputs to have similar outputs (a well-chosen hashing function can have such a property)

Required installations

We need to do some preparation for this recipe. Since we will download a Kaggle dataset, first, we need to install the Kaggle API:

1. Install the library with `pip`:

   ```
   pip install kaggle
   ```

2. If you haven't already done so, create a Kaggle account at `www.kaggle.com`.

3. Go to your profile page and create your API token by clicking **Create New API Token**. This should download a `kaggle.json` file to your computer:

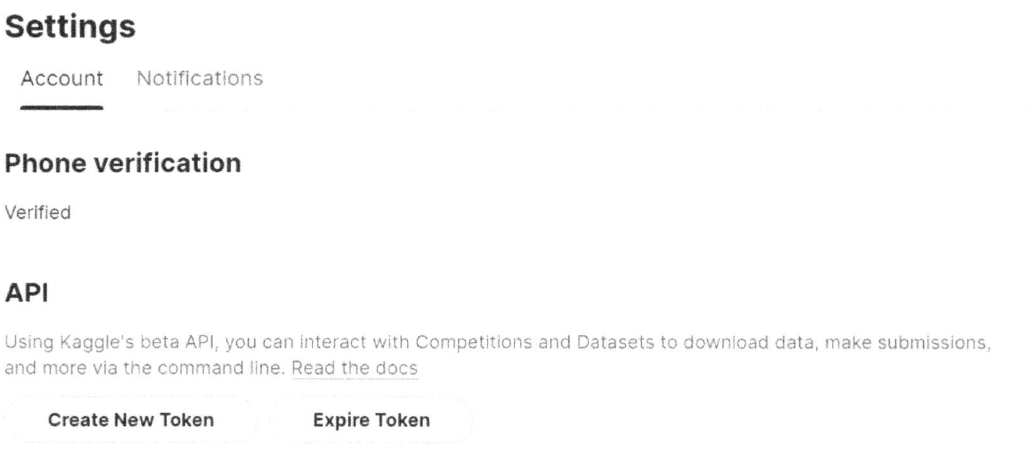

Settings

Account   Notifications

Phone verification

Verified

API

Using Kaggle's beta API, you can interact with Competitions and Datasets to download data, make submissions, and more via the command line. Read the docs

    Create New Token   Expire Token

Figure 5.1 – Screenshot of the Kaggle website

4. You need to move the freshly downloaded kaggle.json file to ~/.kaggle via the following command line:

    ```
    mkdir ~/.kaggle && mv kaggle.json ~/.kaggle/.
    ```

5. You can now download the dataset with the following command line:

    ```
    kaggle datasets download -d reddynitin/aug-train
    ```

6. We should now have a file called aug-train.zip, which contains the data we will use for this recipe. We also need to install the category_encoders, pandas, and sklearn libraries with the following command line:

    ```
    pip install category_encoders pandas scikit-learn
    ```

How to do it...

In this recipe, we will load and quickly prepare the dataset (quickly in the sense that more data preparation could lead to better results), and then apply a logistic regression model to this classification task. On the selected dataset, the `city` feature has 123 possible outcomes, so it can be considered a high cardinality feature. Also, we can fairly assume that the production data could contain more cities, so the hashing trick would make sense here:

1. Import the required modules, functions, and classes: `pandas` for loading the data, `train_test_split` for splitting the data, `StandardScaler` for rescaling quantitative features, `HashingEncoder` for encoding qualitative features, and `LogisticRegression` as the model:

```
Import numpy as np
import pandas as pd
from sklearn.model_selection import train_test_split
from sklearn.preprocessing import StandardScaler, OneHotEncoder
from category_encoders.hashing import HashingEncoder
from sklearn.linear_model import LogisticRegression
```

2. Load the dataset with `pd.read_csv()`. Note that we do not need to unzip the dataset first since the zip only contains one CSV file – `pandas` will take care of that for us:

```
df = pd.read_csv('aug-train.zip')
print('number of unique values for the feature city',
    df['city'].nunique())
```

As we can see, the `city` feature has `123` possible values in the dataset:

number of unique values for the feature city 123

3. Remove any missing data. We take a very brutal policy here: we remove all the features that have a large amount of missing data, and then we remove all the rows with remaining missing data. This is not a recommended approach in general since we lose a lot of potentially useful information. Since dealing with missing data isn't the subject here, we will take this simplistic approach:

```
df = df.drop(columns=['gender', 'major_discipline',
    'company_size', 'company_type'])
df = df.dropna()
```

4. Split the data into training and test sets with the `train_test_split` function:

```
X_train, X_test, y_train, y_test = train_test_split(
    df.drop(columns=['target']), df['target'],
    stratify=df['target'], test_size=0.2,
    random_state=0
)
```

5. Select and rescale any quantitative features. We will use the standard scaler to rescale the selected quantitative features, but any other scaler may work fine too:

```
quanti_feats = ['city_development_index', 'training_hours']
# Instantiate the scaler
scaler = StandardScaler()
# Select quantitative features
X_train_quanti = X_train[quanti_feats]
X_test_quanti = X_test[quanti_feats]
# Rescale quantitative features
X_train_quanti = scaler.fit_transform(X_train_quanti)
X_test_quanti = scaler.transform(X_test_quanti)
```

6. Select and prepare "regular" qualitative features. Here, we will use the one-hot encoder from `scikit-learn`, though we could also apply the hashing trick to those features:

```
quali_feats = ['relevent_experience',
    'enrolled_university', 'education_level',
    'experience', 'last_new_job']
quali_feats = ['last_new_job']
# Instantiate the one hot encoder
encoder = OneHotEncoder()
# Select qualitative features to one hot encode
X_train_quali = X_train[quali_feats]
X_test_quali = X_test[quali_feats]
# Encode those features
X_train_quali = encoder.fit_transform(
    X_train_quali).toarray()
X_test_quali = encoder.transform(
    X_test_quali).toarray()
```

7. Encode the high cardinality `'city'` feature with hashing. Since there are currently 123 possible values for this feature, we could use only 7 bits to encode the whole space of possibilities. This is what is denoted by the `n_components=7` parameter. For safety, we could set it to 8 or more bits, to consider a growing set of possible cities in the data:

```
high_cardinality_feature = ['city']
# Instantiate the hashing encoder
hasher = HashingEncoder(n_components=7)
# Encode the city feature with hashing
X_train_hash = hasher.fit_transform(
    X_train[high_cardinality_feature])
X_test_hash = hasher.fit_transform(
```

```
    X_test[high_cardinality_feature])
# Display the result on the training set
X_train_hash.head()
```

The output will look something like this:

col_0	col_1	col_2	col_3	col_4	col_5	col_6	
18031	1	0	0	0	0	0	0
16295	0	0	0	1	0	0	0
7679	0	0	0	0	0	1	0
18154	0	0	1	0	0	0	0
10843	0	0	0	0	0	1	0

> **Note**
> As we can see, all the values are encoded in seven columns, spanning $2^7 = 128$ possible values.

8. Concatenate all the prepared data:

```
X_train = np.concatenate([X_train_quali,
    X_train_quanti, X_train_hash], 1)
X_test = np.concatenate([X_test_quali,
    X_test_quanti, X_test_hash], 1)
```

9. Instantiate and train the logistic regression model. Here, we will use the default hyperparameters proposed by `scikit-learn` for logistic regression:

```
lr = LogisticRegression()
lr.fit(X_train, y_train)
```

10. Print the accuracy for both the training and test sets using the `.score()` method:

```
print('Accuracy train set:', lr.score(X_train,
    y_train))
print('Accuracy test set:', lr.score(X_test, y_test))
```

The output will look something like this:

```
Accuracy train set: 0.7812087988342239
Accuracy test set: 0.7826810990840966
```

As we can see, we have an accuracy of about 78% on the test set, with no apparent overfitting.

> **Note**
> It is possible that adding some features (for example, with feature engineering) could help improve the model since the model in itself seems to have no room for much improvement based on the fact there is no overfitting.

See also

- The official documentation of the category encoders library: `https://contrib.scikit-learn.org/category_encoders/`
- The category encoders page about hashing: `https://contrib.scikit-learn.org/category_encoders/hashing.html`

Aggregating features

When you're looking at high cardinality features, one possible solution is to reduce the actual cardinality of that feature. Here, aggregating is one possible solution, and it may work very well in some cases. In this recipe, we will explain what aggregating is and discuss when we should use it. Once we've done that, we will apply it.

Getting ready

When dealing with high cardinality features, one-hot encoding leads to high-dimensionality datasets. Because of the so-called curse of dimensionality, the ability for models to generalize properly can be a real issue for one-hot encoded high cardinality features, even with very large training datasets. Thus, aggregating is a way to lower the dimensionality of the one-hot encoding, and then lower the risk of overfitting.

There are several ways to aggregate. Let's, for example, assume that we have a database of clients that contains the "phone model" feature, which consists of many of the possible phone models (that is, hundreds). There could be at least two ways of aggregating such a feature:

- **By occurrence probability**: Any model appearing less than X% in the data is considered as "others"
- **By a given similarity**: We could gather models by generation, brand, or even price

These methods have their pros and cons:

- **Pros**: Aggregating by occurrence is simple, works all the time, and does not require any subject matter knowledge
- **Cons**: Aggregating by a given similarity can be more relevant but requires knowledge about the feature that may not be available, or it could take too long (for example, if there are millions of values)

> **Note**
> Aggregating is also sometimes useful when there is a long tail distribution of values for a feature, which means that some values appear a lot, while many others appear only a small fraction of the time.

In this recipe, we will apply aggregating to a dataset that contains many cities as features but with no information about the city names. This will leave us with the only option being to aggregate by occurrence. We will reuse the same dataset as in the previous recipe, so we will require the Kaggle API. For that, please refer to the previous recipe. Using the Kaggle API, the dataset can be downloaded with the following command:

```
kaggle datasets download -d reddynitin/aug-train
```

We will also need the `pandas` and `scikit-learn` libraries, which can be installed with the following command:

```
pip install pandas scikit-learn.
```

How to do it...

We will use the same dataset as in the previous recipe. To prepare for this recipe, we will aggregate the cities based on a given threshold on their occurrences in the dataset, and then train and evaluate a model on this data:

1. Import the required modules, functions, and classes: `pandas` for loading the data, `train_test_split` for splitting the data, `StandardScaler` for rescaling quantitative features, `OneHotEncoder` for encoding qualitative features, and `LogisticRegression` as the model:

    ```
    Import numpy as np
    import pandas as pd
    from sklearn.model_selection import train_test_split
    from sklearn.preprocessing import OneHotEncoder, StandardScaler
    from sklearn.linear_model import LogisticRegression
    ```

2. Load the dataset with `pandas`. There's no need to unzip the file first – this is all handled by `pandas`:

    ```
    df = pd.read_csv('aug-train.zip')
    ```

3. Remove any missing data. As we did in the previous recipe, we will use a simple policy, remove all features that have a large amount of missing data, and then remove the rows with missing data:

    ```
    df = df.drop(columns=['gender', 'major_discipline',
        'company_size', 'company_type'])
    df = df.dropna()
    ```

4. Split the data into training and test sets with the `train_test_split` function:

    ```
    X_train, X_test, y_train, y_test = train_test_split(
        df.drop(columns=['target']), df['target'],
        stratify=df['target'], test_size=0.2,
        random_state=0
    )
    ```

5. Rescale any quantitative features with the standard scaler provided by `scikit-learn`:

    ```
    quanti_feats = ['city_development_index',
        'training_hours']
    scaler = StandardScaler()
    X_train_quanti = X_train[quanti_feats]
    X_test_quanti = X_test[quanti_feats]
    X_train_quanti = scaler.fit_transform(X_train_quanti)
    X_test_quanti = scaler.transform(X_test_quanti)
    ```

6. Now, we must aggregate the `city` feature:

    ```
    # Get only cities above threshold
    threshold = 0.1
    kept_cities = X_train['city'].value_counts(
        normalize=True)[X_train['city'].value_counts(
        normalize=True) > threshold].index
    # Update all cities below threshold as 'other'
    X_train.loc[~X_train['city'].isin(kept_cities),
        'city'] = 'other'
    X_test.loc[~X_test['city'].isin(kept_cities),
        'city'] = 'other'
    ```

7. Prepare the qualitative features with one-hot encoding, including the newly aggregated `city` feature:

    ```
    # Get qualitative features
    quali_feats = ['city', 'relevent_experience',
        'enrolled_university', 'education_level',
        'experience', 'last_new_job']
    X_train_quali = X_train[quali_feats]
    X_test_quali = X_test[quali_feats]
    # Instantiate the one hot encoder
    encoder = OneHotEncoder()
    # Apply one hot encoding
    X_train_quali = encoder.fit_transform(
    ```

```
        X_train_quali).toarray()
    X_test_quali = encoder.transform(
        X_test_quali).toarray()
```

8. Concatenate the quantitative and qualitative features back together:

```
    X_train = np.concatenate([X_train_quali,
        X_train_quanti], 1)
    X_test = np.concatenate([X_test_quali, X_test_quanti], 1)
```

9. Instantiate and train the logistic regression model. Here, we will just keep the default hyperparameters of the model:

```
    lr = LogisticRegression()
    lr.fit(X_train, y_train)
```

10. Compute and print the model's accuracy on both the training and test sets:

```
    print('Accuracy train set:', lr.score(X_train, y_train))
    print('Accuracy test set:', lr.score(X_test, y_test))
```

The output will look like this:

```
Accuracy train set: 0.7805842759003538

Accuracy test set: 0.774909797391063
```

> **Note**
>
> For this specific case, aggregating did not seem to help much in terms of giving us more robust results, but at the very least it has helped the model be less unpredictable and robust to new cities.

There's more...

Since the aggregation code may have looked complicated, let's take a look at what we did.

So, we have the `city` feature, which has many possible values; there's a frequency for each value in the training set. These can be computed with the `.value_counts(normalize=True)` method:

```
df['city'].value_counts(normalize=True)
```

This will give us the following output:

```
city_103        0.232819
city_21         0.136227
city_16         0.081659
city_114        0.069613
city_160        0.045354
                                   ...
city_111        0.000167
city_129        0.000111
city_8          0.000111
city_140        0.000056
city_171        0.000056
Name: city, Length: 123, dtype: float64
```

It appears that, in the entire dataset, `city_103` is the value more than 23% of the time, while other values such as `city_111` appear less than 1% of the time. We will just apply a threshold to those values so that we get the list of cities appearing more than the given threshold:

```
df['city'].value_counts(normalize=True) > 0.05
```

This will give us the following output:

```
city_103        True
city_21         True
city_16         True
city_114        True
city_160        False
                                   ...
city_111        False
city_129        False
city_8          False
city_140        False
city_171        False
Name: city, Length: 123, dtype: bool
```

Now, all we have to do is get the index (that is, the city name) of all the true values. This is exactly what we can do with the following full line:

```
kept_cities = df['city'].value_counts(normalize=True)[
    df['city'].value_counts(normalize=True) > 0.05].index
kept_cities
```

This shows the following output:

```
Index(['city_103', 'city_21', 'city_16', 'city_114'], dtype='object')
```

As expected, this returns list of cities that occur more than the threshold.

Undersampling an imbalanced dataset

A typical case in machine learning is what we call an imbalanced dataset. An imbalanced dataset simply means that for a given class, some occurrences are much more likely than others, hence the lack of balance. There are plenty of cases of imbalanced datasets: rare diseases in medicine, customer behavior, and more.

In this recipe, we will propose one possible way to handle imbalanced datasets: undersampling. After explaining this process, we will apply it to a credit card fraud detection dataset.

Getting ready

The problem with imbalanced data is that it may bias the results of a machine learning model. Let's assume we're undertaking a classification task of detecting rare diseases present in only 1% of a dataset. A common pitfall with such data is to have a model predicting as always healthy as it would still have 99% accuracy. So, it would be very likely for a machine learning model to minimize its losses.

> **Note**
> In such situations, other metrics such as the F1-score or **ROC Area Under Curve (ROC AUC)** are usually more relevant.

One way to prevent this from happening is to undersample the dataset. More specifically, we can undersample the overrepresented class by removing some samples of it:

- We keep all the samples of the underrepresented class
- We keep only a subsample of the overrepresented class

By doing this, we can artificially balance the dataset and avoid the pitfalls of an imbalanced dataset. For example, let's say we have a dataset composed of the following attributes:

- 100 samples with disease
- 9,900 samples with no disease

A perfectly balanced undersampling would give us the following results in the dataset:

- 100 samples with disease

- 100 randomly selected samples with no disease

Of course, the drawback is that we lose a lot of data in the process.

For this recipe, we first need to download the dataset. To do so, we will use the Kaggle API (refer to the *Hashing high cardinality features* recipe to learn how to install it). The dataset can be downloaded with the following command line:

```
kaggle datasets download -d mlg-ulb/creditcardfraud
```

The following libraries are also needed: `pandas` for loading the data, `scikit-learn` for modeling, `matplotlib` for displaying the data, and `imbalanced-learn` for undersampling. They can be installed with the following command line:

```
pip install pandas scikit-learn matplotlib imbalanced-learn
```

How to do it...

In this recipe, we will apply undersampling to a credit card fraud dataset. This is a rather extreme case of an imbalanced dataset since only about 0.18% of the samples are positive:

1. Import the required modules, classes, and functions:

    - `pandas` for data loading and manipulation

    - `train_test_split` for data splitting

    - `StandardScaler` for data rescaling (the dataset holds only quantitative features)

    - `RandomUnderSampler` for undersampling

    - `LogisticRegression` for modeling

    - `roc_auc_score` for displaying the ROC and ROC AUC computations:

        ```
        import pandas as pd
        import matplotlib.pyplot as plt
        from sklearn.model_selection import train_test_split
        from sklearn.preprocessing import StandardScaler
        from imblearn.under_sampling import RandomUnderSampler
        from sklearn.linear_model import LogisticRegression
        from sklearn.metrics import roc_auc_score
        ```

2. Load the data with `pandas`. We can load the ZIP file directly. We will also display the relative amount of each label: we have around 99.8% of regular transactions compared to less than 0.18% of fraudulent transactions:

```
df = pd.read_csv('creditcardfraud.zip')
df['Class'].value_counts(normalize=True)
```

The output will look like this:

```
0        0.998273
1        0.001727
Name: Class, dtype: float64
```

3. Split the data into training and test sets. The need to stratify the label in such cases can be critical:

```
X_train, X_test, y_train, y_test = train_test_split(
    df.drop(columns=['Class']), df['Class'],
    test_size=0.2, random_state=0,
    stratify=df['Class'])
```

4. Apply random undersampling, up to a 10% sampling strategy. This means we must undersample the overrepresented class until there is a 10 to 1 ratio in the class balance. We could go up to a 1 to 1 ratio, but this would be at the cost of even more dropped data. This ratio is defined by the `sampling_strategy=0.1` parameter. We must also set the random state for reproducibility:

```
# Instantiate the object with a 10% strategy
rus = RandomUnderSampler(sampling_strategy=0.1,
    random_state=0)
# Undersample the train dataset
X_train, y_train = rus.fit_resample(X_train, y_train)
# Check the balance
y_train.value_counts()
```

This gives us the following output:

```
0        3940
1         394
Name: Class, dtype: int64
```

After undersampling, we end up with 3940 regular transaction samples compared to 394 fraudulent transactions.

5. Rescale the data using a standard scaler:

```
# Scale the data
scaler = StandardScaler()
X_train = scaler.fit_transform(X_train)
X_test = scaler.transform(X_test)
```

> **Note**
> Arguably, we could apply rescaling before resampling. This would give more weight to the overrepresented class when rescaling but would not be considered data leakage.

6. Instantiate and train the logistic regression model on the training set:

```
lr = LogisticRegression()
lr.fit(X_train, y_train)
```

7. Compute the ROC AUC on both the training and test sets. To do so, we need the predicted probabilities for each sample, which we can get with the `predict_proba()` method, as well as the imported `roc_auc_score()` function:

```
# Get the probas
y_train_proba = lr.predict_proba(X_train)[:, 1]
y_test_proba = lr.predict_proba(X_test)[:, 1]
# Display the ROC AUC
print('ROC AUC training set:', roc_auc_score(y_train,
    y_train_proba))
print('ROC AUC test set:', roc_auc_score(y_test,
    y_test_proba))
```

This returns the following:

```
ROC AUC training set: 0.9875041871730784
ROC AUC test set: 0.9731067071595099
```

We got a ROC AUC of about 97% on the test set and close to 99% on the training set.

There's more...

Optionally, we can plot the ROC curves for both the training and test sets. To do so, we can use the `roc_curve()` function from `scikit-learn`:

```
import matplotlib.pyplot as plt
from sklearn.metrics import roc_curve
# Display the ROC curve
fpr_test, tpr_test, _ = roc_curve(y_test, y_test_proba)
fpr_train, tpr_train, _ = roc_curve(y_train, y_train_proba)
plt.plot(fpr_test, tpr_test, label='test')
plt.plot(fpr_train, tpr_train, label='train')
plt.xlabel('False positive rate')
plt.ylabel('True positive rate')
plt.legend()
plt.show()
```

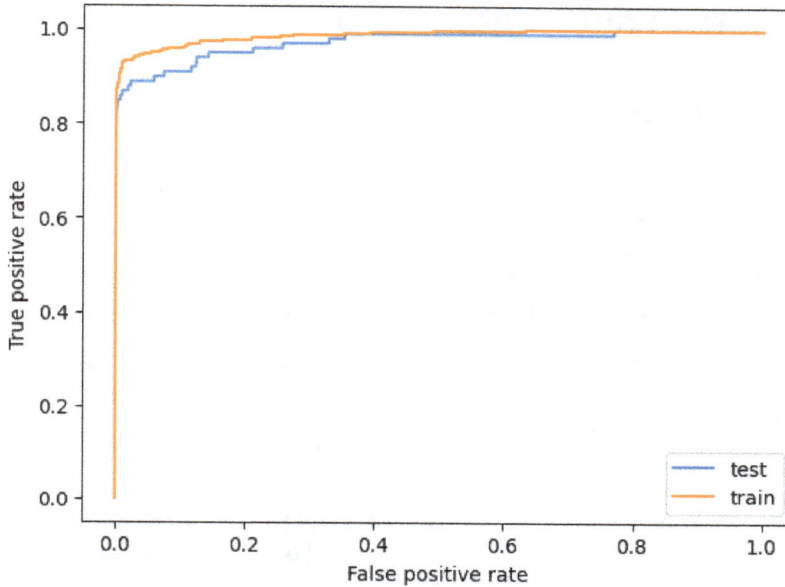

Figure 5.2 – ROC curve for the train and test sets. Plot produced by the code

As we can see, while the ROC AUC is very similar for the training and test sets, the curve for the test set is a bit lower. This means that, as expected, the model is overfitting slightly.

Note that fine-tuning `sampling_strategy` might be helpful to get better results.

> **Note**
> To optimize the sampling strategy and the model hyperparameters at the same time, you can use scikit-learn's `Pipeline` class.

See also

- The documentation for `RandomUnderSampler`: `https://imbalaced-learn.org/stable/references/generated/imblearn.under_sampling.RandomUnderSampler.html`

- The documentation for `Pipeline`: `https://scikit-learn.org/stable/modules/generated/sklearn.pipeline.Pipeline.html`

- Here's a great code example of a two-step pipeline hyperparameter optimization: `https://scikit-learn.org/stable/tutorial/statistical_inference/putting_together.html`

Oversampling an imbalanced dataset

Another solution when dealing with imbalanced datasets is random oversampling. This is the opposite of random undersampling. In this recipe, we'll learn how to use it on the credit card fraud detection dataset.

Getting ready

Random oversampling can be seen as the opposite of random undersampling: the idea is to duplicate samples of the underrepresented dataset to rebalance the dataset.

As for the previous recipe, let's assume a 1%-99% imbalanced dataset that contains the following:

- 100 samples with disease
- 9,900 samples with no disease

To apply oversampling to this dataset using a 1/1 strategy (so, a perfectly balanced dataset), we would need to have 99 duplicates of each sample of the disease class. So, the oversampled dataset would need to contain the following:

- 9,900 samples with disease (100 original samples duplicated 99 times on average)
- 9,900 samples with no disease

We can easily guess the pros and cons of this method:

- **Pro**: Unlike undersampling, we do not waste any data from the overrepresented class, which means our model can be trained on the full picture of the data we have
- **Con**: We may have a lot of duplicates of the underrepresented class, leading to potential overfitting on this data

Fortunately, we can choose a rebalancing strategy below 1/1 so that the underrepresented data duplication can be limited.

For this recipe, we need to download the dataset. If you completed the *Undersampling an imbalanced dataset* recipe, you don't need to do anything else.

Otherwise, using the Kaggle API (refer to the *Hashing high cardinality features* recipe to learn how to install it), we need to download the dataset via the following command line:

```
kaggle datasets download -d mlg-ulb/creditcardfraud
```

The following libraries are also needed: `pandas` for loading the data, `scikit-learn` for modeling, `matplotlib` for displaying the data, and `imbalanced-learn` for the oversampling part. They can be installed via the following command line:

```
pip install pandas scikit-learn matplotlib imbalanced-learn.
```

How to do it...

In this recipe, we will apply oversampling to the credit card fraud dataset:

1. Import the required modules, classes, and functions:

   - `pandas` for data loading and manipulation
   - `train_test_split` for data splitting
   - `StandardScaler` for data rescaling (the dataset only contains quantitative features)
   - `RandomOverSampler` for oversampling
   - `LogisticRegression` for modeling
   - `roc_auc_score` for displaying the ROC and ROC AUC computations:

     ```
     import pandas as pd
     import matplotlib.pyplot as plt
     from sklearn.model_selection import train_test_split
     from sklearn.preprocessing import StandardScaler
     from imblearn.over_sampling import RandomOverSampler
     from sklearn.linear_model import LogisticRegression
     from sklearn.metrics import roc_auc_score
     ```

2. Load the data with pandas. We can load the ZIP file directly. As we did in the previous recipe, we will display the relative amount of each label to remind us that we have about 99.8% of regular transactions and less than 0.18% of fraudulent transactions:

   ```
   df = pd.read_csv('creditcardfraud.zip')
   df['Class'].value_counts(normalize=True)
   ```

   This prints the following:
   ```
   0          0.998273
   1          0.001727
   Name: Class, dtype: float64
   ```

3. Split the data into training and test sets. We must specify stratification on the labels to make sure the balance is still the same:

```
X_train, X_test, y_train, y_test = train_test_split(
    df.drop(columns=['Class']), df['Class'],
    test_size=0.2, random_state=0,
    stratify=df['Class'])
```

4. Apply random oversampling with a 10% sampling strategy. This means that we oversample the underrepresented class until there is a 10 to 1 ratio in the class balance. This ratio is defined by the `sampling_strategy=0.1` parameter. We must also set the random state for reproducibility:

```
# Instantiate the oversampler with a 10% strategy
ros = RandomOverSampler(sampling_strategy=0.1,
    random_state=0)
# Overersample the train dataset
X_train, y_train = ros.fit_resample(X_train, y_train)
# Check the balance
y_train.value_counts()
```

This outputs the following:

```
0           227451
1            22745
Name: Class, dtype: int64
```

After oversampling, we now have `227451` regular transactions in the training set (which is unchanged) versus `22745` fraudulent transactions.

> **Note**
>
> It is possible to change the sampling strategy. As usual, it is a matter of doing a tradeoff: a greater sampling strategy means more duplicated samples for more balance, while a smaller sampling strategy means fewer duplicated samples but less balance.

5. Rescale the data using a standard scaler:

```
# Scale the data
scaler = StandardScaler()
X_train = scaler.fit_transform(X_train)
X_test = scaler.transform(X_test)
```

6. Instantiate and train the logistic regression model on the training set:

```
lr = LogisticRegression()
lr.fit(X_train, y_train)
```

7. Compute the ROC AUC on both the training and test sets. To do so, we need the predicted probabilities for each sample, which we can get by using the predict_proba() method, as well as the imported roc_auc_score() function:

```
# Get the probas
y_train_proba = lr.predict_proba(X_train)[:, 1]
y_test_proba = lr.predict_proba(X_test)[:, 1]
# Display the ROC AUC
print('ROC AUC training set:', roc_auc_score(y_train,
    y_train_proba))
print('ROC AUC test set:', roc_auc_score(y_test,
    y_test_proba))
```

This returns the following:

```
ROC AUC training set: 0.9884952360756659
ROC AUC test set: 0.9721115830969416
```

The results are quite comparable to the ones we obtained with undersampling. However, this does not mean that these two techniques are always equal.

There's more...

Optionally, just like we did in the *Undersampling an imbalanced dataset* recipe, we can plot the ROC curves for both the training and test sets using the roc_curve() function from scikit-learn:

```
import matplotlib.pyplot as plt
from sklearn.metrics import roc_curve
# Display the ROC curve
fpr_test, tpr_test, _ = roc_curve(y_test, y_test_proba)
fpr_train, tpr_train, _ = roc_curve(y_train, y_train_proba)
plt.plot(fpr_test, tpr_test, label='test')
plt.plot(fpr_train, tpr_train, label='train')
plt.xlabel('False positive rate')
plt.ylabel('True positive rate')
plt.legend()
plt.show()
```

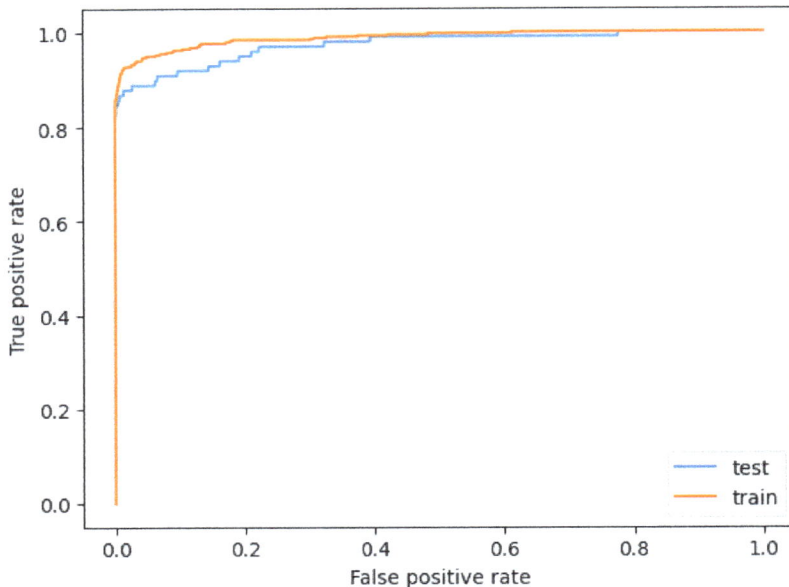

Figure 5.3 – ROC curve for the train and test sets. Plot produced by the code

In this case, the ROC AUC curve for the test set is clearly below the one for the training set, which means that the model is overfitting slightly.

See also

The documentation for `RandomUnderSampler` is available at `https://imbalanced-learn.org/stable/references/generated/imblearn.under_sampling.RandomUnderSampler.html`.

Resampling imbalanced data with SMOTE

Finally, a more complex solution for dealing with imbalanced datasets is a method called SMOTE. After explaining the SMOTE algorithm, we will apply this method to the credit card fraud detection dataset.

Getting ready

SMOTE stands for **Synthetic Minority Oversampling TEchnique**. As its name suggests, it creates synthetic samples for an underrepresented class. But how exactly does it create synthetic data?

This method uses the k-NN algorithm on the underrepresented class. The SMOTE algorithm can be summarized with the following steps:

1. Randomly pick a sample, x_i, in the minority class.

2. Using k-NN, randomly pick one of the k-nearest neighbors of x_i in the minority class. Let's call this sample x_{zi}.

3. Compute the new synthetic sample, $x_{new} = x_i + \lambda(x_{zi} - x_i)$, with λ being randomly drawn in the [0, 1] range:

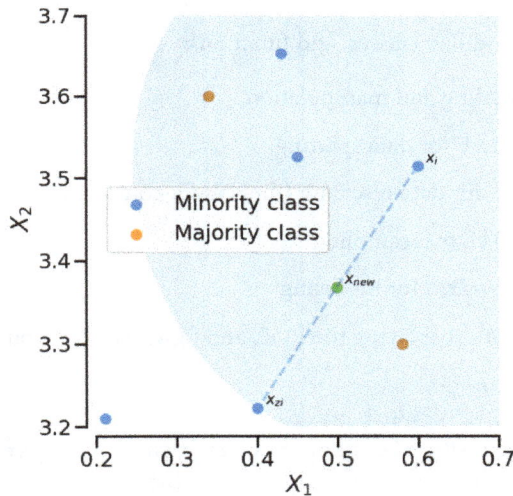

Figure 5.4 – Visual representation of SMOTE

Compared to random oversampling, this method is more complex since it has one hyperparameter: the number of nearest neighbors, k, to consider. This method also comes with pros and cons:

- **Pro**: Unlike random oversampling, it limits the risks of overfitting on the underrepresented class since samples are not duplicated

- **Con**: Creating synthetic data is a risky bet; nothing assures you that it has a meaning and would ever be likely on real data

To complete this recipe, you will need to download the credit card fraud dataset if you haven't done so already (check out the *Undersampling an imbalanced dataset* or *Oversampling an imbalanced dataset* recipe to learn how to do this).

Using the Kaggle API (refer to the *Hashing high cardinality features* recipe to learn how to install it), we have to download the dataset via the following command line:

```
kaggle datasets download -d mlg-ulb/creditcardfraud
```

The following libraries are needed: `pandas` for loading the data, `scikit-learn` for modeling, `matplotlib` for displaying the data, and `imbalanced-learn` for undersampling. They can be installed via the following command line:

```
pip install pandas scikit-learn matplotlib imbalanced-learn.
```

How to do it...

In this recipe, we will apply SMOTE to the credit card fraud dataset:

1. Import the required modules, classes, and functions:

    -   `pandas` for data loading and manipulation

    -   `train_test_split` for data splitting

    -   `StandardScaler` for data rescaling (the dataset contains only quantitative features)

    -   SMOTE for the SMOTE oversampling

    -   `LogisticRegression` for modeling

    -   `roc_auc_score` for displaying the ROC and ROC AUC computations:

        ```
        import pandas as pd
        import matplotlib.pyplot as plt
        from sklearn.model_selection import train_test_split
        from sklearn.preprocessing import StandardScaler
        from imblearn.over_sampling import SMOTE
        from sklearn.linear_model import LogisticRegression
        from sklearn.metrics import roc_auc_score
        ```

2. Load the data with `pandas`. We can load the ZIP file directly. As we did in the previous two recipes, we will display the relative amount of each label. Again, we will have about 99.8% of regular transactions and less than 0.18% of fraudulent transactions:

    ```
    df = pd.read_csv('creditcardfraud.zip')
    df['Class'].value_counts(normalize=True)
    ```

    The output will look like this:

    ```
    0           0.998273
    1           0.001727
    Name: Class, dtype: float64
    ```

3. Split the data into training and test sets. We must specify stratification on the labels to make sure the balance is still the same:

```
X_train, X_test, y_train, y_test = train_test_split(
    df.drop(columns=['Class']), df['Class'],
    test_size=0.2, random_state=0,
    stratify=df['Class'])
```

4. Apply SMOTE with a 10% sampling strategy with the `sampling_strategy=0.1` parameter. By doing this, we will generate synthetic data of the underrepresented class until there is a 10 to 1 ratio in the class balance. We must also set the random state for reproducibility:

```
# Instantiate the SLOT with a 10% strategy
smote = SMOTE(sampling_strategy=0.1, random_state=0)
# Overersample the train dataset
X_train, y_train = smote.fit_resample(X_train,
    y_train)
# Check the balance
y_train.value_counts()
```

With this, we will get the following output:

```
0           227451
1            22745
Name: Class, dtype: int64
```

After oversampling, we now have 227451 regular transactions in the training set (which is unchanged) versus 22745 fraudulent transactions, including many synthetically generated samples.

5. Rescale the data using a standard scaler:

```
# Scale the data
scaler = StandardScaler()
X_train = scaler.fit_transform(X_train)
X_test = scaler.transform(X_test)
```

6. Instantiate and train the logistic regression model on the training set:

```
lr = LogisticRegression()
lr.fit(X_train, y_train)
```

7. Compute the ROC AUC on both the training and test sets. To do so, we need the predicted probabilities for each sample, which we can get by using the `predict_proba()` method, as well as the imported `roc_auc_score()` function:

```
# Get the probas
y_train_proba = lr.predict_proba(X_train)[:, 1]
y_test_proba = lr.predict_proba(X_test)[:, 1]
```

```
# Display the ROC AUC
print('ROC AUC training set:', roc_auc_score(y_train,
    y_train_proba))
print('ROC AUC test set:', roc_auc_score(y_test,
    y_test_proba))
```

Now, the output should be as follows:

ROC AUC training set: 0.9968657635906649
ROC AUC test set: 0.9711737923925902

The results are slightly different from the ones we got for random undersampling and oversampling. While the performances on the test set are quite similar, there seems to be more overfitting in this case. There are several possible explanations for such results, with one of them being that the synthetic samples were not very helpful for the model.

> **Note**
>
> The results that we got for the resampling strategies on this dataset are not necessarily representative of the results we would get on any other dataset. Moreover, we had to fine-tune the sampling strategies and models to get a proper performance comparison.

There's more...

Optionally, we can plot the ROC curves for both the training and test sets using the `roc_curve()` function from `scikit-learn`:

```
import matplotlib.pyplot as plt
from sklearn.metrics import roc_curve
# Display the ROC curve
fpr_test, tpr_test, _ = roc_curve(y_test, y_test_proba)
fpr_train, tpr_train, _ = roc_curve(y_train, y_train_proba)
plt.plot(fpr_test, tpr_test, label='test')
plt.plot(fpr_train, tpr_train, label='train')
plt.xlabel('False positive rate')
plt.ylabel('True positive rate')
plt.legend()
plt.show()
```

Here is the plot for it:

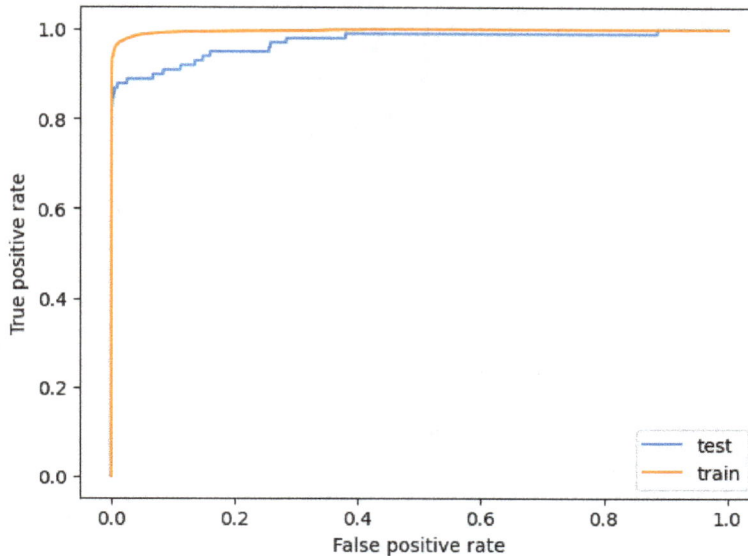

Figure 5.5 – ROC curve for the train and test sets after using SMOTE

Compared to random undersampling and oversampling, overfitting appears to be even clearer here.

See also

The official documentation for SMOTE can be found at `https://imbalanced-learn.org/stable/references/generated/imblearn.over_sampling.SMOTE.html`.

It is not recommended that you apply this implementation to categorical features as it assumes that a feature value for a sample can be any linear combination of values of other samples. This is not true for categorical features.

Working implementations for categorical features have also been proposed, including the following:

- **SMOTENC**: For working with datasets that contain both categorical and non-categorical features: `https://imbalanced-learn.org/stable/references/generated/imblearn.over_sampling.SMOTENC.html`

- **SMOTEN**: For working with datasets that contain only categorical features: `https://imbalanced-learn.org/stable/references/generated/imblearn.over_sampling.SMOTEN.html`

6

Deep Learning Reminders

Deep learning is the specific domain of machine learning based on neural networks. Deep learning is known to be particularly powerful with unstructured data, such as text, audio, and image, but can be useful for time series and structured data too. In this chapter, we will review the basics of deep learning, from a perceptron to training a neural network. We will provide recipes for training neural networks for three main use cases: regression, binary classification, and multiclass classification.

In this chapter, we'll cover the following recipes:

- Training a perceptron
- Training a neural network for regression
- Training a neural network for binary classification
- Training a multiclass classification neural network

Technical requirements

In this chapter, you will train a perceptron, as well as several neural networks. To do so, the following libraries are required:

- NumPy
- pandas
- scikit-learn
- PyTorch
- torchvision

Training a perceptron

The perceptron is arguably the building block of deep learning. Even if the perceptron is not directly used in production systems, understanding what it is can be an asset for building a strong foundation in deep learning.

In this recipe, we will review what a perceptron is and then train one using scikit-learn on the Iris dataset.

Getting started

The perceptron is a machine learning method first proposed to mimic a biological neuron. It was first proposed in the 1940s and then implemented in the 1950s.

From a high-level point of view, a neuron can be described as a cell that receives input signals and fires a signal itself when the sum of the input signals is above a given threshold. This is exactly what a perceptron does; all you have to do is the following:

- Replace the input signals with features
- Apply a weighted sum to those features and apply an activation function to it
- Replace the output signal with a prediction

More formally, assuming n input features   and n weights w^i, the output $\hat{y}$ of a perceptron is the following:

$$\hat{y} = g\left(\sum_i w_i x_i + w_0\right)$$

Where w_0 is the bias and g is the activation function; historically, this is the step function, which returns 1 for a positive input value, 0 otherwise. So, at the end, for n input features, a perceptron is made of $n+1$ parameters: one parameter per feature plus the bias.

> **Tip**
> The step function is also called the **Heaviside function** and is widely used in other fields, such as physics.

The perceptron forward computation is summarized in *Figure 6.1*. As you can see, given a list of features x_i and weights w_i, the forward computation is just the weighted sum, to which an activation function is applied.

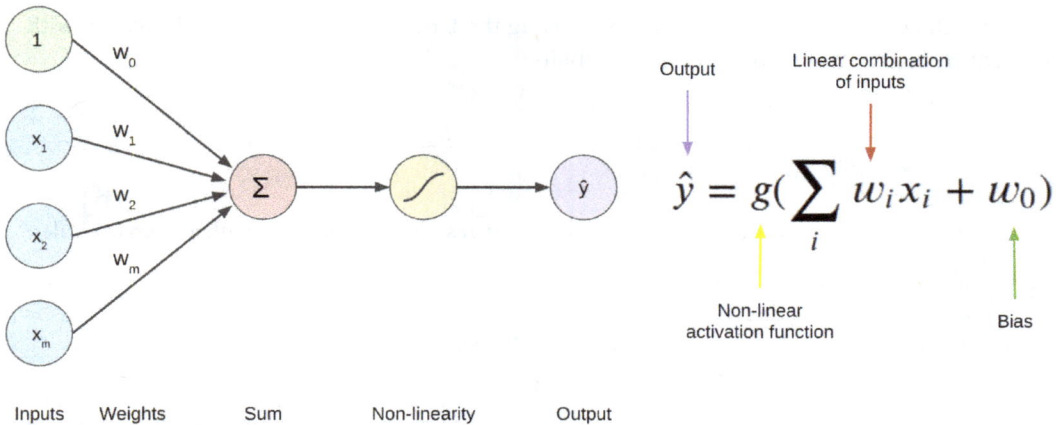

$$\hat{y} = g(\sum_i w_i x_i + w_0)$$

Figure 6.1 – A mathematical representation of a perceptron: from input features to output, through weights and the activation function

On the practical side, scikit-learn is the only thing required for installation for this recipe. It can be installed with the `pip install scikit-learn` command.

How to do it...

We will use the Iris dataset again since the perceptron does not really perform well on complex classification tasks:

1. Make the required imports from scikit-learn:

   - `load_iris`: A function to load the dataset
   - `train_test_split`: A function to split the data
   - `StandardScaler`: A class allowing us to rescale the data
   - `Perceptron`: The class containing the implementation of the perceptron:

     ```
     from sklearn.datasets import load_iris
     from sklearn.model_selection import train_test_split
     from sklearn.preprocessing import StandardScaler
     from sklearn.linear_model import Perceptron
     ```

2. Load the Iris dataset:

   ```
   # Load the Iris dataset
   X, y = load_iris(return_X_y=True)
   ```

3. Split the data into training and test sets using the `train_test_split` function, with `random_state` set to 0 for reproducibility:

```
# Split the data
X_train, X_test, y_train, y_test = train_test_split(
    X, y, random_state=0)
```

4. Since all the features are quantitative here, we simply rescale all the features with a standard scaler:

```
# Rescale the data
scaler = StandardScaler()
X_train = scaler.fit_transform(X_train)
X_test = scaler.transform(X_test)
```

5. Instantiate the model with the default parameters and fit it on the training set with the `.fit()` method:

```
perc = Perceptron()perc.fit(X_train, y_train)
```

6. Evaluate the model on both the training and test sets with the `.score()` method of the `LinearRegression` class, providing the accuracy score:

```
# Print the R2-score on train and test
print('R2-score on train set:',
    perc.score(X_train, y_train))
print('R2-score on test set:',
    perc.score(X_test, y_test))
```

Here is the output:

```
R2-score on train set: 0.9285714285714286
R2-score on test set: 0.8421052631578947
```

7. Out of curiosity, we can have a look at the weights in `.coef_` and the bias in `.intercept_`.

```
print('weights:', perc.coef_)
print('bias:', perc.intercept_)
```

Here is the output:

```
weights: [[-0.49201984  2.77164495 -3.07208498 -2.51124259]
 [ 0.41482008 -1.94508614  3.2852582  -2.60994774]
 [-0.32320969  0.48524348  5.73376173  4.93525738]] bias:
[-2. -3. -6.]
```

> **Important note**
> There are three sets of four weights and one bias, since scikit-learn handles on its own the One-vs-Rest multiclass classification, so we have one perceptron per class.

There's more...

The perceptron is not only a machine learning model. It can be used to simulate logical gates: OR, AND, NOR, NAND, and XOR. Let's have a look.

We can easily implement a forward propagation for the perceptron with the following code:

```
import numpy as np
class LogicalGatePerceptron:
    def __init__(self, weights: np.array, bias: float):
        self.weights = weights
        self.bias = bias
    def forward(self, X: np.array) -> int:
        return (np.dot(
            X, self.weights) + self.bias > 0).astype(int)
```

This code does not consider many edge cases, but is used here simply to explain and demonstrate simple concepts.

The AND gate has the inputs and expected outputs defined in the following truth table:

Input 1	Input 2	Output
0	0	0
0	1	0
1	0	0
1	1	1

Table 6.1 – AND gate truth table

Let's reproduce this data with an array X that has two features (input 1 and input 2) and four samples, and an array y with the expected outputs:

```
# Define X and y
X = np.array([[0, 0], [0, 1], [1, 0], [1, 1]])
y = [0, 0, 0, 1]
```

We can now find a set of weights and bias that will allow the perceptron to act as an AND gate, and check the results to see whether it's actually working:

```
gate = LogicalGatePerceptron(np.array([1, 1]), -1)
y_pred = gate.forward(X)
print('Error:', (y - y_pred).sum())
```

Here is the output:

```
Error: 0
```

With the same logic, most basic logic gates can be created out of a perceptron:

- AND gate: weights [1, 1] and bias -1

- OR gate: weights [1, 1] and bias 0

- NOR gate: weights [-1, -1] and bias 1

- NAND gate: weights [-1, -1] and bias 2

- XOR gate: this requires two perceptrons

> **Tip**
> You can guess the weights and bias with a trial-and-error approach. But you can also use the truth table of a logic gate to make an educated guess or even to solve a set of equations.

This means that using perceptrons, any logic function can be computed.

See also

The official documentation of the scikit-learn implementation: https://scikit-learn.org/stable/modules/generated/sklearn.linear_model.Perceptron.html.

Training a neural network for regression

A perceptron is not a powerful and commonly used machine learning model. But having many perceptrons employed together in a neural network can become a powerful machine learning model. In this recipe, we will review a simple neural network, sometimes called a **multi-layer perceptron** or **vanilla neural network**. And we will then train such a neural network on a regression task on the California housing dataset with PyTorch, a widely used framework in deep learning.

Getting started

Let's start by reviewing what a neural network is, and how to feed forward a neural network from input features.

A neural network can be divided into three parts:

- **The input layer**, containing the input features
- **The hidden layers**, which can be any number of layers and units
- **The output layer**, which is defined by the expected output of the neural network

In both the hidden and output layers, we consider each unit (or neuron) to be a perceptron, with its own weights and bias.

These three parts are well represented in *Figure 6.2*.

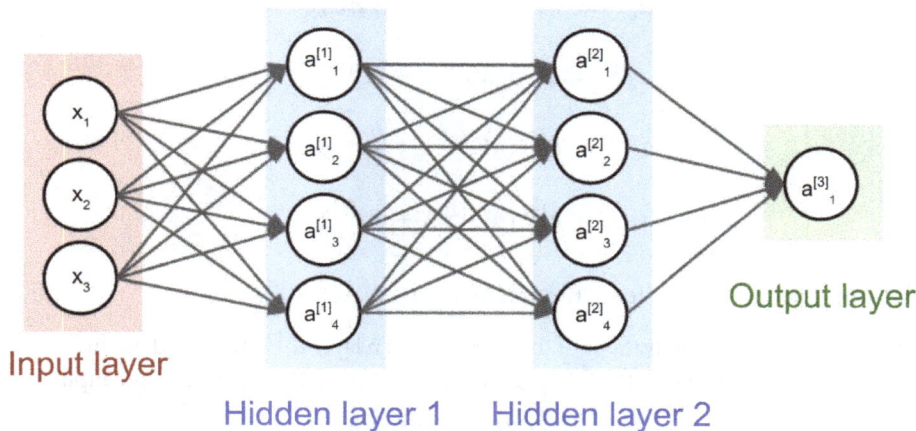

Figure 6.2 – A typical representation of a neural network: on the left the input layer, in the middle the hidden layers, on the right the output layer

We will note the input features $X = x_1, x_2,.... x_N, a_i^{[l]}$, the activation of the unit i of the layer l, and $W_i^{[l]}$, the weights of the unit i of the layer l.

> **Tip**
>
> We consider a neural network to involve deep learning if there is at least one hidden layer.

Training a neural network in regression is not so different from training a linear regression. It is made of the same ingredients:

- Forward propagation, from input features and weights to a prediction
- A loss function to minimize
- An algorithm to update the weights

Let's have a look at those ingredients.

Forward propagation

The forward propagation is what allows to compute and output from the input features. It must be computed from left to right, from the input layer (the input features) to the output layer (the output prediction). Each unit being a perceptron, the first hidden layer is fairly easy to compute:

$$a_1^{[1]} = g^{[1]} (W_1^{[1]} X + b_1^{[1]})$$

$$a_2^{[1]} = g^{[1]} (W_2^{[1]} X + b_2^{[1]})$$

$$a_3^{[1]} = g^{[1]} (W_3^{[1]} X + b_3^{[1]})$$

$$a_4^{[1]} = g^{[1]} (W_4^{[1]} X + b_4^{[1]})$$

Where $bi^{[1]}$ is the bias, and $g^{[1]}$ is the activation function of layer 1.

Now, if we want to compute the activations of the second hidden layer $a_i^{[2]}$, we would use the exact same formulas, but with the activations of the first hidden layer as input ($a^{[1]}$), instead of the input features:

$$a_1^{[2]} = g^{[2]} (W_1^{[2]} . a^{[1]} + b_1^{[2]})$$

$$a_2^{[2]} = g^{[2]} (W_2^{[2]} . a^{[1]} + b_2^{[2]})$$

$$a_3^{[2]} = g^{[2]} (W_3^{[2]} . a^{[1]} + b_3^{[2]})$$

$$a_4^{[2]} = g^{[2]} (W_4^{[2]} . a^{[1]} + b_4^{[2]})$$

> Tip
>
> You can easily generalize to any number of hidden layers and any number of units per layer – the principle remains the same.

Finally, the output layer would be computed in exactly the same way, except in this case we have only one output neuron:

$$a_1^{[3]} = g^{[3]} (W_1^{[3]} \cdot a^{[2]} + b_1^{[3]})$$

One interesting thing to underline: the activation function is also layer dependent, meaning that each layer can have a different activation function. This is particularly critical for the output layer, which needs a specific output function depending on the task and expected output.

For a regression task, it is common to have a linear activation function, allowing the output values of the neural network to be any number.

> **Tip**
>
> The activation function plays a decisive role in neural networks: it adds non-linearity. If we have only linear activation functions for hidden layers, no matter the number of layers, it is equivalent to having no hidden layer.

Loss function

The loss function in a regression task can be the same as in a linear regression: the mean squared error. In our example, if we consider the prediction $\hat{y}$ to be our output value $\hat{y} = a_1^{[3]}$, the loss L is simply the following:

$$L = \frac{1}{2m} \sum_j (y^{(j)} - \hat{y}^{(j)})^2$$

Assuming j is the sample index.

Updating the weights

Updating the weights is done by trying to minimize the loss function. Again, this is almost the same as in a linear regression. The tricky part is that, unlike linear regression, we have several layers of units with weights and biases that all need to be updated. This is where the so-called backpropagation allows the updating of each layer step by step, from the rightmost to the leftmost (following the convention in *Figure 6.2*).

The details of the backpropagation, although useful and interesting, are beyond this book's scope.

Also, just like there are several algorithms to optimize the weights in a logistic regression (the `solver` parameter in scikit-learn's `LogisticRegression`), there are several algorithms to train a neural network. They are commonly called **optimizers**. Among the most frequently used are the **Stochastic Gradient Descent (SGD)** and **Adaptive momentum (Adam)**.

PyTorch

PyTorch is a widely used framework for deep learning, allowing us to easily train and reuse deep learning models.

It is fairly easy to use and can be easily installed with the following command:

```
pip install torch
```

For this recipe, we will also need scikit-learn and matplotlib, which can be installed with `pip install scikit-learn matplotlib`.

How to do it...

In this recipe, we will build and train a neural network on the California housing dataset:

1. First, we need the required imports. Among the imports are some from scikit-learn that we have already used in this book:

    *   `fetch_california_housing` to load the dataset
    *   `train_test_split` to split the data into training and test sets
    *   `StandardScaler` to rescale the quantitative data
    *   `r2_score` to evaluate the model

2. For display purposes, we also import matplotlib:

    ```
    from sklearn.datasets import fetch_california_housing
    from sklearn.model_selection import train_test_split
    from sklearn.preprocessing import StandardScaler
    from sklearn.metrics import r2_score
    import matplotlib.pyplot as plt
    ```

3. We also need some imports from torch:

    *   `torch` itself for some functions at the lower level of the library
    *   `torch.nn` containing many useful classes for building a neural network
    *   `torch.nn.functional` for some useful functions
    *   `Dataset` and `DataLoader` for handling the data operations:

    ```
    import torch
    import torch.nn as nn
    import torch.nn.functional as F
    from torch.utils.data import Dataset, DataLoader
    ```

4. We need to load the data using the `fetch_california_housing` function and return the features and labels:

```
X, y = fetch_california_housing(return_X_y=True)
```

5. We can then split the data into training and test sets using the `train_test_split` function. We set a test size of 20% and a random state for reproducibility:

```
X_train, X_test, y_train, y_test = train_test_split(
    X.astype(np.float32), y.astype(np.float32),
        test_size=0.2, random_state=0)
```

6. We can now rescale the data with a standard scaler:

```
scaler = StandardScaler()
X_train = scaler.fit_transform(X_train)
X_test = scaler.transform(X_test)
```

> **Important note**
>
> Note that we convert the *X* and *y* variables to float32 variables. This is necessary to prevent later troubles with PyTorch not properly handling float64 variables.

7. For PyTorch, we need to create the dataset class. Nothing complicated here though; this class requires only the following to work properly:

   - It has to inherit from the `Dataset` class (imported earlier)
   - It has to have a constructor (`__init__` method) that deals with (and optionally prepares) the data
   - It has to have a `__len__` method, so that the number of samples can be fetched
   - It has to have a `__getitem__` method, in order to get X and y for any given index

   Let's implement this for the California dataset, and let's call our class `CaliforniaDataset`:

```
class CaliforniaDataset(Dataset):
    def __init__(self, X: np.array, y: np.array):
        self.X = torch.from_numpy(X)
        self.y = torch.from_numpy(y)
    def __len__(self) -> int:
        return len(self.X)
    def __getitem__(self, idx: int) -> tuple[torch.Tensor]:
        return self.X[idx], self.y[idx]
```

If we break this class down, we have the following functions:

- The `init` constructor simply converts X and y to torch tensors with the `torch.from_numpy` function and stores the results as class attributes

- The `len` method just returns the length of the X attribute; it would work equally using the length of the y attribute

- The `getitem` method simply returns a tuple with the given item `idx` of the X and y tensors

This is quite straightforward, and will then allow `pytorch` to know what the data is, how many samples are in the dataset, and what the sample `i` is. For that, we will need to instantiate a `DataLoader` class.

> **Tip**
> Rescaling can also be computed in this `CaliforniaDataset` class, as well as any preprocessing.

8. Now we instantiate the `CaliforniaDataset` objects for the training and test datasets. Then we instantiate the associated loaders using the imported `DataLoader` class:

```
# Instantiate datasets
training_data = CaliforniaDataset(X_train, y_train)
test_data = CaliforniaDataset(X_test, y_test)
# Instantiate data loaders
train_dataloader = DataLoader(training_data,
    batch_size=64, shuffle=True)
test_dataloader = DataLoader(test_data, batch_size=64,
    shuffle=True)
```

The data loader instances have a couple of options available. Here, we specify the following:

- `batch_size`: The batch size for training. It may have an impact on the final results.

- `shuffle`: Determines whether to shuffle the data at each epoch.

9. We can finally create the neural network model class. For this class, we only need to fill in two methods:

- The constructor with whatever is useful, such as parameters and attributes

- The `forward` method that computes the forward propagation:

```
class Net(nn.Module):
    def __init__(self, input_shape: int,
        hidden_units: int = 24):
            super(Net, self).__init__()
            self.hidden_units = hidden_units
            self.fc1 = nn.Linear(input_shape,
```

```
                        self.hidden_units)
            self.fc2 = nn.Linear(self.hidden_units,
                self.hidden_units)
            self.output = nn.Linear(self.hidden_units,
                1)
    def forward(self,
        x: torch.Tensor) -> torch.Tensor:
            x = self.fc1(x)
            x = F.relu(x)
            x = self.fc2(x)
            x = F.relu(x)
            output = self.output(x)
            return output
```

If we break it down, we have designed a class that takes two input parameters:

- input_shape is the input shape of the neural networks – this is basically the number of features in the dataset

- hidden_units is the number of units in the hidden layers, which defaults to 24

The neural network itself comprises the following:

- Two hidden layers of hidden_units units with ReLU activation functions

- One output layer of one unit, since we need to predict one value

> **Important note**
>
> More context about ReLU and other activation functions will be given in the next *There's more* subsection.

10. We can now instantiate a neural network and test it on random data of the expected shape to check whether the forward method is working properly:

```
# Instantiate the network
net = Net(X_train.shape[1])
# Generate one random sample of 8 features
random_data = torch.rand((1, X_train.shape[1]))
# Compute the forward
propagationprint(net(random_data))
```

We'll get this output:

```
tensor([[-0.0003]], grad_fn=<AddmmBackward0>)
```

As we can see, the computation of the forward propagation on the random data worked well and returns one single value, as expected. Any error in that step would mean we did something wrong.

11. Before being able to train the neural network on the data, we need to define the loss function and the optimizer. Fortunately, the mean squared error is already implemented and available as nn.MSELoss(). There are plenty of optimizers available; we decided to use Adam here, but other optimizers can also be tested:

```
criterion = nn.MSELoss()
optimizer = torch.optim.Adam(net.parameters(), lr=0.001)
```

> **Important note**
> The optimizer needs the network parameters as input to its constructor.

12. Finally, we can train the neural networks on 10 epochs with the following piece of code:

```
losses = []
# Loop over the dataset multiple times
for epoch in range(10):
    # Reset the loss for this epoch
    running_loss = 0.
    For I, data in enumerate(train_dataloader, 0):
        # Get the inputs per batch: data is a list of [inputs,
labels]
        inputs, labels = data
        # Zero the parameter gradients
        optimizer.zero_grad()
        # Forward propagate + backward + optimize
        outputs = net(inputs)
        # Unsqueeze for dimension matching
        labels = labels.unsqueeze(1)
        # Compute the loss
        Loss = criterion(outputs, labels)
        # Backpropagate and update the weights
        loss.backward()
        optimizer.step()
        # Add this loss to the running loss
        running_loss += loss.item()
     # Compute the loss for this epoch and add it to the list
    epoch_loss = running_loss / len(
        train_dataloader)
    losses.append(epoch_loss)
    # Print the epoch and training loss
    print(f'[epoch {epoch + 1}] loss: {
        epoch_loss:.3f}')print('Finished Training')
```

Hopefully, the comments are self-explanatory. Basically, there are two nested loops:

- One outer loop over the epochs: the number of times the model is trained over the whole dataset
- One inner loop over the samples: for each step, a batch of `batch_size` samples is used to train the model

For each step in the inner loop, we have the following main steps:

- Get a batch of the data: both features and labels
- Forward propagate on this data and get output predictions
- Compute the loss: the mean squared error between predictions and labels
- Update the weights of the network with backpropagation

At the end of each step, we print the loss, which hopefully decreases with each epoch.

13. We can plot the loss as a function of the epoch. This is quite visual and lets us ensure the network is learning if the loss is decreasing:

```
plt.plot(losses)
plt.xlabel('epoch')
plt.ylabel('loss (MSE)')plt.show()
```

Here is the resulting graph:

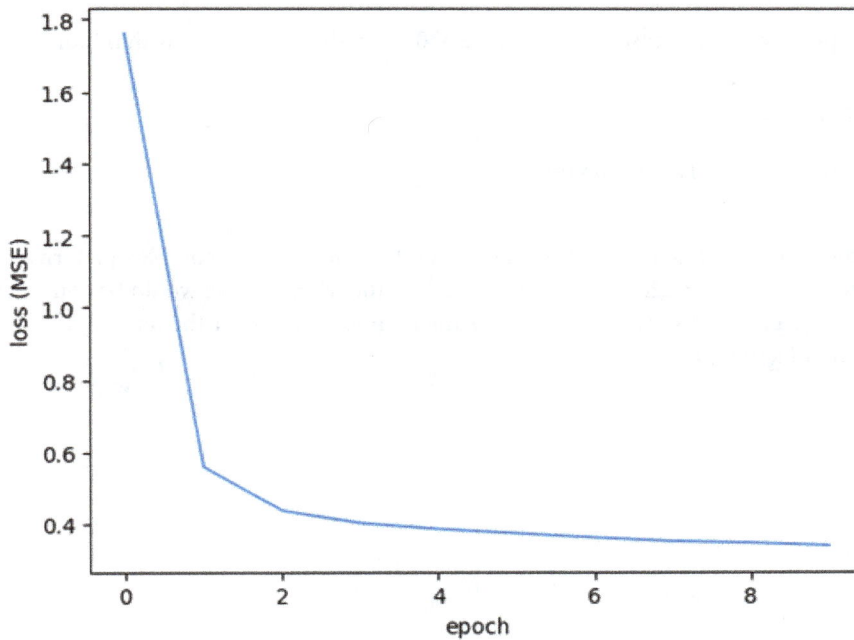

Figure 6.3 – Resulting MSE loss as a function of the epoch

> **Important note**
> We could also keep track of the loss on the test set and display it at this step for more information. We will do that in the next recipe to avoid being drowned in too much information.

14. We can finally evaluate the model on both the training and test sets. As we did previously in this book with regression tasks, we will use the R2-score. Any other relevant metric can be used too:

```
# Compute the predictions with the trained neural
Network
y_train_pred = net(torch.tensor((
    X_train))).detach().numpy()
y_test_pred = net(torch.tensor((
    X_test))).detach().numpy()
# Compute the R2-score
print('R2-score on training set:',
    r2_score(y_train, y_train_pred))
print('R2-score on test set:',
    r2_score(y_test, y_test_pred))
```

Here's the output:

```
R2-score on training set: 0.7542622050620708 R2-score on test
set: 0.7401526252651656
```

As we can see here, we have a reasonable R2-score of 0.74 on the training set, with minor overfitting.

There's more...

In this recipe, we mentioned activation functions without really explaining what they are or why they are needed.

Put simply, they add non-linearities, allowing the model to learn more complex patterns. Indeed, if we had a neural network with no activation function, the whole model would be equivalent to a linear model (e.g., a linear regression), no matter the number of layers or the number of units. This is summarized in *Figure 6.4*.

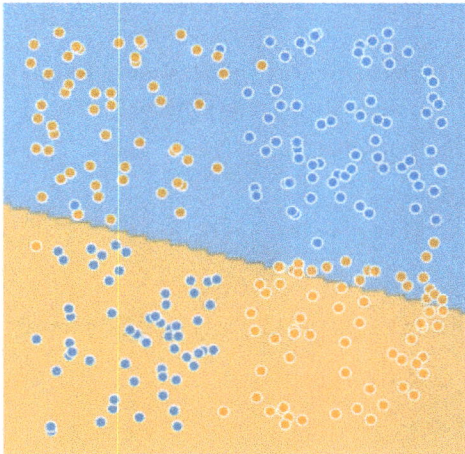

Linear activation function   Sigmoid activation function

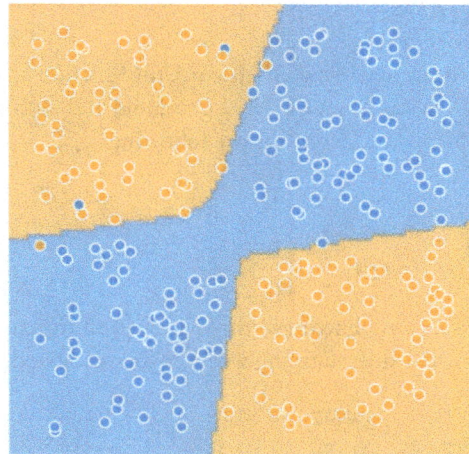

Figure 6.4 – On the left, a neural network with no activation functions can
only learn linearly separable decision functions. On the right, a neural network
with activation functions can learn complex decision functions

There are many available activation functions, but some of the most common ones for hidden layers
are sigmoid, ReLU, and tanh.

Sigmoid

The sigmoid function is the same as that used in logistic regression:

$$sigmoid\ (x) = \frac{1}{1 + e^{-x}}$$

This function's values range from 0 to 1, and outputs 0.5 if $x = 0$.

tanh

The tanh or hyperbolic tangent function ranges from -1 to 1, with a value of 0 if x is 0:

$$tanh(x) = \frac{e^x - e^{-x}}{e^x + e^{-x}}$$

ReLU

The **ReLU** or **Rectified Linear Unit** function just returns 0 for any input negative value, and x for any positive input value x. Unlike sigmoid and tanh, it does not plateau and thus limits vanishing gradient problems. Its formula is the following:

$$ReLU(x) = max(0, x)$$

Visualization

We can visualize these three activation functions (sigmoid, tanh, and ReLU) together for a more intuitive understanding with the following code:

```
import numpy as np
import matplotlib.pyplot as plt
x = np.arange(-2, 2, 0.02)
sigmoid = 1./(1+np.exp(-x))
tanh = (np.exp(x)-np.exp(-x))/(np.exp(x)+np.exp(-x))
relu = np.max([np.zeros(len(x)), x], axis=0)
plt.plot(x, sigmoid)
plt.plot(x, tanh)
plt.plot(x, relu)plt.grid()
plt.xlabel('x')
plt.ylabel('activation')
plt.legend(['sigmoid', 'tanh', 'relu'])
plt.show()
```

You'll get this output upon running the previous code, which computes the output values of these functions in the [-2, 2] range:

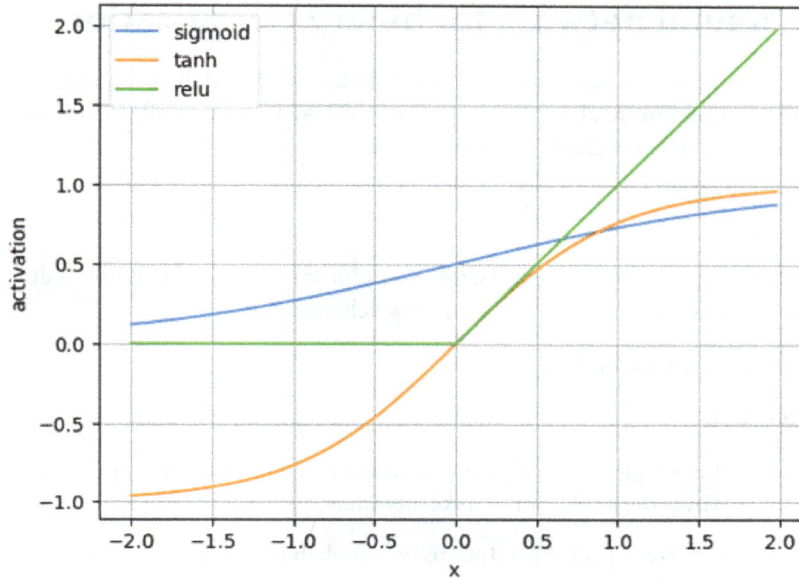

Figure 6.5 – Resulting plot of the sigmoid, tanh, and ReLU activation functions in the [-2, 2] input range

For more about the available activation functions in PyTorch, have a look at the following link: `https://pytorch.org/docs/stable/nn.html#non-linear-activations-weighted-sum-nonlinearity`.

See also

Here are several links to PyTorch tutorials that can be helpful for gaining familiarity with it, along with a deeper understanding of how it works:

- `https://pytorch.org/tutorials/recipes/recipes/defining_a_neural_network.html`

- `https://pytorch.org/tutorials/beginner/blitz/neural_networks_tutorial.html`

- `https://pytorch.org/tutorials/beginner/blitz/cifar10_tutorial.html`

- `https://pytorch.org/tutorials/beginner/basics/data_tutorial.html?highlight=dataset`

And the following link is for a very well-written website about deep learning, for those who wish to have a better understanding of neural networks, gradient descent, and backpropagation: `http://neuralnetworksanddeeplearning.com/`.

Training a neural network for binary classification

In this recipe, let's train our first neural network for a binary classification task on the breast cancer dataset. We will also learn more about the impact of the learning rate and the optimizer on the optimization, as well as how to evaluate the model against the test set.

Getting ready

As we will see in this recipe, training a neural network for binary classification is not so different from training a neural network for regression. Primarily, two changes have to be made:

- The output layer's activation function
- The loss function

In the previous recipe for a regression task, the output layer had no activation function. Indeed, for a regression, one can expect the prediction to take any value.

For a binary classification, we expect the output to be a probability, so a value between 0 and 1, just like the logistic regression. This is why when doing a binary classification, the output layer's activation function is usually the sigmoid function. The resulting predictions will be just like those of a logistic regression: a number on which to apply a threshold (e.g., 0.5) above which we consider the prediction to be class 1.

As the labels are 0s and 1s, and the predictions are values between 0 and 1, the mean squared error is no longer suited to train such a model. So, just like for a logistic regression, we would use binary cross-entropy loss.

The required libraries for this recipe are matplotlib, scikit-learn, and PyTorch, they can be installed with `pip install matplotlib scikit-learn torch`.

How to do it...

We will train a simple neural network with two hidden layers on a binary classification task on the breast cancer dataset. Even though this dataset is not really suited for deep learning since it is a small dataset, it allows us to easily understand all the steps involved in training a neural network for binary classification:

1. We have the following required imports from scikit-learn:

    - `load_breast_cancer` to load the dataset
    - `train_test_split` to split the data into training and test sets
    - `StandardScaler` to rescale the quantitative data
    - `accuracy_score` to evaluate the model

We also need matplotlib for display, and we need the following from torch:

- `torch` itself
- `torch.nn` containing required classes for building a neural network
- `torch.nn.functional` for activation functions such as ReLU
- `Dataset` and `DataLoader` for handling the data

```
from sklearn.datasets import load_breast_cancer
from sklearn.model_selection import train_test_split
from sklearn.preprocessing import StandardScaler
from sklearn.metrics import accuracy_score
import matplotlib.pyplot as plt
import torchimport torch.nn as nn
import torch.nn.functional as F
from torch.utils.data import Dataset, DataLoader
```

2. Load the features and labels with the `load_breast_cancer` function:

```
X, y = load_breast_cancer(return_X_y=True)
```

3. Split the data into training and test sets, specifying the random state for reproducibility. Also cast the features and labels for `float32` for later compatibility with PyTorch:

```
X_train, X_test, y_train, y_test = train_test_split(
    X.astype(np.float32), y.astype(np.float32),
    test_size=0.2, random_state=0)
```

4. Create the `Dataset` class for handling the data. Note that in this recipe we integrate the data rescaling in this step, unlike in the previous recipe:

```
class BreastCancerDataset(Dataset):
    def __init__(self, X: np.array, y: np.array,
        x_scaler: StandardScaler = None):
            if x_scaler is None:
                self.x_scaler = StandardScaler()
                X = self.x_scaler.fit_transform(X)
            else:
                self.x_scaler = x_scaler
                X = self.x_scaler.transform(X)
            self.X = torch.from_numpy(X)
            self.y = torch.from_numpy(y)
    def __len__(self) -> int:
        return len(self.X)
    def __getitem__(self, idx: int) -> tuple[torch.Tensor]:
        return self.X[idx], self.y[idx]
```

> **Important note**
> Having the scaler in the class has pros and cons, where a pro is properly handling data leakage between train and test sets.

5. Instantiate the training and test sets and loaders. Note that no scaler is provided to the training dataset, while the test dataset is given the training set scaler to ensure that all the data is processed the same with no data leakage:

```
training_data = BreastCancerDataset(X_train, y_train)
test_data = BreastCancerDataset(X_test, y_test,
    training_data.x_scaler)
train_dataloader = DataLoader(training_data,
    batch_size=64, shuffle=True)
test_dataloader = DataLoader(test_data, batch_size=64,
    shuffle=True)
```

6. Build the neural network. Here, we build a neural network with two hidden layers. In the `forward` method, the `torch.sigmoid()` function is applied to the output layer before returning the value, ensuring we have a prediction between 0 and 1. The only parameter needed to instantiate the model is the input shape, which is simply the number of features here:

```
class Net(nn.Module):
    def __init__(self, input_shape: int,
        hidden_units: int = 24):
            super(Net, self).__init__()
                self.hidden_units = hidden_units
                self.fc1 = nn.Linear(input_shape,
                    self.hidden_units)
                self.fc2 = nn.Linear(
                    self.hidden_units,
                    self.hidden_units)
                self.output = nn.Linear(
                    self.hidden_units, 1)

    def forward(self, x: torch.Tensor) -> torch.Tensor:
        x = self.fc1(x)
        x = F.relu(x)
        x = self.fc2(x)
        x = F.relu(x)
        output = torch.sigmoid(self.output(x))
        return output
```

7. We can now instantiate the model with the right input shape and check that the forward propagation works properly on a given random tensor:

```
# Instantiate the network
net = Net(X_train.shape[1])
# Generate one random sample
random_data = torch.rand((1, X_train.shape[1]))
# Compute the forward propagation
print(net(random_data))
```

After running the previous code, we get the following output:

```
tensor([[0.4487]], grad_fn=<SigmoidBackward0>)
```

8. Define the loss function and the optimizer. As stated, we will use the binary cross-entropy loss, available as nn.BCELoss() in PyTorch. The chosen optimizer is Adam, but other optimizers can be tested too:

```
criterion = nn.BCELoss()
optimizer = torch.optim.Adam(net.parameters(),
    lr=0.001)
```

> **Important note**
> More explanations about the optimizer are provided in the next *There's more* subsection.

9. We can now train the neural network for 50 epochs. We also compute both the training and test set loss at each epoch, so we can plot them afterward. To do so, we need to switch mode for the model:

    *   Before training on the training set, switch to train mode with model.train()

    *   Before evaluating the test set, switch to the eval model with model.eval()

```
train_losses = []
test_losses = []
# Loop over the dataset 50 times
for epoch in range(50):
    ## Train the model on the training set
    running_train_loss = 0.
    # Switch to train mode
    net.train()
    # Loop over the batches in train set
    for i, data in enumerate(train_dataloader, 0):
        # Get the inputs: data is a list of [inputs, labels]
        inputs, labels = data
```

```
                # Zero the parameter gradients
                optimizer.zero_grad()
                # Forward + backward + optimize
                outputs = net(inputs)
                loss = criterion(outputs, labels.unsqueeze(1))
                loss.backward()
                optimizer.step()
                # Add current loss to running loss
                running_train_loss += loss.item()
        # Once epoch is over, compute and store the epoch loss
        train_epoch_loss = running_train_loss / len(
            train_dataloader)
        train_losses.append(train_epoch_loss)
        ## Evaluate the model on the test set
        running_test_loss = 0.
        # Switch to eval model
        net.eval()
        with torch.no_grad():
            # Loop over the batches in test set
            for i, data in enumerate(test_dataloader, 0):
                # Get the inputs
                inputs, labels = data
                # Compute forward propagation
                outputs = net(inputs)
                # Compute loss
                loss = criterion(outputs,
                    labels.unsqueeze(1))
                # Add to running loss
                running_test_loss += loss.item()
                # Compute and store the epoch loss
                test_epoch_loss = running_test_loss / len(
                    test_dataloader)
                test_losses.append(test_epoch_loss)
        # Print stats
        print(f'[epoch {epoch + 1}] Training loss: {
            train_epoch_loss:.3f} | Test loss: {
                test_epoch_loss:.3f}')
    print('Finished Training')
```

> **Important note**
>
> Notice the use of `with torch.no_grad()` around the evaluation part. This line of code allows us to deactivate the autograd engine and speed up processing.

10. Now we plot the losses for the train and test sets as a function of the epoch, using the two computed lists in the previous step:

```
plt.plot(train_losses, label='train')
plt.plot(test_losses, label='test')
plt.xlabel('epoch')plt.ylabel('loss (BCE)')
plt.legend()
plt.show()
```

Here's the output:

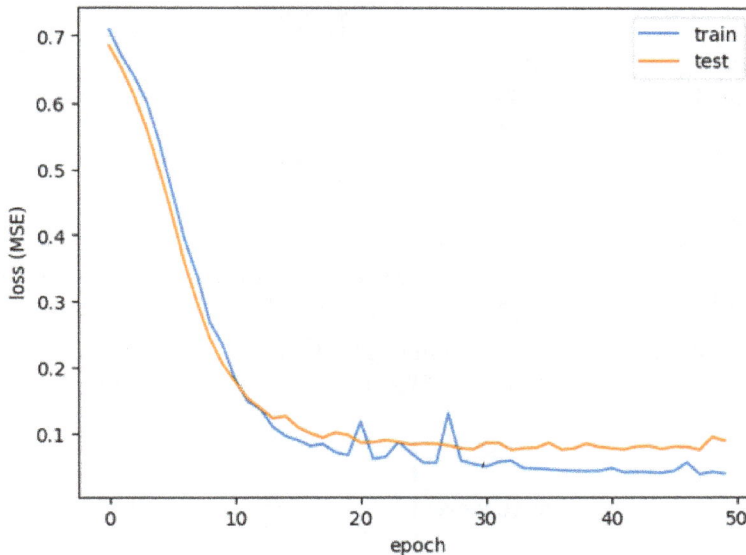

Figure 6.6 – Resulting MSE loss for the train and test sets

As we can see, both losses are decreasing. At first, the train and test losses are almost equal, but after 10 epochs, the train loss keeps decreasing while the test does not, meaning the model has overfit on the training set.

11. It is possible to evaluate the model using the accuracy scores from both the training and test sets, via the `accuracy_score` function of scikit-learn. It requires a few more steps to compute the predictions, since we have to do the following operations before getting actual class predictions:

- Rescale the data with the scaler used for training, available in the `training_data.x_scaler` attribute

- Cast the NumPy data to the torch tensor with `torch.tensor()`

- Apply forward propagation to the model

- Cast the output torch tensor back to NumPy with `.detach().numpy()`
- Apply a threshold to convert a probability prediction (between 0 and 1) to a class prediction with `> 0.5`

```
# Compute the predictions with the trained neural network
y_train_pred = net(torch.tensor((
    training_data.x_scaler.transform(
        X_train)))).detach().numpy() > 0.5
y_test_pred = net(torch.tensor((
    training_data.x_scaler.transform(
        X_test)))).detach().numpy() > 0.5
# Compute the accuracy score
print('Accuracy on training set:', accuracy_score(
    y_train, y_train_pred))
print('Accuracy on test set:', accuracy_score(y_test,
    y_test_pred))
```

Here is the output of the preceding code:

```
Accuracy on training set: 0.9912087912087912 Accuracy on test
set: 0.9649122807017544
```

We get an accuracy of 99% on the training and 96% on the test set, proving there is indeed overfitting, as expected from the curve of the train and test losses as a function of the epoch.

There's more...

As we have seen here, the loss is decreasing over time, meaning the model is actually learning.

> **Important note**
> Even if it's sometimes a bit noisy with bumps in the loss, as long as the overall trend remains good, there is nothing to worry about.

There are two important notions that may alter these results, somewhat related to each other: the learning rate and the optimizer. As with logistic regression or linear regression, the optimizer's goal is to find the parameters that provide the lowest possible loss value. Therefore, this is a minimization problem and can be represented as in *Figure 6.7*: we seek to find the set of parameters that give the lowest possible value.

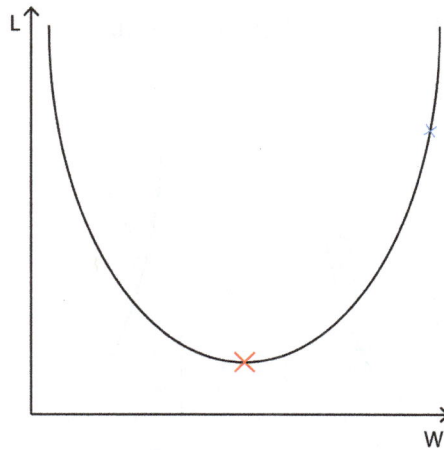

Figure 6.7 – Representation of the loss L as a function of the parameters w. The red cross
is the optimal point, while the blue cross is a random arbitrary set of weights

Let's see how the learning rate can impact the learning curve.

Learning rate

The learning rate is set in PyTorch when instantiating the optimizer, with the *lr=0.001* parameter for
example. Arguably, we can have four main cases for the learning rate value, as presented in *Figure 6.8*,
from a low learning rate to a very high learning rate.

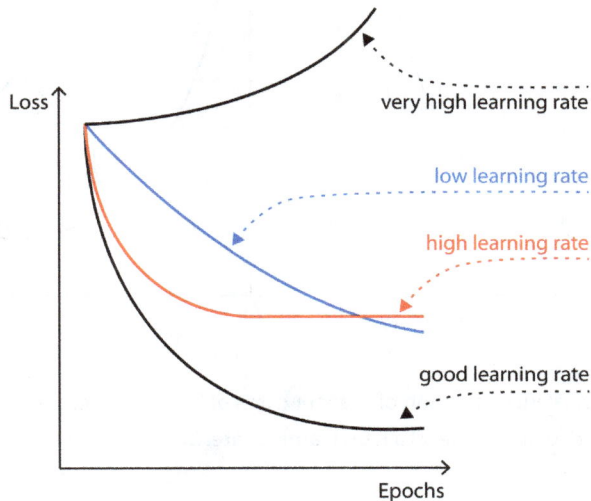

Figure 6.8 – The four main categories of learning rate: too low learning rate, good
learning rate, high learning rate, very high learning rate (diverging loss)

In terms of loss, the rates can be intuited from *Figure 6.9*, presenting the evolution of the weights and the loss for several epochs.

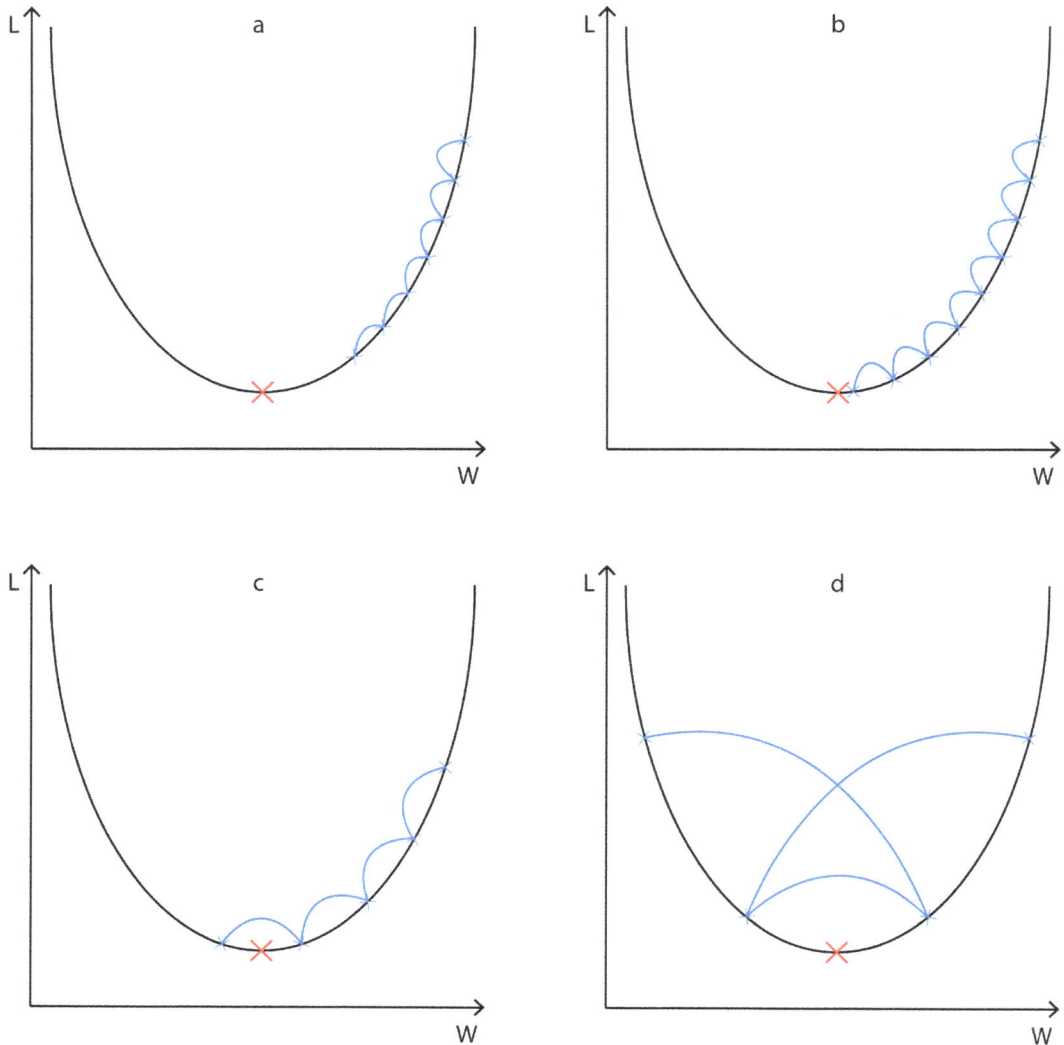

Figure 6.9 – A visual interpretation of the four cases of learning rate: a) a low learning rate, b) a good learning rate, c) a high learning rate, d) a very high learning rate

Figure 6.9 can be further explained with the following:

- **A low learning rate (a)**: The loss will decrease over the epochs but too slowly and may take a very long time to converge. It may also get the model stuck in a local minimum.

- **A good learning rate (b)**: The loss will decrease steadily until it gets close enough to the global minimum.

- **A slightly too large learning rate (c)**: The loss will decrease steeply at first but may soon jump over the global minimum without being able to ever reach it.

- **A very high learning rate (d)**: The loss will rapidly diverge, taking learning steps that are way too large.

Tuning the learning rate may sometimes help to produce the best results. Several techniques, such as the so-called learning rate decay, decrease the learning rate over time to hopefully more accurately catch the global minimum.

Optimizer

There are many robust and useful optimizers in deep learning besides the arguably most famous ones (the stochastic gradient descent and Adam). Without getting into the details of those optimizers, let's just give some insight into how they work and their differences, summarized in *Figure 6.10*:

- **Stochastic gradient descent** simply computes the gradients from the loss for each batch, without any further sophistication. It means that sometimes, the optimization of one batch may be almost in the opposite direction of another batch.

- **Adam** uses momentum, meaning that for each batch, not only the gradient from this batch is used, but also the momentum of the previously computed gradients. This allows Adam to keep an overall more consistent direction, and hopefully to converge faster.

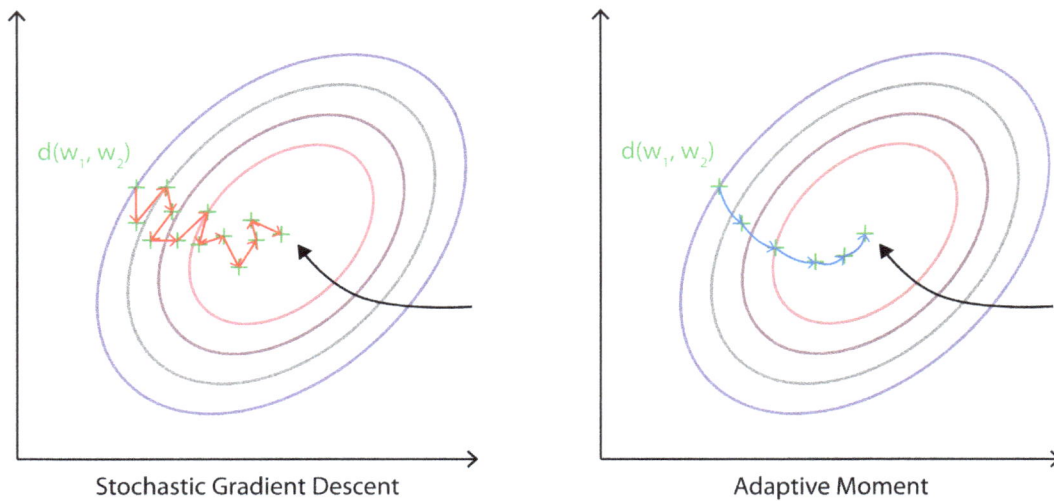

Figure 6.10 – A visual representation of training towards a global minimum, on the left with stochastic gradient descent; on the right with Adam keeping the momentum of previous steps

The optimization process can be summarized with quite a simple metaphor. This is like hiking somewhere up a mountain, and then trying to go down (to the global minimum) while surrounded by fog. You can either go down with stochastic gradient descent or Adam:

- With stochastic gradient descent, you look around you, choose the direction with the steepest slope downward, and take a step in that direction. And then do it again.

- With Adam, you do the same as stochastic gradient descent but running. You quickly look around you, see the direction with the steepest slope downward, and try to take a step in that direction while keeping the inertia from your previous steps since you're running. And then do it again.

Note that in this analogy, the step size would be the learning rate.

See also

A list of available optimizers on PyTorch: `https://pytorch.org/docs/stable/optim.html#algorithms`

Training a multiclass classification neural network

In this recipe, we will have a look at another very common task: multiclass classification with neural networks, in this instance using PyTorch. We will work on a very iconic dataset in deep learning: **MNIST handwritten digit recognition**. This dataset is a set of small grayscale images of 28x28 pixels, depicting handwritten digits between 0 and 9, having thus 10 classes.

Getting ready

In classical machine learning, multiclass classification is usually not handled natively. For example, when training logistic regression with scikit-learn on a three-class task (e.g., the Iris dataset), scikit-learn will automatically train three models, using the one-versus-the-rest method.

In deep learning, it is possible for the model to natively handle more than two classes. To do so, only a few changes are required compared to binary classification:

- The output layer has as many units as classes: this way, each unit will be responsible for predicting the probability of one class

- The output layer's activation function is the softmax function, a function such that the sum of the units is equal to 1, allowing us to consider it as a probability

- The loss function is the cross-entropy loss, considering multiple classes, unlike the binary cross entropy

In our case, we will need a few other changes in the code that are specific to the data itself. Since the input is now an image, some transformations are required:

- The image, a 2D (or 3D if RGB color image) array, must be flattened to be 1D

- The data must be normalized, just like rescaling for quantitative data

To do that, we will require the following libraries: torch, torchvision (for dataset loading and image transformation), and matplotlib for visualization. They can be installed with `pip install torch torchvision matplotlib`.

How to do it...

In this recipe, we will reuse the same pattern as previously in this chapter: we will train a two-hidden-layer neural network. But a few things will change, though:

- The input data is a grayscale image from the MNIST handwritten digits dataset, so it's a 2D array that needs to be flattened

- The output layer will have not one, but ten units for the ten classes of the dataset; the loss will change accordingly

- We will not only compute the training and test losses in the training loop, but also the accuracy

Let's see how to do that in practice now:

1. Import the required libraries. As in previous recipes, we import several useful torch modules and functions:

    -   `torch` itself

    -   `torch.nn` containing the required classes for building a neural network

    -   `torch.nn.functional` for activation functions such as ReLU

    -   `DataLoader` for handling the data

    We also need some imports from torchvision:

    -   `MNIST` for loading the dataset

    -   `transforms` for transforming the dataset, both rescaling and flattening the data:

    ```
    import torch
    import torch.nn as nn
    import torch.nn.functional as F
    from torch.utils.data import DataLoader
    from torchvision.datasets import MNIST
    import torchvision.transforms as transforms
    import matplotlib.pyplot as plt
    ```

2. Instantiate the transformations. We use the `Compose` class, allowing us to compose two or more transformations. Here, we compose three transformations:

    -   `transforms.ToTensor()`: Converts the input image to `torch.Tensor` format.

    -   `transforms.Normalize()`: Normalizes the image with a mean value and standard deviation. It will subtract the mean (i.e., 0.1307) and then divide by the standard deviation (i.e., 0.3081) for each pixel value.

    -   `transforms.Lambda(torch.flatten)`: Flattens the 2D tensor to a 1D tensor:

    ```
    transform = transforms.Compose([transforms.ToTensor(),
        transforms.Normalize((0.1307), (0.3081)),
        transforms.Lambda(torch.flatten)])
    ```

Important note

Images are commonly normalized with a mean and standard deviation of 0.5. We normalize with the specific values used in the preceding code block because the dataset is made with specific images, but 0.5 would work fine too. Check the *See also* subsection of this recipe for an explanation.

---

3. Load the train and test sets, as well as the train and data loaders. Using the MNIST class, we both get the train and test sets using the train parameter as True and False, respectively. We directly apply the previously defined transformations while loading the data with the MNIST class too. Then we instantiate the data loaders with a batch size of 64:

```python
trainset = MNIST('./data', train=True, download=True,
    transform=transform)
train_dataloader = DataLoader(trainset, batch_size=64,
    shuffle=True)
testset = MNIST('./data', train=False, download=True,
    transform=transform)
test_dataloader = DataLoader(testset, batch_size=64,
    shuffle=True)
```

4. Define the neural network. We define here by default a neural network with 2 hidden layers of 24 units. The output layer has 10 units for the 10 classes of the data (our digits between 0 and 9). Note that the softmax function is applied to the output layer, allowing the sum of the 10 units to be strictly equal to 1:

```python
class Net(nn.Module):
    def __init__(self, input_shape: int,
        hidden_units: int = 24):
            super(Net, self).__init__()
            self.hidden_units = hidden_units
            self.fc1 = nn.Linear(input_shape,
                self.hidden_units)
            self.fc2 = nn.Linear(
                self.hidden_units,
                self.hidden_units)
            self.output = nn.Linear(
                self.hidden_units, 10)
    def forward(self,
        x: torch.Tensor) -> torch.Tensor:
            x = self.fc1(x)
            x = F.relu(x)
            x = self.fc2(x)
            x = F.relu(x)
            output = torch.softmax(self.output(x),
                dim=1)
            return output
```

5. We can now instantiate the model with the right input shape of 784 (28x28 pixels), and check the forward propagation works properly on a given random tensor:

```
# Instantiate the model
net = Net(784)
# Generate randomly one random 28x28 image as a 784 values
tensor
random_data = torch.rand((1, 784))
result = net(random_data)
print('Resulting output tensor:', result)
print('Sum of the output tensor:', result.sum())
```

Here is the output:

```
Resulting output tensor: tensor([[0.0918, 0.0960,
0.0924, 0.0945, 0.0931, 0.0745, 0.1081, 0.1166,
0.1238,              0.1092]], grad_fn=<SoftmaxBackward0>) Sum
of the output tensor: tensor(1.0000, grad_fn=<SumBackward0>)
```

> **Tip**
> Note the output is a tensor of 10 values, with a sum of 1.

6. Define the loss function as the cross-entropy loss, available as nn.CrossEntropyLoss() in PyTorch, and the optimizer as Adam:

```
criterion = nn.CrossEntropyLoss()
optimizer = torch.optim.Adam(net.parameters(),
    lr=0.001)
```

7. Before training, we implement an epoch_step helper function that works for both the train and test sets, allowing us to loop over all the data, compute the loss and the accuracy, and train the model for the training set:

```
def epoch_step(net, dataloader, training_set: bool):
    running_loss = 0.
    Correct = 0.
    For i, data in enumerate(dataloader, 0):
        # Get the inputs: data is a list of [inputs, labels]
        inputs, labels = data
        if training_set:
            # Zero the parameter gradients
            optimizer.zero_grad()
            # Forward + backward + optimize
            outputs = net(inputs)
            loss = criterion(outputs, labels)
            if training_set:
```

```
                    loss.backward()
                    optimizer.step()
            # Add correct predictions for this batch
            correct += (outputs.argmax(
                dim=1) == labels).float().sum()
            # Compute loss for this batch
            running_loss += loss.item()

        return running_loss, correct
```

8. We can now train the neural network on 20 epochs. For each epoch, we also compute the following:

- The loss for both train and test sets

- The accuracy for both train and test sets

As for the previous recipe, before training, the model is switched to train mode with `model.train()`, while before evaluating on the test set, it is switched to `eval` model with `model.eval()`:

```
# Create empty lists to store the losses and accuracies
train_losses = []
test_losses = []
train_accuracy = []
test_accuracy = []
# Loop over the dataset 20 times for 20 epochs
for epoch in range(20):
    ## Train the model on the training set
    net.train()
    running_train_loss, correct = epoch_step(net,
        dataloader=train_dataloader,training_set=True)
    # Compute and store loss and accuracy for this epoch
    train_epoch_loss = running_train_loss / len(
        train_dataloader)
    train_losses.append(train_epoch_loss)
    train_epoch_accuracy = correct / len(trainset)
     rain_accuracy.append(train_epoch_accuracy)


    ## Evaluate the model on the test set
    net.eval()
    with torch.no_grad():
        running_test_loss, correct = epoch_step(net,
            dataloader=test_dataloader,training_set=False)
        test_epoch_loss = running_test_loss / len(
            test_dataloader)
```

```
        test_losses.append(test_epoch_loss)
        test_epoch_accuracy = correct / len(testset)
        test_accuracy.append(test_epoch_accuracy)
    # Print stats
    print(f'[epoch {epoch + 1}] Training: loss={train_epoch_
loss:.3f} accuracy={train_epoch_accuracy:.3f} |\
\t Test: loss={test_epoch_loss:.3f} accuracy={test_epoch_
accuracy:.3f}')
print('Finished Training')
```

9. We can plot the loss for both the train and test sets as a function of the epoch, since we stored those values for each epoch:

```
plt.plot(train_losses, label='train')
plt.plot(test_losses, label='test')
plt.xlabel('epoch')plt.ylabel('loss (CE)')
plt.legend()plt.show()
```

Here is the output:

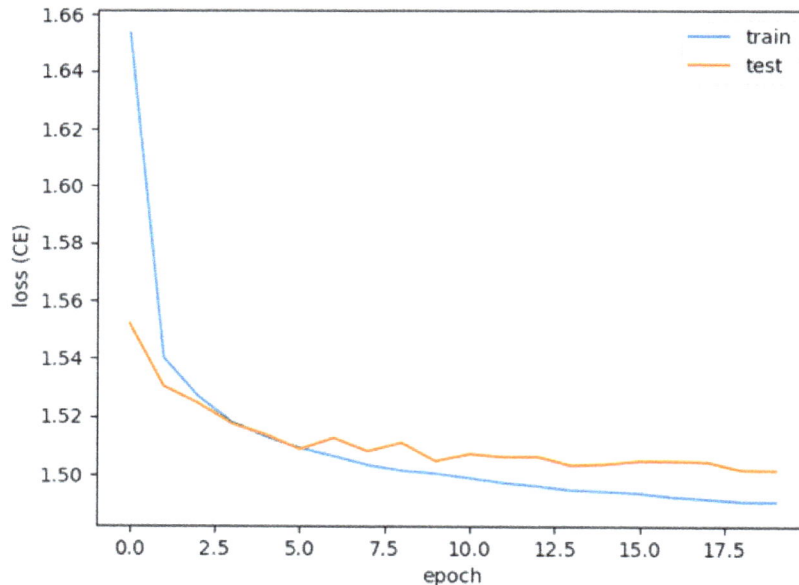

Figure 6.11 – Resulting cross-entropy loss as a function of the epoch, for both the train and test sets

Since the loss seems to keep improving on both the **train** and **test** sets after 20 epochs, it could be interesting in terms of performance to keep training more epochs.

10. It is also possible to do the same with the accuracy score, showing the equivalent results:

```
plt.plot(train_accuracy, label='train')
plt.plot(test_accuracy, label='test')
plt.xlabel('epoch')
plt.ylabel('Accuracy')plt.legend()plt.show()
```

Here are the results:

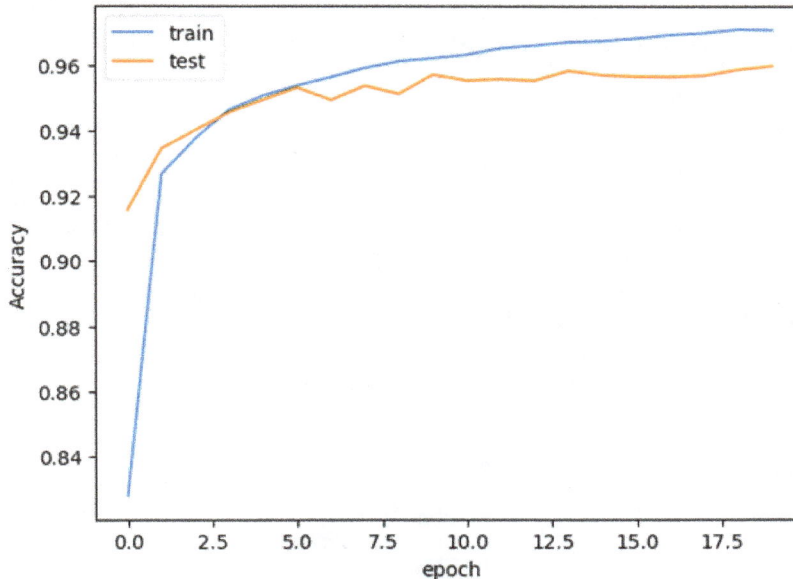

Figure 6.12 – Resulting accuracy as a function of the epoch for the train and test sets

At the end, the accuracy is about 97% on the train set and 96% on the test set.

Once the model has been trained, it is, of course, possible to store it so that it can be used on new data directly. There are several ways to save a model:

- **Saving the state dict**: This saves only the weights, meaning that on loading, the net class must first be instantiated.

- **Saving the entire model**: This saves both the weights and the architecture, meaning only the file needs to be loaded.

- **Saving in torchscript format**: This saves the entire model using a more efficient representation. This method is more suited for deployment and inference at scale.

For now, let's just save the `state` dict, reload it, and then compute inferences on an image:

```
# Save the model's state dict
torch.save(net.state_dict(), 'path_to_model.pt')
# Instantiate a new model
new_model = Net(784)
# Load the model's weights
new_model.load_state_dict(torch.load('path_to_model.pt'))
```

It is now possible to compute inferences using that loaded, already-trained model on a given image:

```
plt.figure(figsize=(12, 8))
for i in range(6):
    plt.subplot(3, 3, i+1)
    # Compute the predicted number
    pred = new_model(
        testset[i][0].unsqueeze(0)).argmax(axis=1)
    # Display the image and predicted number as title
    plt.imshow(testset[i][0].detach().numpy().reshape(
        28, 28), cmap='gray_r')
    plt.title(f'Prediction: {pred.detach().numpy()}')
    plt.axis('off')
```

This is what we get:

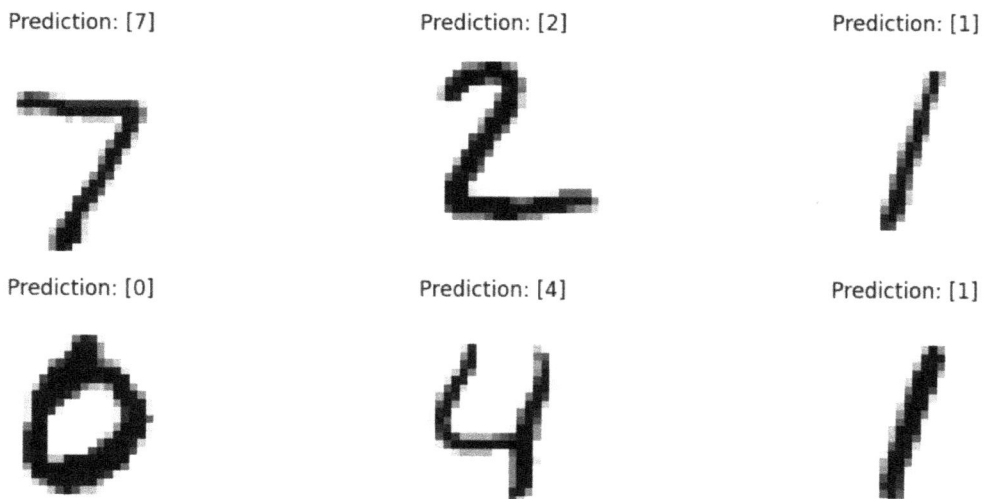

Figure 6.13 – The resulting output with six input images and their predictions from the trained model

As expected, the loaded model can correctly predict the right number on most images.

There's more...

Deep learning is often used in heavily computational tasks requiring a lot of resources. In such cases, the use of a GPU is often a necessity. PyTorch of course allows us to train and infer models on GPUs. Only a few steps are required to do so: declaring a device variable and moving both the model and data to this device. Let's have a quick look at how to do it.

Choosing the device

Declaring a device variable to be the GPU can be done with the following Python code:

```
device = torch.device(
    "cuda" if torch.cuda.is_available() else "cpu") print(device)
```

This line instantiates a `torch.device` object, containing `"cuda"` if CUDA is available, else it contains `"cpu"`. Indeed, if CUDA is not installed, or if there is no GPU on your hardware, the CPU will be used (which is the default behavior).

If the GPU has been correctly detected, the output of `print(device)` is `"cuda"`. Otherwise, the output is `"cpu"`.

Moving the model and data to the GPU

Once the device is correctly set to the GPU, both the model and data have to be moved to the GPU memory. To do so, you only need to call the `.to(device)` method on both the model and the data. For example, the training and evaluation code that we used in this recipe would become the following:

```
train_losses = []
test_losses = []
train_accuracy = []
test_accuracy = []
# Move the model to the GPU
net = net.to(device)
for epoch in range(20):
    running_train_loss = 0.
    correct = 0.
    net.train()
    for i, data in enumerate(train_dataloader, 0):
        inputs, labels = data
        # Move the data to the device
        inputs = inputs.to(device)
        labels = labels.to(device)
    running_test_loss = 0.
    correct = 0.
    net.eval()
```

```
    with torch.no_grad():
        for i, data in enumerate(test_dataloader, 0):
            inputs, labels = data
            # Move the data to the device
            inputs = inputs.to(device)
            labels = labels.to(device)
print('Finished Training')
```

At the beginning, the model is moved to the GPU device once with `net = net.to(device)`.

At each iteration loop for both the training and evaluation, the inputs and labels tensors are moved to the device with `tensor = tensor.to(device)`.

Tip

The data can be either fully loaded on the GPU at loading, or done one batch at a time during training. However, since only rather small datasets can be fully loaded in the GPU memory, we did not present this solution here.

See also

- The official documentation on saving and loading PyTorch models: `https://pytorch.org/tutorials/beginner/saving_loading_models.html`

- The reason the images are transformed with such specific values for the MNIST dataset: `https://discuss.pytorch.org/t/normalization-in-the-mnist-example/457`

7

Deep Learning Regularization

In this chapter, we will cover several tricks and techniques to regularize neural networks. We will reuse the L2 regularization technique, as we did in linear models, for example. But there are other techniques not yet presented in this book, such as early stopping and dropout, which will be covered in this chapter.

In this chapter, we'll look at the following recipes:

- Regularizing a neural network with L2 regularization
- Regularizing a neural network with early stopping
- Regularization with network architecture
- Regularizing with dropout

Technical requirements

In this chapter, we will train neural networks on various tasks. This will require us to use the following libraries:

- NumPy
- Scikit-learn
- Matplotlib
- PyTorch
- torchvision

Regularizing a neural network with L2 regularization

Just like a linear model, whether it be a linear regression or a logistic regression, neural networks have weights. And so, just like a linear model, L2 penalization can be used on those weights to regularize the neural network. In this recipe, we will apply L2 penalization to a neural network on the MNIST handwritten digits dataset.

As a reminder, when training a neural network on this task in *Chapter 6*, there was a small overfitting after 20 epochs, and the results were an accuracy of 97% on the train set and 95% on the test set. Let's try to reduce this overfitting by adding L2 regularization in this recipe.

Getting ready

Just like for linear models, L2 regularization is just adding a new L2 term to the loss. Given the weights W=w1,w2,..., the added term to the loss would be $|w|^2 = \sum w_i^2$. The consequence of this added term to the loss is that the weights are more constrained and must stay close to zero to keep the loss small. As a result, it adds bias to the model and then can help regularize it.

> **Note**
>
> This notation for the weights is simplified here. Actually, there are weights $w_{ij}^{[l]}$ for each unit i, each feature j, and each layer l. But in the end, the L2 term remains the sum of all the squared weights.

For this recipe, only three libraries are needed:

- `matplotlib` for plots
- `pytorch` for deep learning
- `torchvision` for the MNIST dataset

These can be installed with `pip install matplotlib torch torchvision`.

How to do it...

In this recipe, we reuse the exact same code as in the previous chapter when training a multiclass classification model on the MNIST dataset. The only difference will be at *step 6* – feel free to jump there if needed.

The input data is the MNIST handwritten dataset: grayscale images of 28x28 pixels. The data will thus need to be rescaled and flattened before being able to train a custom neural network:

1. Import the required libraries. As in previous recipes, we import several useful `torch` modules and functions:

   - `torch`

   - `torch.nn` containing required classes for building a neural network

   - `torch.nn.functional` for activation functions such as ReLU

   - `DataLoader` for handling the data

   And we have some imports from `torchvision`:

   - `MNIST` for loading the dataset

   - `transforms` for transforming the dataset – both rescaling and flattening the data:

   ```python
   import torch
   import torch.nn as nn
   import torch.nn.functional as F
   from torch.utils.data import DataLoader
   from torchvision.datasets import MNIST
   import torchvision.transforms as transforms
   import matplotlib.pyplot as plt
   ```

2. Instantiate the transformations. The `Compose` class is used here to compose three transformations:

   - `transforms.ToTensor()`: Convert the input image in to `torch.Tensor` format

   - `transforms.Normalize()`: Normalize the image with the mean value and standard deviation. Will subtract the mean (i.e., 0.1307) and then divide it by the standard deviation (i.e., 0.3081) for each pixel value.

   - `transforms.Lambda(torch.flatten)`: Flatten the 2D tensor in to a 1D tensor:

   Here is the code:

   ```python
   transform = transforms.Compose([transforms.ToTensor(),
       transforms.Normalize((0.1307), (0.3081)),
       transforms.Lambda(torch.flatten)])
   ```

> **Note**
>
> Images are commonly normalized with a mean and standard deviation of 0.5. We normalize with those specific values because the dataset is made with specific images, but 0.5 would work fine too.

3. Load the train and test sets, as well as the train and test data loaders. Using the MNIST class, we both get the train and test sets using the train=True and train=False parameters, respectively. We apply the previously defined transformations directly while loading the data with the MNIST class too. Then we instantiate the data loaders with a batch size of 64:

```
trainset = MNIST('./data', train=True, download=True,
    transform=transform)
train_dataloader = DataLoader(trainset, batch_size=64,
    shuffle=True)
testset = MNIST('./data', train=False, download=True,
    transform=transform)
test_dataloader = DataLoader(testset, batch_size=64,
    shuffle=True)
```

4. Define the neural network. We define here, by default, a neural network made of 2 hidden layers of 24 units. The output layer has 10 units since there are 10 classes (digits between 0 and 9). Finally, the softmax function is applied to the output layer, allowing the sum of the 10 units to be strictly equal to 1:

```
class Net(nn.Module):
    def __init__(self, input_shape: int,
    hidden_units: int = 24):
        super(Net, self).__init__()
        self.hidden_units = hidden_units
        self.fc1 = nn.Linear(input_shape,
            self.hidden_units)
        self.fc2 = nn.Linear(self.hidden_units,
            self.hidden_units)
        self.output = nn.Linear(self.hidden_units, 10)

    def forward(self,
    x: torch.Tensor) -> torch.Tensor:
        x = self.fc1(x)
        x = F.relu(x)
        x = self.fc2(x)
        x = F.relu(x)
        output = torch.softmax(self.output(x), dim=1)
        return output
```

5. To check the code, we instantiate the model with the right input shape of 784 (28x28 pixels) and check the forward propagation works properly on a given random tensor:

```
# Instantiate the model
net = Net(784)
# Generate randomly one random 28x28 image as a 784 values
tensor
random_data = torch.rand((1, 784))
result = net(random_data)
print('Resulting output tensor:', result)
print('Sum of the output tensor:', result.sum())
```

The code output would be something like the following (only the sum must be equal to 1; other numbers may be different):

```
Resulting output tensor: tensor([[0.0882, 0.1141, 0.0846,
0.0874, 0.1124, 0.0912, 0.1103, 0.0972, 0.1097,
        0.1048]], grad_fn=<SoftmaxBackward0>)
Sum of the output tensor: tensor(1.0000, grad_fn=<SumBackward0>)
```

6. Define the loss function as the cross-entropy loss, available as nn.CrossEntropyLoss() in pytorch, and the optimizer as Adam. Here we set another parameter to the Adam optimizer: weight_decay=0.001. This parameter is the strength of the L2 penalization. By default, weight_decay is 0, meaning there is no L2 penalization. A higher value means a higher regularization, just like in linear models in scikit-learn:

```
criterion = nn.CrossEntropyLoss()
optimizer = torch.optim.Adam(net.parameters(),
    lr=0.001, weight_decay=0.001)
```

7. Instantiate the epoch_step helper function allowing to compute forward and backward propagation (for the training set only) as well as the loss and accuracy:

```
def epoch_step(net, dataloader, training_set: bool):
    running_loss = 0.
    correct = 0.
    for i, data in enumerate(dataloader, 0):
        # Get the inputs: data is a list of [inputs, labels]
        inputs, labels = data
        if training_set:
            # Zero the parameter gradients
            optimizer.zero_grad()
        # Forward + backward + optimize
        outputs = net(inputs)
        loss = criterion(outputs, labels)
        if training_set:
```

```
        loss.backward()
        optimizer.step()
    # Add correct predictions for this batch
    correct += (outputs.argmax(
        dim=1) == labels).float().sum()
    # Compute loss for this batch
    running_loss += loss.item()

return running_loss, correct
```

8. We can finally train the neural network on 20 epochs and compute the loss and accuracy for each epoch.

Since we both train on the train set and evaluate on the test set, the model is switched to `train` mode with `model.train()` before training, whereas before evaluating on the test set, it is switched to `eval` mode with `model.eval()`:

```
# Create empty lists to store the losses and accuracies
train_losses = []
test_losses = []
train_accuracy = []
test_accuracy = []

# Loop over the dataset 20 times for 20 epochs
for epoch in range(20):
    ## Train the model on the training set
    running_train_loss, correct = epoch_step(net,
        dataloader=train_dataloader,
        training_set=True)
    # Compute and store loss and accuracy for this epoch
    train_epoch_loss = running_train_loss / len(
        train_dataloader)
    train_losses.append(train_epoch_loss)
    train_epoch_accuracy = correct / len(trainset)
    train_accuracy.append(train_epoch_accuracy)

    ## Evaluate the model on the test set
    #running_test_loss = 0.
    #correct = 0.
    net.eval()
    with torch.no_grad():
```

```
        running_test_loss, correct = epoch_step(net,
            dataloader=test_dataloader,
            training_set=False)

        test_epoch_loss = running_test_loss / len(
            test_dataloader)
        test_losses.append(test_epoch_loss)
        test_epoch_accuracy = correct / len(testset)
        test_accuracy.append(test_epoch_accuracy)

    # Print stats
    print(f'[epoch {epoch + 1}] Training: loss={
        train_epoch_loss:.3f}accuracy={
            train_epoch_accuracy:.3f} |\
            \t Test: loss={test_epoch_loss:.3f}
            accuracy={test_epoch_accuracy:.3f}')
print('Finished Training')
```

On the last epoch, the output should look like the following:

```
[epoch 20] Training: loss=1.505 accuracy=0.964 |     Test:
loss=1.505 accuracy=0.962
Finished Training
```

9. For visualization purposes, we can plot the loss for both the train and test sets as a function of the epoch:

```
plt.plot(train_losses, label='train')
plt.plot(test_losses, label='test')
plt.xlabel('epoch')
plt.ylabel('loss (CE)')
plt.legend()
plt.show()
```

Here is the plot for it:

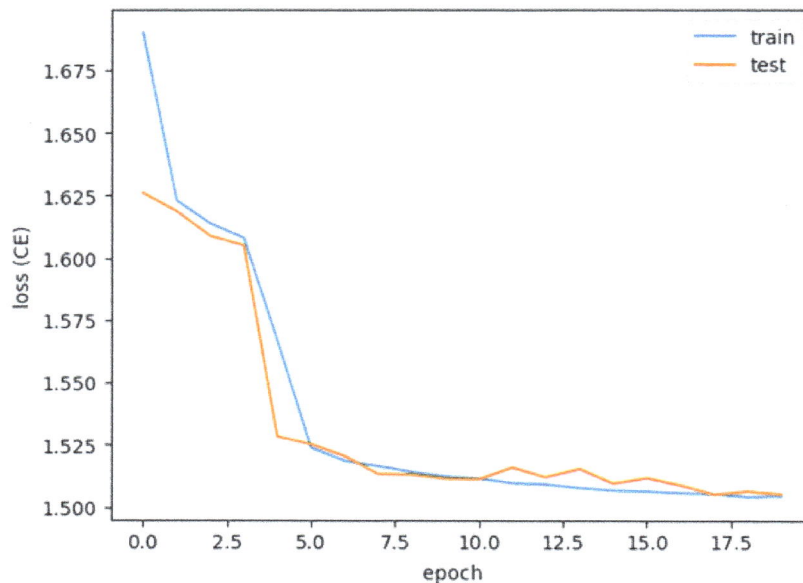

Figure 7.1 – Cross-entropy loss as a function of the epoch; output from the previous code

We can notice that the loss seems to be almost the same for both the training and test set, with no clear divergence. In the previous attempts without L2 penalization, the losses were further apart from each other, meaning we effectively regularized the model.

10. Showing related results, we can do it with accuracy too:

```
plt.plot(train_accuracy, label='train')
plt.plot(test_accuracy, label='test')
plt.xlabel('epoch')
plt.ylabel('Accuracy')
plt.legend()
plt.show()
```

Here is the plot for it:

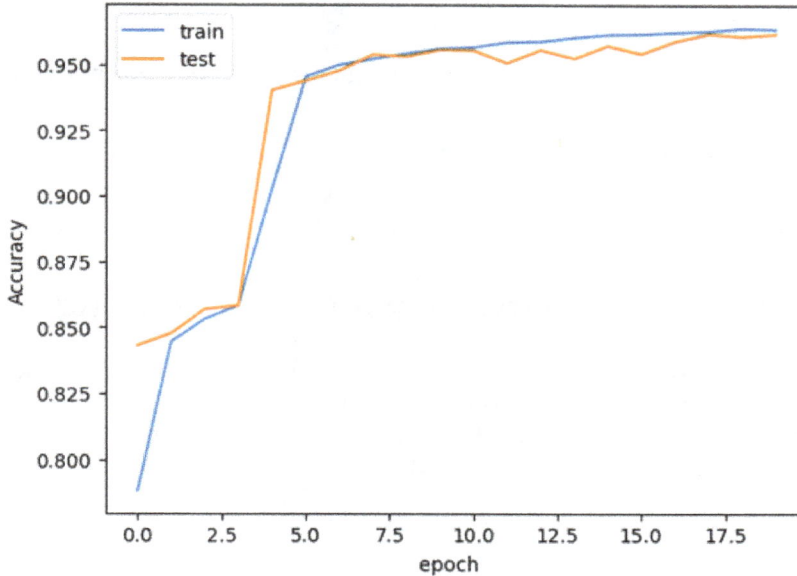

Figure 7.2 – Accuracy as a function of the epoch; output from the previous code

At the end, the accuracy is about 96% for both the train set and the test set, with no significant overfitting.

There's more...

Even if L2 regularization is a quite common technique to regularize linear models such as linear regression and logistic regression, it is not usually the first choice with deep learning. Other methods such as early stopping or dropout are usually preferred.

On another note, in this recipe, we keep mentioning only the train and test sets. But to optimize the `weight_decay` hyperparameter properly, it is required to use a validation set; otherwise, the results will be biased. We have simplified this recipe by having only two sets to keep it concise.

> **Note**
>
> Generally speaking, in deep learning, any other hyperparameter optimization, such as the number of layers, number of units, activation functions, and so on must be optimized for the validation set too, not just for the test set.

See also

It may seem strange to adjust the L2 penalization through the optimizer of the model rather than directly in the loss function, and indeed it is.

Of course, it would be possible to manually add an L2 penalization, but it would probably be suboptimal. See this PyTorch thread for more about this design choice, as well as an example of adding L1 penalization: `https://discuss.pytorch.org/t/simple-l2-regularization/139`.

Regularizing a neural network with early stopping

Early stopping is a commonly employed approach in deep learning to prevent the overfitting of models. The concept is straightforward yet effective: if the model is overfitting due to prolonged training epochs, we terminate the training prematurely to prevent overfitting. We can utilize this technique on the breast cancer dataset.

Getting ready

In a perfect world, there is no need for regularization. What that means is that for both the train and validation sets, the losses are almost perfectly equal, for any number of epochs, as in *Figure 7.3*.

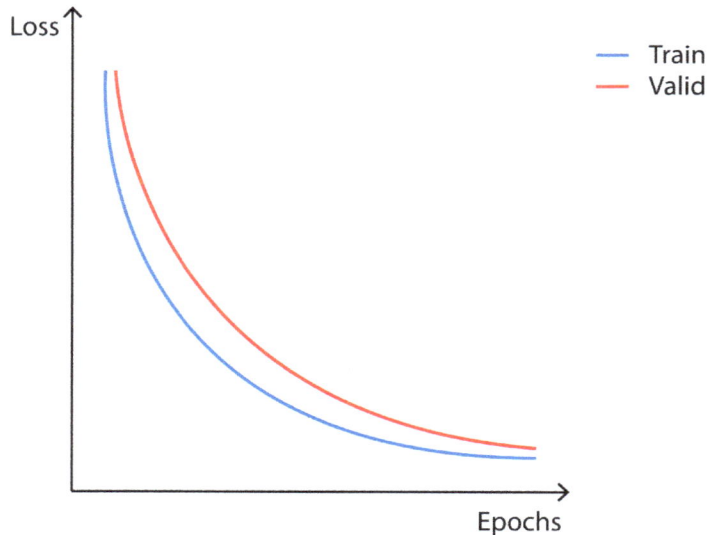

Figure 7.3 – Example with no overfitting of train and valid losses as a function of the number of epochs

But it's not always that perfect. In practice, it may happen that the neural network is learning more and more about the data distribution of the train set at every epoch, at the cost of the generalization to new data. This case is depicted by the example in *Figure 7.4*.

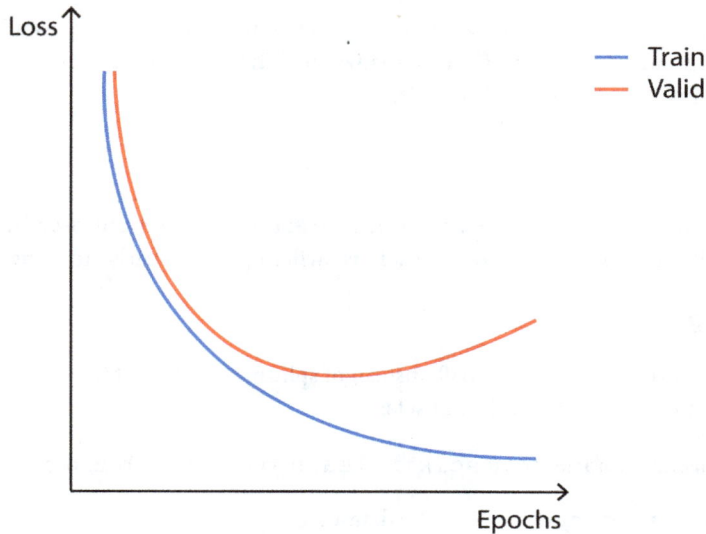

Figure 7.4 – Example with overfitting of train and valid losses as a function of the number of epochs

When dealing with such a scenario, a natural solution would be to halt the training process once the **valid** loss of the model stops decreasing. Once the validation loss of the model stops decreasing, continuing to train the model for additional epochs may cause it to become better at memorizing the training data, rather than improving its ability to make accurate predictions on new, unseen data. This technique is **called early stopping** and allows to prevent a model from overfitting.

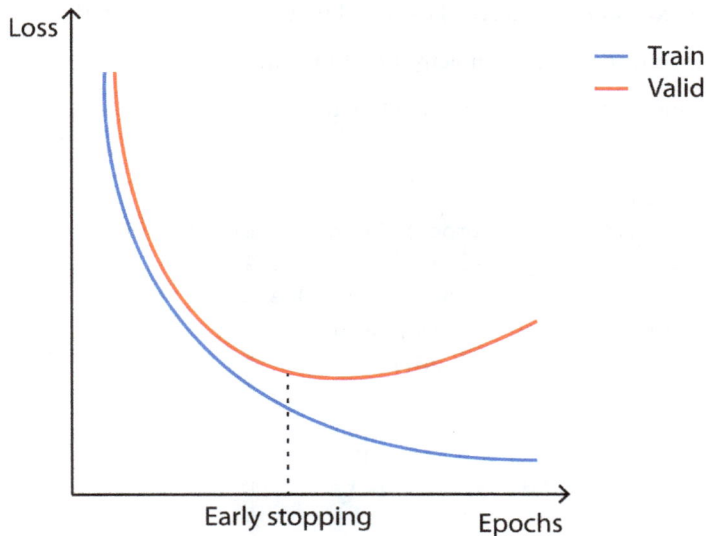

Figure 7.5 – As soon as the valid loss stops decreasing, we can stop the
learning and consider the model fully trained; this is early stopping

Since this recipe will be applied to the breast cancer dataset, scikit-learn must be installed, along with `torch` for the models and `matplotlib` for visualization. These libraries can be installed with `pip install sklearn torch matplotlib`.

How to do it...

In this recipe, we will first train a neural network on the breast cancer dataset and visualize the overfitting effect amplifying with the number of epochs. Then, we will implement early stopping, to regularize.

Regular training

Since the breast cancer dataset is rather small, instead of splitting the dataset into train, valid, and test sets, we will consider only the train and valid sets:

1. Import the needed libraries from `scikit-learn`, `matplotlib`, and `torch`:

    *   `load_breast_cancer` to load the dataset
    *   `train_test_split` to split the data into training and validation sets
    *   `StandardScaler` to rescale the quantitative data
    *   `accuracy_score` to evaluate the model
    *   `matplotlib` for display
    *   `torch` itself
    *   `torch.nn` containing required classes for building a neural network
    *   `torch.nn.functional` for activation functions such as ReLU
    *   `Dataset` and `DataLoader` for handling the data

    Here is the code for it:

    ```
    import numpy as np
    from sklearn.datasets import load_breast_cancer
    from sklearn.model_selection import train_test_split
    from sklearn.preprocessing import StandardScaler
    from sklearn.metrics import accuracy_score
    import matplotlib.pyplot as plt
    import torch
    import torch.nn as nn
    import torch.nn.functional as F
    from torch.utils.data import Dataset, DataLoader
    ```

2. Load the features and labels with the `load_breast_cancer` function:

```
X, y = load_breast_cancer(return_X_y=True)
```

3. Split the data into training and validation sets, specifying the random state for reproducibility, and convert the features and labels in to `float32` for later compatibility with PyTorch:

```
X_train, X_val, y_train, y_val = train_test_split(
    X.astype(np.float32), y.astype(np.float32),
    test_size=0.2, random_state=0)
```

4. Create the `Dataset` class for handling the data. We are simply reusing the class implemented in the previous chapter:

```
class BreastCancerDataset(Dataset):
    def __init__(self, X: np.array, y: np.array,
        x_scaler: StandardScaler = None):
            if x_scaler is None:
                self.x_scaler = StandardScaler()
                X = self.x_scaler.fit_transform(X)
            else:
                self.x_scaler = x_scaler
                X = self.x_scaler.transform(X)
            self.X = torch.from_numpy(X)
            self.y = torch.from_numpy(y)

    def __len__(self) -> int:
        return len(self.X)

    def __getitem__(self, idx: int) -> tuple[torch.Tensor]:
        return self.X[idx], self.y[idx]
```

5. Instantiate the training and validation sets and loaders for PyTorch. Notice that we provide the training scaler when instantiating the validation dataset to make sure the scaler used with both datasets is the one fitted on the training set:

```
training_data = BreastCancerDataset(X_train, y_train)
val_data = BreastCancerDataset(X_val, y_val,
    training_data.x_scaler)
train_dataloader = DataLoader(training_data,
    batch_size=64, shuffle=True)
val_dataloader = DataLoader(val_data, batch_size=64,
    shuffle=True)
```

6. Define the neural network architecture – 2 hidden layers of 36 units and an output layer with 1 unit with a sigmoid activation function since it's a binary classification task:

```
class Net(nn.Module):
    def __init__(self, input_shape: int,
        hidden_units: int = 36):
            super(Net, self).__init__()
            self.hidden_units = hidden_units
            self.fc1 = nn.Linear(input_shape,
                self.hidden_units)
            self.fc2 = nn.Linear(self.hidden_units,
                self.hidden_units)
            self.output = nn.Linear(self.hidden_units,
                1)

    def forward(self, x: torch.Tensor) ->
        torch.Tensor:
            x = self.fc1(x)
            x = F.relu(x)
            x = self.fc2(x)
            x = F.relu(x)
            output = torch.sigmoid(self.output(x))
            return output
```

7. Instantiate the model with the expected input shape (the number of features). Optionally, we can check the forward propagation works properly on a given random tensor:

```
# Instantiate the model
net = Net(X_train.shape[1])
# Generate randomly one random 28x28 image as a 784 values
tensor
random_data = torch.rand((1, X_train.shape[1]))
result = net(random_data)
print('Resulting output tensor:', result)
```

The output of this code is the following (the value itself may change, but will be between 0 and 1 since it's a sigmoid activation function on the last layer):

```
Resulting output tensor: tensor([[0.5674]], grad_
fn=<SigmoidBackward0>)
```

8. Define the loss function as the binary cross entropy loss since this is a binary classification task. Instantiate the optimizer too:

```
criterion = nn.BCELoss()
optimizer = torch.optim.Adam(net.parameters(),
    lr=0.001)
```

9. Implement a helper function, epoch_step, that computes forward propagation, backpropagation (for the training set), loss, and accuracy for one epoch:

```
def epoch_step(net, dataloader, training_set: bool):
    running_loss = 0.
    correct = 0.
    for i, data in enumerate(dataloader, 0):
        # Get the inputs: data is a list of [inputs, labels]
        inputs, labels = data
        labels = labels.unsqueeze(1)
        if training_set:
            # Zero the parameter gradients
            optimizer.zero_grad()
        # Forward + backward + optimize
        outputs = net(inputs)
        loss = criterion(outputs, labels)
        if training_set:
            loss.backward()
            optimizer.step()
        # Add correct predictions for this batch
        correct += ((
            outputs > 0.5) == labels).float().sum()
        # Compute loss for this batch
        running_loss += loss.item()
    return running_loss, correct
```

10. Let's now implement the train_model function allowing us to train a model, with or without patience. This function stores each epoch and then returns the following:

   - The loss and accuracy for the train set

   - The loss and accuracy for the valid set

   Here is the code for the model:

```
def train_model(net, train_dataloader, val_dataloader,
    criterion, optimizer, epochs, patience=None):
    # Create empty lists to store the losses and accuracies
    train_losses = []
    val_losses = []
```

```
    train_accuracy = []
    val_accuracy = []

    best_val_loss = np.inf
    best_val_loss_epoch = 0

    # Loop over the dataset 20 times for 20 epochs
    for epoch in range(500):
        ## If the best epoch was more than the patience, just
stop training
        if patience is not None and epoch - best_val_loss_epoch
> patience:
            break

        ## Train the model on the training set
        net.train()
        running_train_loss, correct = epoch_step(net,
            dataloader=train_dataloader,
            training_set=True)
        # Compute and store loss and accuracy for this epoch
        train_epoch_loss = running_train_loss / len(
            train_dataloader)
        train_losses.append(train_epoch_loss)
        train_epoch_accuracy = correct / len(training_data)
        train_accuracy.append(train_epoch_accuracy)

        ## Evaluate the model on the val set
        net.eval()
        with torch.no_grad():
            running_val_loss, correct = epoch_step(
                net, dataloader=val_dataloader,
                training_set=False)

            val_epoch_loss = running_val_loss / len(
                val_dataloader)
            val_losses.append(val_epoch_loss)
            val_epoch_accuracy = correct / len(val_data)
            val_accuracy.append(val_epoch_accuracy)

        # If the loss is better than the current best,
update it
```

```
        if best_val_loss >= val_epoch_loss:
            best_val_loss = val_epoch_loss
            best_val_loss_epoch = epoch + 1

    # Print stats
    print(f'[epoch {epoch + 1}] Training: loss={
        train_epoch_loss:.3f} accuracy={
        train_epoch_accuracy:.3f} |\
            \t Valid: loss={val_epoch_loss:.3f}
            accuracy={val_epoch_accuracy:.3f}')

return train_losses, val_losses, train_accuracy,
    val_accuracy
```

Let's now train the neural network on 500 epochs reusing the previously implemented `train_model` function. Here is the code for it:

```
train_losses, val_losses, train_accuracy,
    val_accuracy = train_model(
        net, train_dataloader, val_dataloader,
        criterion, optimizer, epochs=500
)
```

After 500 epochs, the code output will be something like the following:

```
[epoch 500] Training: loss=0.000 accuracy=1.000 |    Validation:
loss=0.099 accuracy=0.965
```

11. We can now plot the loss for both training and validation sets, as a function of the epoch, and visualize the overfitting effect increasing with the number of epochs:

```
plt.plot(train_losses, label='train')
plt.plot(val_losses, label='valid')
plt.xlabel('epoch')
plt.ylabel('loss (CE)')
plt.legend()
plt.show()
```

Here is the plot for it:

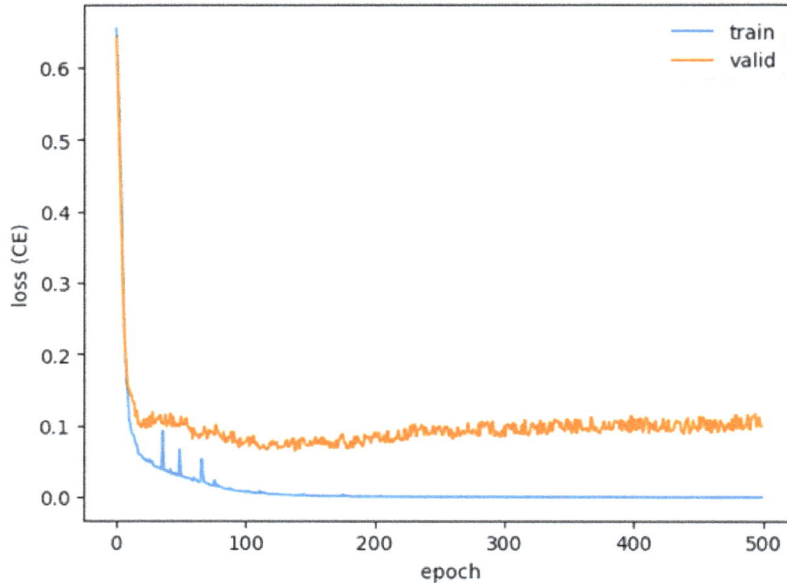

Figure 7.6 – Cross-entropy loss as a function of the epoch. (despite
a few bumps, the training loss keeps decreasing)

We indeed have a training loss that keeps decreasing overall, even reaching a value of zero. On the other hand, the valid loss starts decreasing to reach a minimum somewhere around epoch 100 and then increases slowly over the epochs.

We can implement early stopping to avoid this situation in several ways:

- After the first training, we could retrain the model up to 100 epochs (or any identified optimal validation loss), hopefully having the same results. This would be a waste of CPU time.

- We could save the model at every epoch, and then pick the best one afterward. This solution is sometimes implemented but can be a waste of storage memory, especially for large models.

- We could automatically stop the training after a given number of epochs not improving validation loss. The minimum number of steps without validation loss improvement is usually called patience.

Let's now implement the latter solution.

Note

Using patience is risky too: a too-small patience may get the model stuck in a local minimum, while a too-large patience may miss the actual optimal epoch by stopping too late.

Training with patience and early stopping

Let's now retrain a model using early stopping. We first instantiate a fresh model to avoid training an already trained model:

1. Instantiate a fresh model as well as a fresh optimizer. No need to test it, nor to instantiate the loss again if you are using the same notebook kernel. If you want to run this code separately, *steps 1* to *8* of the previous recipe must be reused:

```
# Instantiate a fresh model
net = Net(X_train.shape[1])
optimizer = torch.optim.Adam(net.parameters(), lr=0.001)
```

2. We now train this model with a patience of 30. After 30 epochs without improving the `val` loss, the training will just stop:

```
train_losses, val_losses, train_accuracy,
    val_accuracy = train_model(
        net, train_dataloader, val_dataloader,
        criterion, optimizer, patience=30, epochs=500
)
```

The code output will be something like the following (the total number of epochs before reaching the early stopping may vary):

```
[epoch 134] Training: loss=0.004 accuracy=1.000 |    Valid:
loss=0.108 accuracy=0.982
```

The training stopped after about 100 epochs (the result may vary since the results are not deterministic by default), with a validation accuracy of about 98%, far better than the 96% that we got after 500 epochs.

3. Let's plot the train and validation losses again as a function of the number of epochs:

```
plt.plot(train_losses, label='train')
plt.plot(val_losses, label='validation')
plt.xlabel('epoch')
plt.ylabel('loss (BCE)')
plt.legend()
plt.show()
```

Here is the plot for it:

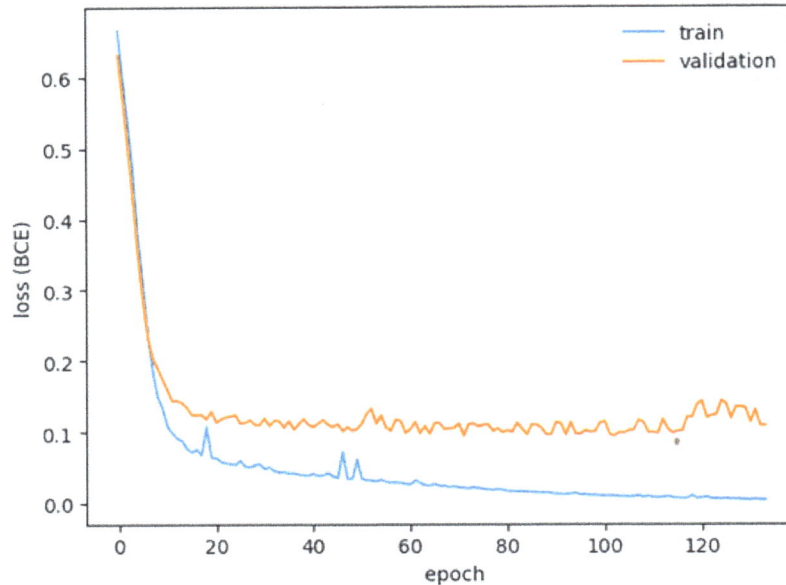

Figure 7.7 – Cross-entropy loss as a function of the epoch

As we can see, the validation loss is already overfitting but did not have time to grow too much, preventing further overfitting.

There's more...

As explained earlier in this recipe, for proper evaluation, it would be necessary to compute the accuracy (or any selected evaluation metric) on a separate test set. Indeed, stopping the training based on the validation set and evaluating the model on this same dataset is a biased approach, and may artificially improve the evaluation.

Regularization with network architecture

In this recipe, we will explore a less popular, but still sometimes useful, regularization method: adapting the neural network architecture. After reviewing why to use this method and when, we will apply it to the California housing dataset, a regression task.

Getting ready

Sometimes, the best way to regularize is not to use any fancy techniques but only common sense. In many cases, it happens that the neural network used is just too large for the input task and dataset. An easy rule of thumb is to have a quick look at the number of parameters in the network (e.g., weights and biases) and compare it to the number of data points: if the ratio is above 1 (i.e., there are more parameters than data points), there is a risk of severe overfitting.

> **Note**
>
> If transfer learning is used, this rule of thumb no longer applies since the network has been trained on a presumably large enough dataset.

If we take a step back and go back to linear models such as linear regression, it is well known that having too many correlated features can deteriorate the model's performance. It can be the same for neural networks: having too many free parameters will do no good to the performances. So, depending on the task, it is not always required to have dozens of layers; just a few may be enough to get the best performances and avoid overfitting. Let's check that in practice on the California dataset.

To do so, the libraries needed are scikit-learn, Matplotlib, and PyTorch. They can be installed with `pip install sklearn matplotlib torch`.

How to do it...

This will be a two-step recipe: first, we will train a large model (compared to the dataset) on the data, to expose the effect of the network on overfitting. Then, we will train another, more adapted model on this same data, hopefully fixing the overfitting issue.

Training a large model

Here are the steps to train a model:

1. The following imports are needed first:

    *   `fetch_california_housing` to load the dataset
    *   `train_test_split` to split the data into training and test sets
    *   `StandardScaler` to rescale the features
    *   `r2_score` to evaluate the model at the end
    *   `matplotlib` to display the loss
    *   `torch` itself for some functions at the lower level of the library
    *   `torch.nn`, which has many useful classes for building a neural network
    *   `torch.nn.functional` for some useful functions
    *   `Dataset` and `DataLoader` for handling the data operations

The following is the code for these `import` statements:

```
import numpy as np
from sklearn.datasets
import fetch_california_housing
from sklearn.model_selection
import train_test_split
from sklearn.preprocessing
import StandardScaler
from sklearn.metrics
import r2_score
import matplotlib.pyplot as plt
import torch
import torch.nn as nn
import torch.nn.functional as F
from torch.utils.data import Dataset, DataLoader
```

2. Load the data using the `fetch_california_housing` function:

```
X, y = fetch_california_housing(return_X_y=True)
```

3. Split the data into training and test sets with a ratio of 80%/20%, using the `train_test_split` function. Set a random state for reproducibility. For `pytorch`, the data is converted in to `float32` variables:

```
X_train, X_test, y_train, y_test = train_test_split(
    X.astype(np.float32), y.astype(np.float32),
    test_size=0.2, random_state=0)
```

4. Rescale the data using the standard scaler:

```
scaler = StandardScaler()
X_train = scaler.fit_transform(X_train)
X_test = scaler.transform(X_test)
```

5. Create the `CaliforniaDataset` class, allowing to handle the data. The only transformation here is the conversion from a numpy array to a `torch` tensor:

```
class CaliforniaDataset(Dataset):
    def __init__(self, X: np.array, y: np.array):
        self.X = torch.from_numpy(X)
        self.y = torch.from_numpy(y)
```

```
def __len__(self) -> int:
    return len(self.X)


def __getitem__(self, idx: int) ->
    tuple[torch.Tensor]: return self.X[idx], self.y[idx]
```

6. Instantiate the datasets for the train and test sets and the data loaders. We define here a batch size of 64 but this can be modified:

```
# Instantiate datasets
training_data = CaliforniaDataset(X_train, y_train)
test_data = CaliforniaDataset(X_test, y_test)
# Instantiate data loaders
train_dataloader = DataLoader(training_data,
    batch_size=64, shuffle=True)
test_dataloader = DataLoader(test_data, batch_size=64,
    shuffle=True)
```

7. Create the neural network architecture. We create a large model here on purpose considering the dataset – 5 hidden layers of 128 units:

```
class Net(nn.Module):
    def __init__(self, input_shape: int,
        hidden_units: int = 128):
            super(Net, self).__init__()
            self.hidden_units = hidden_units
            self.fc1 = nn.Linear(input_shape,
                self.hidden_units)
            self.fc2 = nn.Linear(self.hidden_units,
                self.hidden_units)
            self.fc3 = nn.Linear(self.hidden_units,
                self.hidden_units)
            self.fc4 = nn.Linear(self.hidden_units,
                self.hidden_units)
            self.fc5 = nn.Linear(self.hidden_units,
                self.hidden_units)
            self.output = nn.Linear(self.hidden_units, 1)

    def forward(self, x: torch.Tensor) -> torch.Tensor:
        x = self.fc1(x)
        x = F.relu(x)
        x = self.fc2(x)
        x = F.relu(x)
        x = self.fc3(x)
        x = F.relu(x)
```

```
        x = self.fc4(x)
        x = F.relu(x)
        x = self.fc5(x)
        x = F.relu(x)
        output = self.output(x)
        return output
```

8. Instantiate the model with the given input shape (the number of features). Optionally, we can check the network is correctly created using an input tensor of the expected shape (so here is the number of features):

```
# Instantiate the network
net = Net(X_train.shape[1])
# Generate one random sample of 8 features
random_data = torch.rand((1, X_train.shape[1]))
# Compute the forward propagation
print(net(random_data))
tensor([[0.0674]], grad_fn=<AddmmBackward0>)
```

9. Instantiate the loss to be a mean squared error loss since this is a regression task, and define the optimizer to be Adam, with a learning rate of 0.001:

```
criterion = nn.MSELoss()
optimizer = torch.optim.Adam(net.parameters(), lr=0.001)
```

10. Finally, train the neural network on 500 epochs by using the train_model function. The implementation of this function is similar to previous ones and can be found in the GitHub repository. Again, we purposely chose a large number of epochs; otherwise, the overfitting could be compensated by early stopping. We also store the train and test losses for each epoch, for visualization purposes and information:

```
train_losses, test_losses = train_model(net,
    train_dataloader, test_dataloader, criterion,
    optimizer, 500)
```

After 500 epochs, the final output lines will be like the following:

```
[epoch 500] Training: loss=0.013 | Test: loss=0.312
Finished Training
```

11. Plot the loss for both the train and test set as a function of the epoch:

```
plt.plot(train_losses, label='train')
plt.plot(test_losses, label='test')
plt.xlabel('epoch')
plt.ylabel('loss (MSE)')
plt.legend()
plt.show()
```

Here is the plot for it:

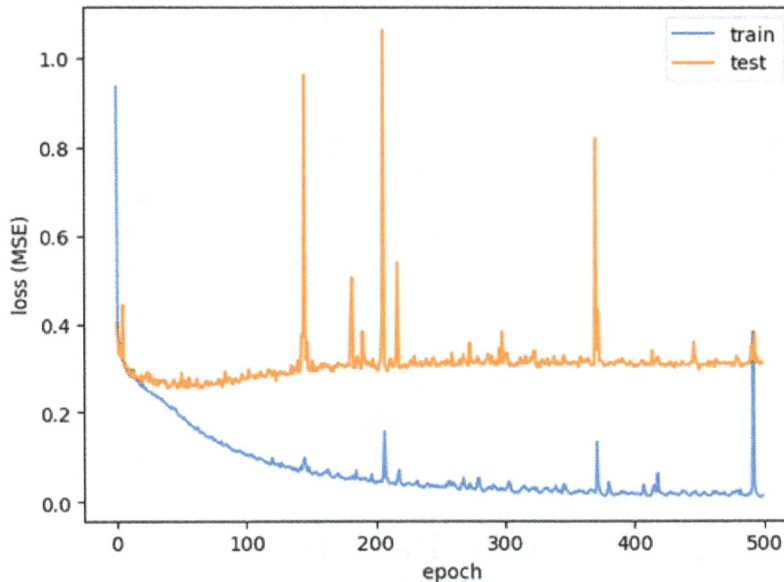

Figure 7.8 – Mean squared error loss as a function of the epoch (note
the clear divergence between the train and test losses)

We can notice that the train loss keeps decreasing over and over, while the test loss soon reaches a plateau before increasing again. This is a clear sign of overfitting. Let's confirm there is overfitting by computing the R2-scores.

12. Finally, let's evaluate the model on both the training and test sets with the R2-score:

```
# Compute the predictions with the trained neural network
y_train_pred = net(
    torch.tensor((X_train))).detach().numpy()
y_test_pred = net(
    torch.tensor((X_test))).detach().numpy()
# Compute the R2-score
print('R2-score on training set:', r2_score(y_train,
    y_train_pred))
print('R2-score on test set:', r2_score(y_test,
    y_test_pred))
```

This code will output values such as the following:

```
R2-score on training set: 0.9922777453770203
R2-score on test set: 0.7610035849523354
```

As expected, we are facing a clear overfitting here, with an almost perfect R2-score on the train set, and an R2-score of about 0.76 on the test set.

> **Note**
>
> This may look like an exaggerated example, but it is fairly easy to choose an architecture that is way too large for the task and dataset.

Regularizing with a smaller network

Let's now train a more reasonable model and see how this impacts overfitting, even with the same number of epochs. The goal is not only to decrease overfitting but also to get better performances on the test set.

If you are using the same kernel, there is no need to redo the first steps. Otherwise, *steps 1 to 6* must be redone:

1. Define the neural network. This time, we only have two hidden layers of 16 units each, so this is much smaller than earlier:

```
class Net(nn.Module):
    def __init__(self, input_shape: int,
        hidden_units: int = 16):
            super(Net, self).__init__()
        self.hidden_units = hidden_units
```

```
        self.fc1 = nn.Linear(input_shape,
            self.hidden_units)
        self.fc2 = nn.Linear(self.hidden_units,
            self.hidden_units)
        self.output = nn.Linear(self.hidden_units, 1)

    def forward(self, x: torch.Tensor) -> torch.Tensor:
        x = self.fc1(x)
        x = F.relu(x)
        x = self.fc2(x)
        x = F.relu(x)
        output = self.output(x)
        return output
```

2. Instantiate the network with the expected number of input features and the optimizer:

```
# Instantiate the network
net = Net(X_train.shape[1])
optimizer = torch.optim.Adam(net.parameters(),
    lr=0.001)
```

3. Train the neural network over 500 epochs so that we have results that we can compare to the previous ones. We will reuse the `train_model` function already used earlier in this recipe:

```
train_losses, test_losses = train_model(net,
    train_dataloader, test_dataloader, criterion,
    optimizer, 500)
[epoch 500] Training: loss=0.248 | Test: loss=0.273
Finished Training
```

4. Plot the loss as a function of the epoch for the train and test sets:

```
plt.plot(train_losses, label='train')
plt.plot(test_losses, label='test')
plt.xlabel('epoch')
plt.ylabel('loss (MSE)')
plt.legend()
plt.show()
```

Here is the plot for it:

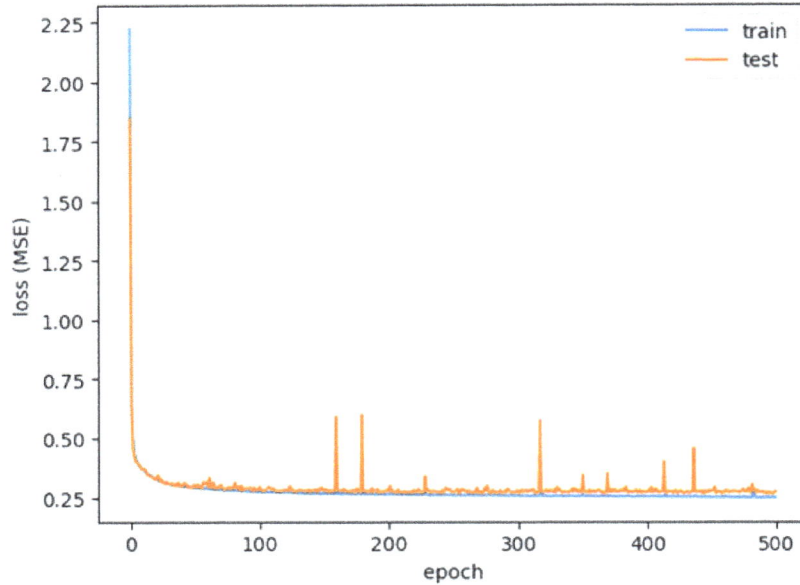

Figure 7.9 – Mean squared error loss as a function of the epoch
(note the train and test sets almost overlapping)

As we can see, this time, even with many epochs, there is no strong overfitting: the train and test losses remain close to each other no matter the number (except for a few noise bumps), even if a small amount of overfitting seems to appear over time.

5. Let's again evaluate the model with the R2-score on the training and test sets:

```
# Compute the predictions with the trained neural network
y_train_pred = net(
    torch.tensor((X_train))).detach().numpy()
y_test_pred = net(
    torch.tensor((X_test))).detach().numpy()
# Compute the R2-score
print('R2-score on training set:', r2_score(y_train,
    y_train_pred))
print('R2-score on test set:', r2_score(y_test,
    y_test_pred))
```

Here is the typical output of this code:

```
R2-score on training set: 0.8161885562733123
R2-score on test set: 0.7906037325658601
```

While the R2-score on the training set decreased from 0.99 to 0.81, the score on the test set increased from 0.76 to 0.79, effectively improving the performance of the model.

Even if it was a rather extreme example, the general idea remains true.

> **Note**
> Early stopping could work well too in this case. The two techniques (early stopping and downsizing the network) are not mutually exclusive and can work well together.

There's more...

The model complexity can arguably be computed using the number of parameters. Even if it's not a direct measure, it remains a good indicator.

For example, the first neural network used in this recipe, with 10 hidden layers of 128 units, had 67,329 trainable parameters. On the other side, the second neural network, with only 2 hidden layers of 16 units, had only 433 trainable parameters.

The number of parameters in a fully connected neural network is based on the number of units and the number of layers: both units and layers do not have to be the same on the number of parameters though.

To compute the number of trainable parameters in the torch network's net, we can use the following code snippet:

```
sum(p.numel() for p in net.parameters() if p.requires_grad)
```

To get an idea, let's take again three examples of neural networks with the same number of neurons, but with a different number of layers. Let's assume they all have 10 input features and 1 unit output layer:

- A neural network with 1 hidden layer of 100 units: 1,201 parameters
- A neural network with 2 hidden layers of 50 units: 3,151 parameters
- A neural network with 10 hidden layers of 10 units: 1,111 parameters

So, there is a trade-off between the number of layers and the number of units per layer to get the most complex neural network for a given number of neurons.

Regularizing with dropout

A widely used method for regularizing is dropout. Dropout is just randomly setting some neurons' activations to zero during the training phase. Let's first review how this works and then apply it to a multiclass classification task, the `sklearn` digits dataset, which is kind of an older and smaller version of the MNIST dataset.

Getting ready

Dropout is a widely adopted regularization approach in deep learning, due to its simplicity and effectiveness. The technique is easy to understand, yet can yield powerful results.

The principle is simple – during training, we randomly ignore some units by setting their activations to zero, as represented in *Figure 7.10*:

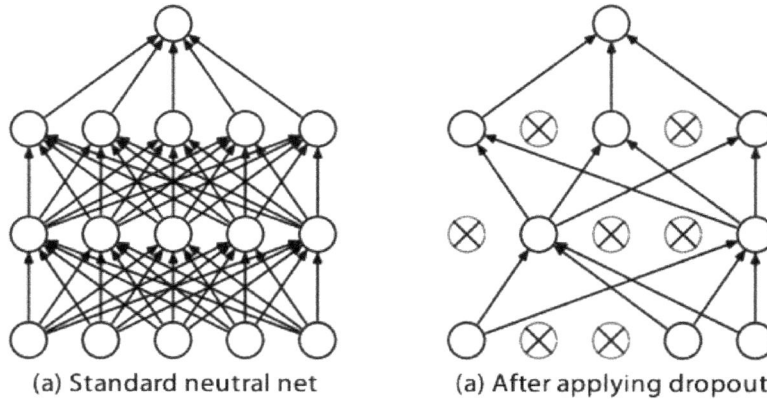

(a) Standard neutral net   (a) After applying dropout

Figure 7.10 – On the left, a standard neural network with its connections, and, on the right, the same neural network with dropout, having, on average, 50% of its neurons ignored at training

Dropout adds one hyperparameter though: the dropout probability. For a 0% probability, there is no dropout. For a 50% probability, about 50% of the neurons will be randomly selected to be ignored. For a 100% probability, well, there is nothing left to learn. The ignored neurons are not always the same: for each new batch size, a new set of units is randomly selected to be ignored.

> **Note**
>
> The remaining activations are consequently scaled to keep a consistent global input for any unit. In practice, for a dropout probability of 1/2, all the neurons that are not ignored are scaled by a factor of 2 (i.e., their activations are multiplied by 2).

Certainly, when evaluating or inferring on new data, dropout is deactivated, causing all neurons to be activated.

But what is the point of doing that? Why would randomly ignoring some neurons help? A formal explanation is beyond the scope of this book, but at least we can provide some intuition. The idea is to avoid confusing the neural network with too much information. As a human, having too much information can hurt more than it helps: sometimes, having less information allows you to make better decisions, preventing you from being flooded by it. This is the idea of dropout: instead of giving the network all the information at once, it is gently trained with less information by turning off a few neurons randomly for a short amount of time. Hopefully, this will help the network make better decisions in the end.

In this recipe, this will be run on the `digits` dataset of scikit-learn, which is just a link to the *Optical Recognition of Handwritten Digits* dataset. A small subset of these images is represented in *Figure 7.11*.

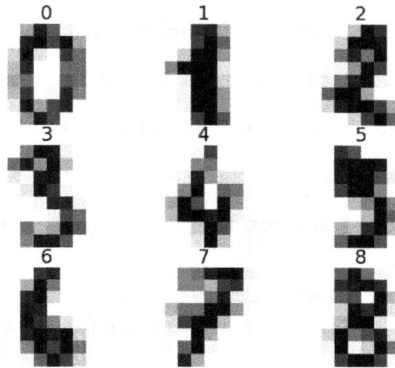

Figure 7.11 – A sample of images from the dataset and their labels: each image is composed of 8x8 pixels

Each image is an 8x8-pixel picture of a handwritten digit. Thus, the dataset is made up of 10 classes, 1 for each digit.

To run the code of this recipe, the required libraries are `sklearn`, `matplotlib`, and `torch`. They can be installed with `pip install sklearn matplotlib torch`.

How to do it...

This recipe will comprise two steps:

1. First, we will train a neural network without dropout with a rather large model, considering the data.
2. Then, we will train the same neural network with dropout, hopefully, to improve the model's performance.

We will use the same data for both configurations, the same batch size, and the same number of epochs, so that we can compare the results.

Without dropout

Here are the steps to regularize without dropout:

1. The following imports must be loaded:

    *   `load_digits` from `sklearn` to load the dataset

    *   `train_test_split` from `sklearn` to split the dataset

    *   `torch`, `torch.nn`, and `torch.nn.functional` for the neural network

    *   `Dataset` and `DataLoader` from `torch` for the dataset loading in `torch`

    *   `matplotlib` for the visualization of the loss

    Here is the code for the `import` statements:

    ```python
    import numpy as np
    from sklearn.datasets import load_digits
    from sklearn.model_selection import train_test_split
    import torch
    import torch.nn as nn
    import torch.nn.functional as F
    from torch.utils.data import Dataset, DataLoader
    import matplotlib.pyplot as plt
    ```

2. Load the data. The dataset is made of 1,797 samples, and the images are already flattened to 64 values between 0 and 16 for the 8x8 pixels:

    ```python
    X, y = load_digits(return_X_y=True)
    ```

3. Split the data into training and test sets, with 80% in the training set and 20% in the test set. The features are converted in to `float32`, while the labels are converted into `int64` to avoid `torch` errors later:

    ```python
    X_train, X_test, y_train, y_test = train_test_split(
        X.astype(np.float32), y.astype(np.int64),
        test_size=0.2, random_state=0)
    ```

4. Create the `DigitsDataset` class for PyTorch. The only transformation to the features, besides converting them into `torch` tensors, is to divide the values by 255 to have a range of features in `[0, 1]`:

    ```python
    class DigitsDataset(Dataset):
        def __init__(self, X: np.array, y: np.array):
            self.X = torch.from_numpy(X/255)
            self.y = torch.from_numpy(y)
    ```

```
def __len__(self) -> int:
    return len(self.X)

def __getitem__(self, idx: int) -> tuple[torch.Tensor]:
    return self.X[idx], self.y[idx]
```

5. Instantiate the datasets for the train and test sets and the data loaders with a batch size of 64:

```
# Instantiate datasets
training_data = DigitsDataset(X_train, y_train)
test_data = DigitsDataset(X_test, y_test)
# Instantiate data loaders
train_dataloader = DataLoader(training_data,
    batch_size=64, shuffle=True)
test_dataloader = DataLoader(test_data, batch_size=64,
    shuffle=True)
```

6. Define the neural network architecture – here, there are 3 hidden layers of 128 units (by default) and a dropout probability set to 25% applied to all the hidden layers:

```
class Net(nn.Module):
    def __init__(self, input_shape: int,
        hidden_units: int = 128,
        dropout: float = 0.25):
            super(Net, self).__init__()
            self.hidden_units = hidden_units
            self.fc1 = nn.Linear(input_shape,
                self.hidden_units)
            self.fc2 = nn.Linear(self.hidden_units,
                self.hidden_units)
            self.fc3 = nn.Linear(self.hidden_units,
                self.hidden_units)
            self.dropout = nn.Dropout(p=dropout)
            self.output = nn.Linear(self.hidden_units, 10)

    def forward(self, x: torch.Tensor) -> torch.Tensor:
        x = self.fc1(x)
        x = F.relu(x)
        x = self.dropout(x)
        x = self.fc2(x)
        x = F.relu(x)
        x = self.dropout(x)
        x = self.fc3(x)
        x = F.relu(x)
        x = self.dropout(x)
```

```
output = torch.softmax(self.output(x), dim=1)
return output
```

Here, dropout is added in two steps:

- Instantiate an `nn.Dropout(p=dropout)` class in the constructor, having the provided dropout probability

- Apply the dropout layer (defined in the constructor) after the activation function for each hidden layer with `x = self.dropout(x)`

Note

In the case of a ReLU activation function, setting the dropout before or after the activation function won't change the output. For other activation functions such as the sigmoid, though, this makes a difference.

---

7. Instantiate the model with the right input shape of 64 (8x8 pixels) and a dropout of 0 since we want to check the results without dropout first. Check the forward propagation works properly on a given random tensor:

```
# Instantiate the model
net = Net(X_train.shape[1], dropout=0)
# Generate randomly one random 28x28 image as a 784 values
tensor
random_data = torch.rand((1, 64))
result = net(random_data)
print('Resulting output tensor:', result)
print('Sum of the output tensor:', result.sum())
```

The output of this code should look like the following:

```
Resulting output tensor: tensor([[0.0964, 0.0908, 0.1043,
0.1083, 0.0927, 0.1047, 0.0949, 0.0991, 0.1012,
        0.1076]], grad_fn=<SoftmaxBackward0>)
Sum of the output tensor: tensor(1., grad_fn=<SumBackward0>)
```

8. Define the loss function as the cross-entropy loss and the optimizer as Adam:

```
criterion = nn.CrossEntropyLoss()
optimizer = torch.optim.Adam(net.parameters(), lr=0.001)
```

9. Train the neural network on 500 epochs using the `train_model` function available in the GitHub repository. For each epoch, we store and compute the loss and the accuracy for both the training and test sets:

```
train_losses, test_losses, train_accuracy,
    test_accuracy = train_model(
        net, train_dataloader, test_dataloader,
        criterion, optimizer, epochs=500
    )
```

After 500 epochs, you should get an output like this:

```
[epoch 500] Training: loss=1.475 accuracy=0.985 |          Test:
loss=1.513 accuracy=0.947
Finished Training
```

10. Plot the cross-entropy loss for both the training and test sets as a function of the epoch number:

```
plt.plot(train_losses, label='train')
plt.plot(test_losses, label='test')
plt.xlabel('epoch')
plt.ylabel('loss (CE)')
plt.legend()
plt.show()
```

Here is the plot for it:

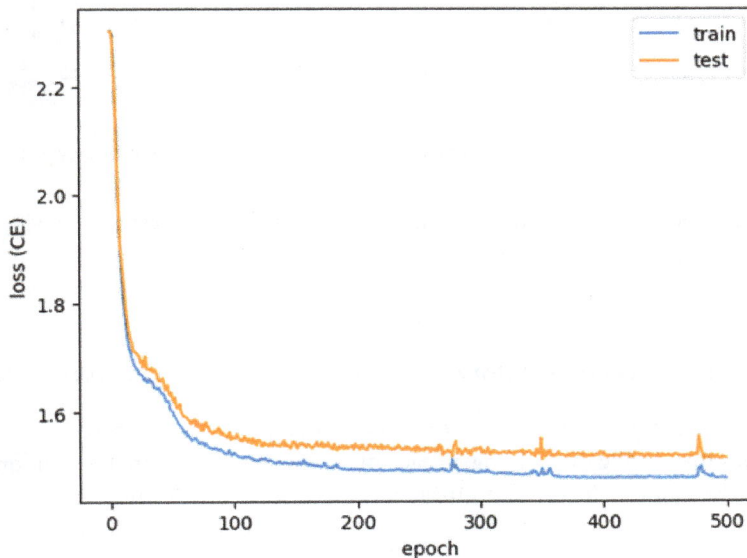

Figure 7.12 – Cross-entropy loss as a function of the epoch (note the slight divergence between train and test sets)

11. Plotting the accuracy will show the equivalent results:

```
plt.plot(train_accuracy, label='train')
plt.plot(test_accuracy, label='test')
plt.xlabel('epoch')
plt.ylabel('Accuracy')
plt.legend()
plt.show()
```

Here is the plot for it:

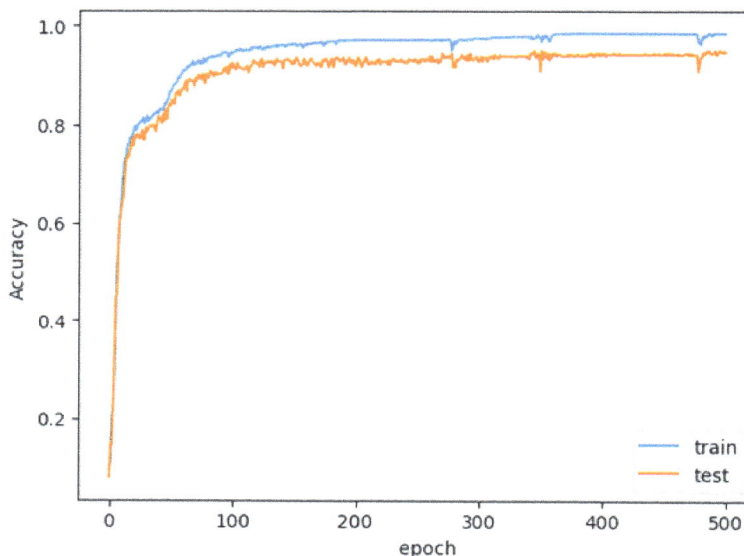

Figure 7.13 – Accuracy as a function of the epoch; we can again notice the overfitting

The final accuracy is about 98% on the train set and only about 95% on the test set, showing overfitting. Let's try now to add dropout to reduce this overfitting.

With dropout

In this part, we will simply restart from *step 7*, but with dropout, and then compare the results:

1. Instantiate the model with an input share of 64 and a dropout probability of 25%. A probability of 25% means that during the training, in each of the hidden layers, about 32 randomly selected neurons will be ignored. Instantiate a fresh optimizer, still using Adam:

```
# Instantiate the model
net = Net(X_train.shape[1], dropout=0.25)
optimizer = torch.optim.Adam(net.parameters(), lr=0.001)
```

2. Train the neural network again for 500 epochs, while storing the train and test loss and accuracy:

```
train_losses, test_losses, train_accuracy, test_accuracy =
train_model(
    net, train_dataloader, test_dataloader, criterion,
        optimizer, epochs=500
)
[epoch 500] Training: loss=1.472 accuracy=0.990 |         Test:
loss=1.488 accuracy=0.975
Finished Training
```

3. Plot the train and test losses again as a function of the epoch:

```
plt.plot(train_losses, label='train')
plt.plot(test_losses, label='test')
plt.xlabel('epoch')
plt.ylabel('loss (CE)')
plt.legend()
plt.show()
```

Here is the plot for it:

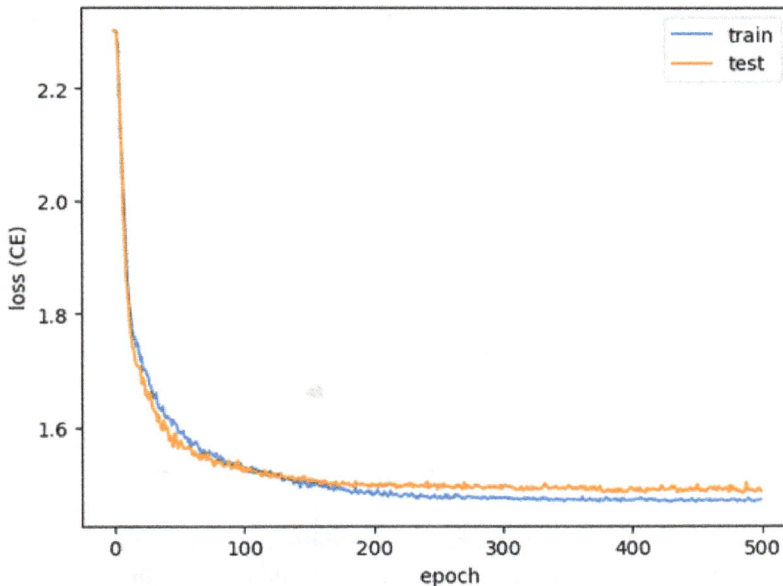

Figure 7.14 – Cross-entropy loss as a function of the epoch, with reduced divergence thanks to dropout

We face a different behavior here than seen previously. The train and test losses do not seem to grow apart too much with the epochs. During the initial 100 epochs, the test loss is marginally lower than the train loss, but afterward, the train loss decreases further, indicating slight overfitting of the model.

4. Finally, plot the train and test accuracy as a function of the epoch:

```
plt.plot(train_accuracy, label='train')
plt.plot(test_accuracy, label='test')
plt.xlabel('epoch')
plt.ylabel('Accuracy')
plt.legend()
plt.show()
```

Here is the plot for it:

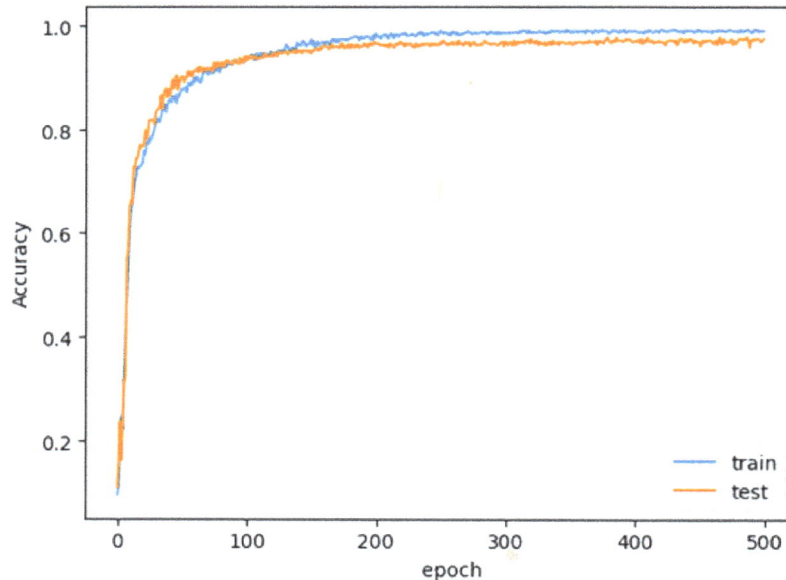

Figure 7.15 – Accuracy as a function of the epoch (the overfitting is largely reduced thanks to dropout)

We have a train accuracy of 99% against the 98% seen previously. More interestingly, the test accuracy climbed to 97%, from 95% previously, effectively regularizing and reducing the overfitting.

There's more...

Although dropout is not always foolproof, it has been demonstrated to be an effective regularization technique, particularly when training large networks on small datasets. More about this can be found in the publication *Improving neural networks by preventing co-adaptation of feature detectors*, by Hinton et al. This publication can be found here on `arxiv`: `https://arxiv.org/abs/1207.0580`.

See also

- The official location of the `digits` dataset: `https://archive.ics.uci.edu/ml/datasets/Optical+Recognition+of+Handwritten+Digits`
- The PyTorch documentation about dropout: `https://pytorch.org/docs/stable/generated/torch.nn.Dropout.html`

8

Regularization with Recurrent Neural Networks

In this chapter, we will work with **Recurrent Neural Networks** (**RNNs**). As we will see, they are well suited for **Natural Language Processing** (**NLP**) tasks, even if they also apply well to time series tasks. After learning how to train RNNs, we will apply several regularization methods, such as using dropout and the sequence maximum length. This will allow you to gain foundational knowledge that can be applied to NLP or time series-related tasks. This will also give you the necessary knowledge to understand more advanced techniques covered in the next chapter.

In this chapter, we'll cover the following recipes:

- Training an RNN
- Training a **Gated Recurrent Unit** (**GRU**)
- Regularizing with dropout
- Regularizing with a maximum sequence length

Technical requirements

In this chapter, we will train RNNs on various tasks using the following libraries:

- NumPy
- pandas
- scikit-learn
- Matplotlib
- PyTorch
- Transformers

Training an RNN

In NLP, input data is commonly textual data. Since a text is usually nothing but a sequence of words, using RNNs is sometimes a good solution. Indeed, RNNs, unlike fully connected networks, consider data's sequential information.

In this recipe, we will train an RNN on tweets to predict whether they are positive, negative, or neutral.

Getting started

In NLP, we usually manipulate textual data, which is unstructured. To handle it properly, this is usually a multi-step process – first, convert the text into numbers, and then only train a model on those numbers.

There are several ways to convert text into numbers. In this recipe, we will use a simple approach called **tokenization**. Tokenization is just converting a sentence into tokens. A token can be as simple as a word, so that a sentence like *"The dog is out"* would be tokenized as `['the', 'dog', 'is', 'out']`. There is usually one more step in the tokenization process – once the sentence is converted into a list of words, it must be converted to a number. Each word is assigned to a number so that the sentence *"The dog is out"* could be tokenized as `[3, 198, 50, 3027]`.

> **Tip**
>
> This is a quite simplistic explanation of tokenization. Check the *See also* subsection for more resources.

In this recipe, we will train an RNN on tweets for a multiclass classification task. However, how does RNN work? An RNN takes as input a sequence of features, as represented in *Figure 8.1*.

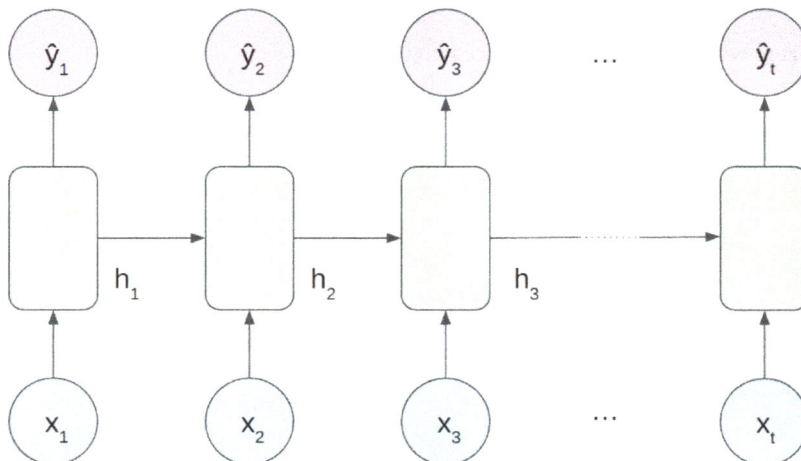

Figure 8.1 – A representation of an RNN. At the bottom level is the input features,
in the middle is the hidden layers, and at the top level is the output layer

In *Figure 8.1*, the hidden layer of an RNN has two inputs and two outputs:

- **Inputs:** The features at the current step, x_t, and the hidden state of the previous step, h_{t-1}
- **Outputs:** The hidden state, h_t (fed to the next step), and this step's activation output a_t

In the case of a one-layer RNN, the activation function is simply the output $\hat{y}_t$.

Going back to our example, the input features are the tokens. So, at each sequence step, one or more layers of neural networks take as input both the features that are at this sequence step and the hidden state of the previous step.

> **Important note**
>
> RNNs can be used in other contexts, such as forecasting, where the input features can be both quantitative and qualitative features.

RNNs also have several sets of weights. As represented in *Figure 8.2*, there are three sets of weights:

- W_{hh} : The weights applied to the hidden state of the previous step, for the current hidden state computation
- W_{xh}: The weights applied to the input features, for the current hidden state computation
- W_{hy}: The weights applied to the current hidden state, for the current output

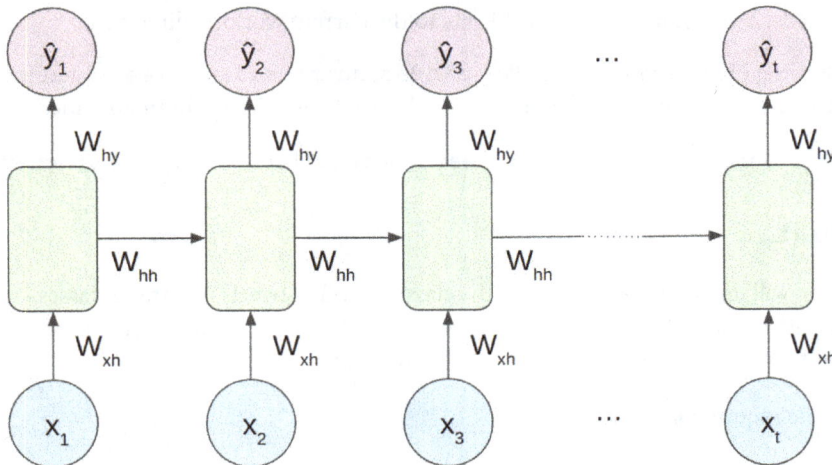

Figure 8.2 – A representation of an RNN with different sets of weights

Overall, considering all of this, the computation of the hidden state and the activation output can be computed as follows:

$$h_t = g(W_{hh}h_{t-1} + W_{xh}X_t + b_h)$$

$$\hat{y}_t = softmax\,(W_{hy}h_t + b_y)$$

Here, g is the activation function, and b_h and b_y are biases. We use *softmax* here for the output computation, assuming it is a multiclass classification, but any activation function is possible depending on the task.

Finally, the loss can be computed easily, as for any other machine-learning task (for example, for a classification task, a cross-entropy loss can be computed between the ground truth and the neural network's output). Backpropagation on such neural networks, called **backpropagation through time**, is beyond the scope of this book.

On a practical side, for this recipe, we will need a Kaggle dataset. To get this dataset, once the Kaggle API has been set, the following command lines can be used to get the dataset in the current working directory:

```
kaggle datasets download -d crowdflower/twitter-airline-sentiment
--unzip
```

This line should download a `.zip` file and unzip its content, and then a file named `Tweets.csv` should be available. You can move or copy this file to the current working directory.

Finally, the following libraries must be installed: `pandas`, `numpy`, `scikit-learn`, `matplotlib`, `torch`, and `transformers`. They can be installed with the following command line:

```
pip install pandas numpy scikit-learn matplotlib torch transformers
```

How to do it...

In this recipe, we will use an RNN to perform the classification of tweets into three classes – negative, neutral, and positive. As explained in the previous section, this will be a multi-step process – first, a tokenization of the tweet's texts, and then just model training:

1. Import the required libraries:

    *   `torch` and some related modules and classes for the neural network
    *   `train_test_split` and `LabelEncoder` from scikit-learn for preprocessing
    *   `AutoTokenizer` from Transformers to tokenize the tweets

- pandas to load the dataset

- matplotlib for visualization:

```
import torch import torch.nn as nn
import torch.optim as optim from torch.utils.data
import DataLoader, Dataset from sklearn.model_selection
import train_test_split from sklearn.preprocessing
import LabelEncoder from transformers
import AutoTokenizer
import pandas as pd
import matplotlib.pyplot as plt
```

2. Load the data from the .csv file with pandas:

```
# Load data
data = pd.read_csv('Tweets.csv')
data[['airline_sentiment', 'text']].head()
```

The output will be the following:

	airline_sentiment	Text
0	Neutral	@VirginAmerica What @dhepburn said.
1	Positive	@VirginAmerica plus you've added commercials t...
2	Neutral	@VirginAmerica I didn't today... Must mean I n...
3	Negative	@VirginAmerica it's really aggressive to blast...
4	Negative	@VirginAmerica and it's a really big bad thing...

Table 8.1 – Output with the data classified

The data we will use is made of labels from the airline_sentiment column (either negative, neutral, or positive) and their associated raw tweets texts from the text column.

3. Split the data into train and test sets, using the train_test_split function, with a test size of 20% and a specified random state for reproducibility:

```
# Split data into train and test sets
train_data, test_data = train_test_split(data,
    test_size=0.2, random_state=0)
```

4. Implement the `TextClassificationDataset` dataset class, handling the data. At instance creation, this class will so the following:

    -   Instantiate `AutoTokenizer` from Transformers

    -   Tokenize the tweets with that previously instantiated tokenizer and store the results

    -   Encode the labels and store them:

    ```python
    # Define dataset class class TextClassificationDataset(Dataset):
    def __init__(self, data, max_length):
        self.data = data
        self.tokenizer = AutoTokenizer.from_pretrained(
            'bert-base-uncased')
        self.tokens = self.tokenizer(
            data['text'].to_list(), padding=True,
            truncation=True, max_length=max_length,
            return_tensors='pt')['input_ids']
        le = LabelEncoder()
        self.labels = torch.tensor(le.fit_transform(
            data['airline_sentiment']))
    def __len__(self):
        return len(self.data)
    def __getitem__(self, index):
        return self.tokens[index], self.labels[index]
    ```

Several options are specified with this tokenizer:

-   It is instantiated with the `'bert-base-uncased'` tokenizer, a tokenizer used for BERT models

-   The tokenization is made, with a maximum length provided as a constructor argument

-   The padding is set to `True`, meaning that if a tweet has less than the maximum length, it will be filled with zeros to match that length

-   The truncation is set to `True`, meaning that if a tweet has more than the maximum length, the remaining tokens will be ignored

-   The return tensor is specified as `'pt'` so that it returns a PyTorch tensor

---

Tip

See the *There's more...* subsection for more details about what the tokenizer does.

5. Instantiate the `TextClassificationDataset` objects for the train and test sets, as well as the related data loaders. We specify here a maximum number of words of 24 and a batch size of 64. This means that each tweet will be converted into a sequence of exactly 24 tokens:

```
batch_size = 64 max_length = 24
# Initialize datasets and dataloaders
train_dataset = TextClassificationDataset(train_data,
    max_length)
test_dataset = TextClassificationDataset(test_data,
    max_length)
train_dataloader = DataLoader(train_dataset,
    batch_size=batch_size, shuffle=True)
test_dataloader = DataLoader(test_dataset,
    batch_size=batch_size, shuffle=True)
```

6. Implement the RNN model:

```
# Define RNN model
class RNNClassifier(nn.Module):
    def __init__(self, vocab_size, embedding_dim,
        hidden_size, output_size, num_layers=3):
            super(RNNClassifier, self).__init__()
            self.num_layers = num_layers
            self.hidden_size = hidden_size
            self.embedding = nn.Embedding(
                num_embeddings=vocab_size,
                embedding_dim=embedding_dim)
            self.rnn = nn.RNN(
                input_size=embedding_dim,
                hidden_size=hidden_size,
                num_layers=num_layers,
                nonlinearity='relu',
                batch_first=True)
            self.fc = nn.Linear(hidden_size, output_size)
    def forward(self, inputs):
        batch_size = inputs.size(0)
        zero_hidden = torch.zeros(self.num_layers,
            batch_size, self.hidden_size)
        embedded = self.embedding(inputs)
        output, hidden = self.rnn(embedded, zero_hidden)
        output = torch.softmax(self.fc(output[:, -1]),
            dim=1)
        return output
```

The RNN model defined here can be described in several steps:

- An embedding that takes the tokens as input, with the input the size of the vocabulary and the output the size of the given embedding dimension

- Three layers of RNN that take as input the embedding output, with the given number of layers, a hidden size, and a ReLU activation function

- Finally, an embedding that takes the tokens as input, with the input the size of the vocabulary and the output the size of the given embedding dimension; note that the output is computed only for the last sequence step (that is, `output[:, -1]`), and a softmax activation function is applied

> **Important note**
>
> The output is not necessarily computed only for the last sequence step. Depending on the task, it can be useful to output a value at each step alike (for example, forecasting) or only a final value (for example, classification task).

7. Instantiate and test the model. The vocabulary size is given by the tokenizer and the output size of three is defined by the task; there are three classes (negative, neutral, positive). The other arguments are hyperparameters; here, the following values are chosen:

- An embedding dimension of 64

- A hidden dimension of 64

Other values can, of course, be tested:

```
vocab_size = train_dataset.tokenizer.vocab_size
embedding_dim = 64
hidden_dim = 64
output_size = 3
model = RNNClassifier(
    vocab_size=vocab_size,
    embedding_dim=embedding_dim,
    hidden_size=hidden_dim,
    output_size=output_size, )
random_data = torch.randint(0, vocab_size,
    size=(batch_size, max_length))
result = model(random_data)
print('Resulting output tensor:', result.shape) print('Sum of
the output tensor:', result.sum())
```

The code will output the following:

```
Resulting output tensor: torch.Size([64, 3]) Sum of the output
tensor: tensor(64.0000, grad_fn=<SumBackward0>)
```

8. Instantiate the optimizer; here, we will use an Adam optimizer, with a learning rate of 0.001. The loss is the cross-entropy loss, since this is a multiclass classification task:

```
optimizer = optim.Adam(model.parameters(), lr=0.001)
criterion = nn.CrossEntropyLoss()
```

9. Let's define two helper functions to train the model.

    epoch_step_tweet will compute the loss and accuracy for one epoch, as well as update the weights for the training set:

```
def epoch_step_tweet(model, dataloader,
    training_set: bool):
        running_loss = 0
        correct = 0.
    for i, data in enumerate(dataloader, 0):
        # Get the inputs: data is a list of [inputs, labels]
        inputs, labels = data
        if training_set:
            # Zero the parameter gradients
            optimizer.zero_grad()
        # Forward + backward + optimize
        outputs = model(inputs)
        loss = criterion(outputs, labels) .long()
        if training_set:
            loss.backward()
            optimizer.step()
        # Add correct predictions for this batch
        correct += (outputs.argmax(
            dim=1) == labels).float().sum()
        # Compute loss for this batch
        running_loss += loss.item()

    return running_loss, correct
```

    train_tweet_classification will use loop over the epochs and use epoch_step_tweet to compute and store the loss and accuracy:

```
def train_tweet_classification(model,
    train_dataloader, test_dataloader, criterion,
    epochs: int = 20):
        # Train the model
```

```
            train_losses = []
            test_losses = []
            train_accuracy = []
            test_accuracy = []

        for epoch in range(20):
            running_train_loss = 0.
            correct = 0.
            model.train()
            running_train_loss,
            correct = epoch_step_tweet(model,
                dataloader=train_dataloader,
                training_set=True)
            # Compute and store loss and accuracy for this epoch
            train_epoch_loss = running_train_loss / len(
                train_dataloader)
            train_losses.append(train_epoch_loss)
            train_epoch_accuracy = correct / len(
                train_dataset)
            train_accuracy.append(train_epoch_accuracy)

            ## Evaluate the model on the test set
            running_test_loss = 0.
            correct = 0.
            model.eval()
            with torch.no_grad():
                running_test_loss,
                correct = epoch_step_tweet(model,
                    dataloader=test_dataloader,
                    training_set=False)

                test_epoch_loss = running_test_loss / len(
                    test_dataloader)
                test_losses.append(test_epoch_loss)
                test_epoch_accuracy = correct / len(
                test_dataset)
                test_accuracy.append(test_epoch_accuracy)

            # Print stats
            print(f'[epoch {epoch + 1}] Training: loss={train_epoch_
        loss:.3f} accuracy={train_epoch_accuracy:.3f} |\
            \t Test: loss={test_epoch_loss:.3f} accuracy={test_epoch_
        accuracy:.3f}')
```

```
    return train_losses, test_losses, train_accuracy,
    test_accuracy
```

10. Reusing the helper functions, we can now train the model on 20 epochs. Here, we will compute and store the accuracy and the loss for both the train and test sets at each epoch, to plot them afterward:

```
train_losses, test_losses, train_accuracy, test_accuracy =
train_tweet_classification(model,
    train_dataloader, test_dataloader, criterion,
    epochs=20)
```

After 20 epochs, the output should be something like the following:

```
[epoch 20] Training: loss=0.727 accuracy=0.824 |    Test:
loss=0.810 accuracy=0.738
```

11. Plot the loss as a function of the epoch number, for both the train and test sets:

```
plt.plot(train_losses, label='train')
plt.plot(test_losses, label='test')
plt.xlabel('epoch') plt.ylabel('loss (CE)')
plt.legend() plt.show()
```

Here is the resulting graph:

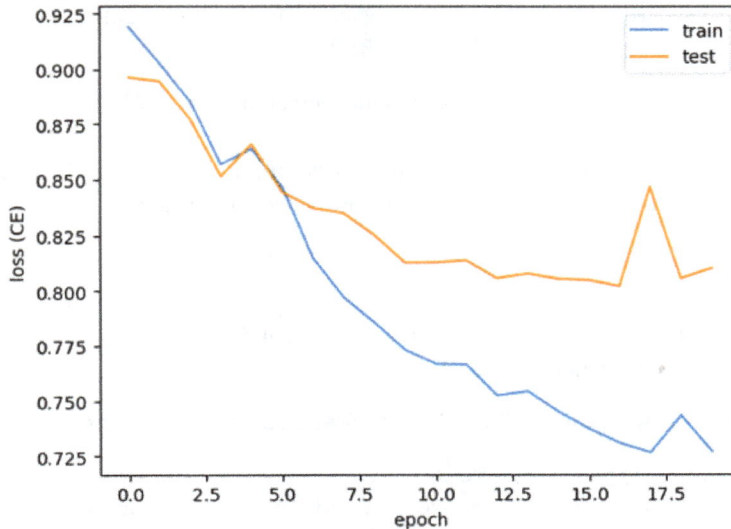

Figure 8.3 – Cross-entropy loss as a function of the epoch

We can see some overfitting as early as the fifth epoch, since the train loss keeps decreasing while the test loss reaches a plateau.

12. Similarly, plot the accuracy as a function of the epoch number of both the train and test sets:

```
plt.plot(train_accuracy, label='train')
plt.plot(test_accuracy, label='test')
plt.xlabel('epoch') plt.ylabel('Accuracy')
plt.legend() plt.show()
```

We then get this graph:

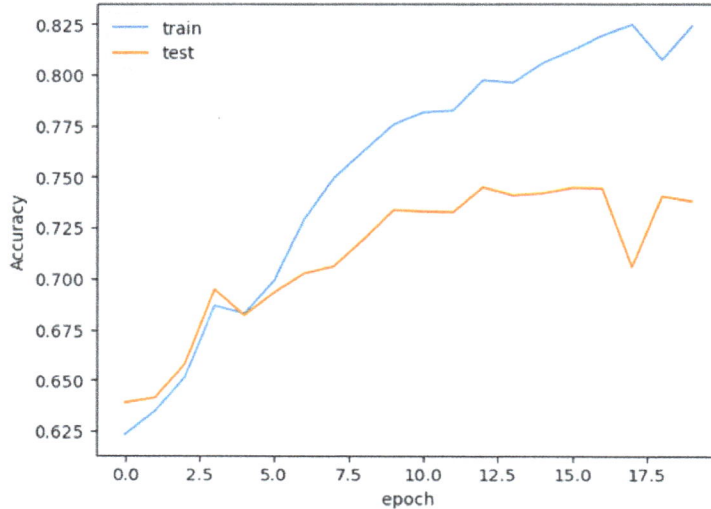

Figure 8.4 – Accuracy as a function of the epoch

After 20 epochs, the accuracy of the train set is about 82%, but it is only about 74% on the test set, meaning there might be room for improvement with proper regularization.

There's more...

In this recipe, we used the HuggingFace tokenizer, but what does it actually do? Let's have a look at a text example to fully understand what it is.

First, let's define a brand-new tokenizer with the AutoTokenizer class, specifying the BERT tokenizer:

```
from transformers import AutoTokenizer
tokenizer = AutoTokenizer.from_pretrained('bert-base-uncased')
```

> **Important note**
>
> There are many tokenizers that have different methods and, thus, different outputs for the same given text. `'bert-base-uncased'` is quite a common one, but many others can be used.

Let's now apply this tokenizer to a given text, using the `tokenize` method, to see what the output is:

```
tokenizer.tokenize("Let's use regularization in ML. Regularization
should help to improve model robustness")
```

The code output is the following:

```
['let', "'", 's', 'use', 'regular', '##ization', 'in',
 'ml', '.', 'regular', '##ization', 'should', 'help',
 'to', 'improve', 'model', 'robust', '##ness']
```

So, the tokenization can be described as splitting a sentence into smaller chunks. Other tokenizers can have different chunks (or tokens) at the end, but the process remains essentially the same.

Now, if we just apply the tokenization in this same sentence, we can get the token numbers with `'input_ids'`:

```
tokenizer("Let's use regularization in ML. Regularization should help
to improve model robustness")['input_ids']
```

The code output is now the following:

```
[101, 2292, 1005, 1055, 2224, 3180, 3989, 1999, 19875,
 1012, 3180, 3989, 2323, 2393, 2000, 5335, 2944, 15873,
 2791, 102]
```

> **Important note**
>
> Note that the 3180 and 3989 tokens are present twice. Indeed, the word `regularization` (tokenized as two separate tokens) is present twice.

For a given tokenizer, the vocabulary size is just the number of existing tokens. This is stored in the `vocab_size` attribute. In this case, the vocabulary size is 30522.

> **Tip**
>
> If you're curious, you can also directly have a look at the whole vocabulary, stored in the `.vocab` attribute as a dictionary.

See also

- This is great content about tokenizers by HuggingFace: `https://huggingface.co/docs/transformers/tokenizer_summary`

- The official documentation about `AutoTokenizer`: `https://huggingface.co/docs/transformers/v4.27.2/en/model_doc/auto#transformers.AutoTokenizer`

- The official documentation about RNNs: `https://pytorch.org/docs/stable/generated/torch.nn.RNN.html`

Training a GRU

In this recipe, we will keep exploring RNNs with the **GRU** – what it is, how it works, and how to train such a model.

Getting started

One of the main limitations of RNNs is the memory of the network throughout their steps. GRUs try to overcome this limit by adding a memory gate.

If we take a step back and describe an RNN cell with a simple diagram, it could look like *Figure 8.5*.

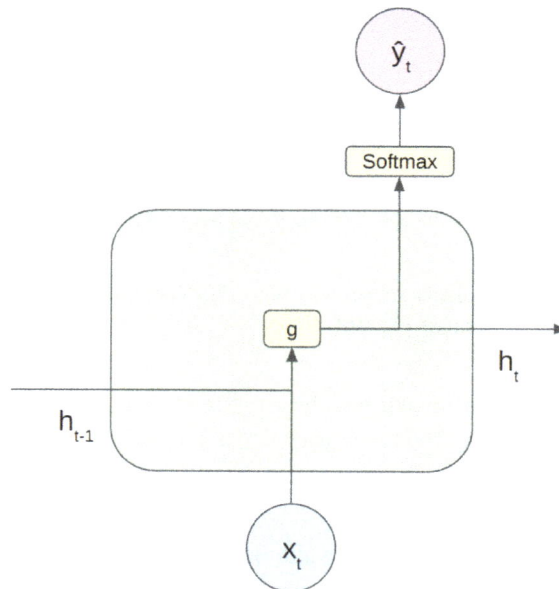

Figure 8.5 – A diagram of an RNN cell

So basically, at each step t, there are both a hidden state h_{t-1} and a set of features x_t. They are concatenated, then weights are applied and an activation function g resulting in a new hidden state h_t. Optionally, an output $\hat{y}_t$ is computed from this hidden state, and so on.

But what if this step of features x_t is not relevant? Or what if it would be useful for the network to just remember fully this hidden state h_{t-1} from time to time? This is exactly what a GRU does, by adding a new set of parameters through what is called a gate.

A **gate** is learned through backpropagation too, using a new set of weights, and allows a network to learn more complex patterns, as well as remember relevant past information.

A GRU is made up of two gates:

- Γ_u: the update gate, responsible for learning whether to update the hidden state
- Γ_r: the relevance gate, responsible for learning how relevant the hidden state is

In the end, a simplified diagram of the GRU unit looks like the one shown in *Figure 8.6*.

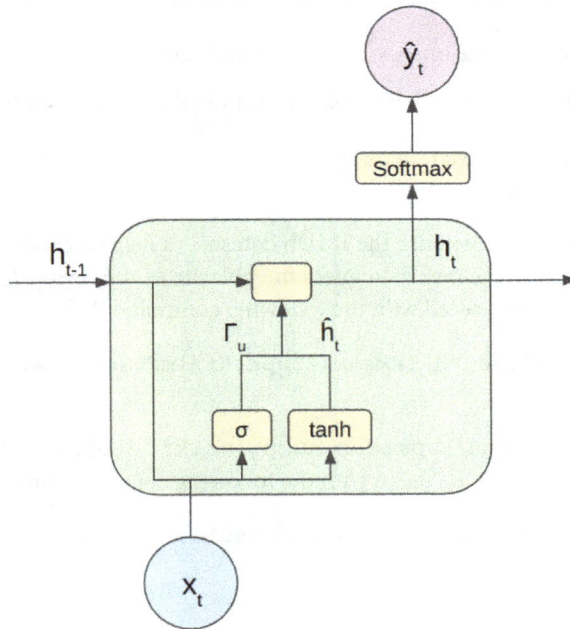

Figure 8.6 – A diagram of a GRU cell. The relevance gate is omitted for clarity

The forward computation is now slightly more complicated than for a simple RNN cell, and it can be described with the following set of formulas:

$$h_t = g(W_{hh}h_{t-1} + W_{xh}X_t + b_h)$$

$$\hat{y}_t = softmax\ (W_{hy}h_t + b_y)$$

These equations can be simply described in a few words. Compared to a simple RNN, there are three major differences:

- Two gates, Γ_u and Γ_r, are computed with associated weights
- The relevance gate Γ_r is used to compute the intermediate hidden state $\hat{h}_t$
- The final hidden state h_t is a linear combination of the previous hidden state and the current intermediate hidden state, with the update gate Γ_u as the weight

The major trick here is the use of the update gate, which can be interpreted in extreme cases as follows:

- If Γ_u is only made of ones, the previous hidden state is forgotten
- If Γ_u is only made of zeros, the new hidden state is not taken into account

Although the concepts can be quite complex at first, the GRU is fortunately super easy to use with PyTorch, as we will see in this recipe.

To run the code in this recipe, we will use the IMDb dataset – a dataset containing movie reviews, and positive or negative labels. The task is to guess the polarity of the review (positive or negative) based on the text. It can be downloaded with the following command line:

```
kaggle datasets download -d lakshmi25npathi/imdb-dataset-of-50k-
moviereviews --unzip
```

We will also need the following libraries: pandas, numpy, scikit-learn, matplotlib, torch, and transformers. They can be installed with the following command line:

```
pip install pandas numpy scikit-learn matplotlib torch transformers
```

How to do it...

In this recipe, we will train a GRU on the same IMDb dataset for a binary classification task. As we will see, the code to train a GRU is almost the same as that to train a simple RNN:

1. Import the required libraries:

    *   torch and some related modules and classes for the neural network
    *   train_test_split and LabelEncoder from scikit-learn for preprocessing
    *   AutoTokenizer from Transformers to tokenize the reviews
    *   pandas to load the dataset
    *   matplotlib for visualization:

    ```
    import torch
    import torch.nn as nn
    import torch.optim as optim from torch.utils.data
    import DataLoader,Dataset from sklearn.model_selection
    import train_test_split from sklearn.preprocessing
    import LabelEncoder from transformers
    import AutoTokenizer
    import pandas as pd
    import numpy as np
    import matplotlib.pyplot as plt
    ```

2. Load the data from the .csv file with pandas. This is a 50,000-row dataset, with textual reviews and labels:

    ```
    # Load data
    data = pd.read_csv('IMDB Dataset.csv')
    data.head()
    ```

    The code output is the following:

    ```
                                        review    sentiment 0  One of
    the other reviewers has mentioned that ...      positive
    1  A wonderful little production. <br /><br />The... positive
    2  I thought this was a wonderful way to spend
    ti...      positive
    3  Basically there's a family where a little boy
    ...        negative
    4  Petter Mattei's "Love in the Time of Money"
    is...      positive
    ```

3. Split the data into train and test sets, using the `train_test_split` function, with a test size of 20% and a specified random state for reproducibility:

```
# Split data into train and test sets
Train_data, test_data = train_test_split(data,
    test_size=0.2, random_state=0)
```

4. Implement the `TextClassificationDataset` dataset class, handling the data. At instance creation, this class will do the following:

   • Instantiate `AutoTokenizer` from the transformers library, using the `bert-base-uncased` tokenizer

   • Tokenize the tweets with the previously instantiated tokenizer, along with the provided maximum length, padding, and truncation

   • Encode the labels and store them:

```
# Define dataset class
class TextClassificationDataset(Dataset):
    def __init__(self, data, max_length):
        self.data = data
        self.tokenizer = AutoTokenizer.from_pretrained('bert-
base-uncased')
        self.tokens = self.tokenizer(
            data['review'].to_list(), padding=True,
            truncation=True, max_length=max_length,
            return_tensors='pt')['input_ids']
        le = LabelEncoder()
        self.labels = torch.tensor(le.fit_transform(
            data['sentiment']).astype(np.float32))
    def __len__(self):
        return len(self.data)
    def __getitem__(self, index):
        return self.tokens[index],self.labels[index]
```

5. Instantiate the `TextClassificationDataset` objects for the train and test sets, as well as the related data loaders, with a maximum number of words of 64 and a batch size of 64. This means that each movie review will be converted as a sequence of exactly 64 tokens:

```
batch_size = 64
max_words = 64
# Initialize datasets and dataloaders
train_dataset = TextClassificationDataset(train_data,
    max_words)
test_dataset = TextClassificationDataset(test_data,
```

```
        max_words)
    train_dataloader = DataLoader(train_dataset,
        batch_size=batch_size, shuffle=True)
    test_dataloader = DataLoader(test_dataset,
        batch_size=batch_size, shuffle=True)
```

6. Implement the GRU classifier model. It is made up of the following elements:

- An embedding layer (taking a zero vector as the first input)

- Three layers of GRU

- A fully connected layer on the last sequence step, with a sigmoid activation function, since it's a binary classification:

```
# Define GRU model
class GRUClassifier(nn.Module):
    def __init__(self, vocab_size, embedding_dim,
        hidden_size, output_size, num_layers=3):
            super(GRUClassifier, self).__init__()
            self.num_layers = num_layers
            self.hidden_size = hidden_size
            self.embedding = nn.Embedding(
                num_embeddings=vocab_size,
                embedding_dim=embedding_dim)
            self.gru = nn.GRU(
                input_size=embedding_dim,
                hidden_size=hidden_size,
                num_layers=num_layers,
                batch_first=True)
            self.fc = nn.Linear(hidden_size,
                output_size)
    def forward(self, inputs):
        batch_size = inputs.size(0)
        zero_hidden = torch.zeros(
            self.num_layers, batch_size,
            self.hidden_size).to(device)
        embedded = self.embedding(inputs)
        output, hidden = self.gru(embedded,
            zero_hidden)
        output = torch.sigmoid(self.fc(output[:, -1]))
        return output
```

7. Instantiate the GRU model, with an embedding dimension and a hidden dimension of 32:

```
vocab_size = train_dataset.tokenizer.vocab_size
embedding_dim = 32
hidden_dim = 32
output_size = 1
# Optionally, set the device to GPU if you have one device =
torch.device(
    'cuda' if torch.cuda.is_available() else 'cpu')
model = GRUClassifier(
    vocab_size=vocab_size,
    embedding_dim=embedding_dim,
    hidden_size=hidden_dim,
    output_size=output_size,
).to(device)
random_data = torch.randint(0, vocab_size,
    size=(batch_size, max_words)).to(device)
result = model(random_data)
print('Resulting output tensor:', result.shape)
print('Sum of the output tensor:', result.sum())
```

The code output is the following:

```
Resulting output tensor: torch.Size([64, 1]) Sum of the output
tensor: tensor(31.0246, device='cuda:0', grad_fn=<SumBackward0>)
```

8. Instantiate the optimizer as an Adam optimizer, with a learning rate of 0.001. The loss is defined as the binary cross-entropy loss because this is a binary classification task:

```
optimizer = optim.Adam(model.parameters(), lr=0.001)
criterion = nn.BCELoss()
```

9. Let's now implement two helper functions.

    `epoch_step_IMDB` updates the weights on the train set and computes the binary cross-entropy loss and accuracy for a given epoch:

```
def epoch_step_IMDB(model, dataloader, device,
    training_set: bool):
        running_loss = 0.
        correct = 0.
        for i, data in enumerate(dataloader, 0):
    # Get the inputs: data is a list of [inputs, labels]
            inputs, labels = data
            inputs = inputs.to(device)
            labels = labels.unsqueeze(1).to(device)
            if training_set:
```

```
                        # Zero the parameter gradients
                        optimizer.zero_grad()
                        # Forward + backward + optimize
                        outputs = model(inputs)
                        loss = criterion(outputs, labels)
                   if training_set:
                        loss.backward()
                        optimizer.step()
                    # Add correct predictions for this batch
                        correct += (
                    (outputs > 0.5) == labels).float().sum()
                        # Compute loss for this batch
                        running_loss += loss.item()


        return running_loss, correct
```

train_IMDB_classification loops over the epochs, trains a model, and stores the accuracy and loss for the train and test sets:

```
def train_IMDB_classification(model, train_dataloader,
    test_dataloader, criterion, device,
    epochs: int = 20):
        # Train the model
        train_losses = []
        test_losses = []
        train_accuracy = []
        test_accuracy = []

    for epoch in range(20):
        running_train_loss = 0.
        correct = 0.
        model.train()
        running_train_loss, correct = epoch_step_IMDB(
            model, train_dataloader, device,
                training_set=True
        )

        # Compute and store loss and accuracy for this epoch
        train_epoch_loss = running_train_loss / len(
            train_dataloader)
        train_losses.append(train_epoch_loss)
        train_epoch_accuracy = correct / len(
            train_dataset)
        train_accuracy.append(
```

```
                        train_epoch_accuracy.cpu().numpy())

            ## Evaluate the model on the test set
            running_test_loss = 0.
            correct = 0.
            model.eval()
            with torch.no_grad():
                running_test_loss,
                correct = epoch_step_IMDB(
                    model, test_dataloader, device,
                    training_set=False
                )

                test_epoch_loss = running_test_loss / len(
                    test_dataloader)
                test_losses.append(test_epoch_loss)
                test_epoch_accuracy = correct / len(
                    test_dataset)
                test_accuracy.append(
                    test_epoch_accuracy.cpu().numpy())

            # Print stats
            print(f'[epoch {epoch + 1}] Training: loss={train_epoch_
        loss:.3f} accuracy={train_epoch_accuracy:.3f} |\
            \t Test: loss={test_epoch_loss:.3f} accuracy={test_epoch_
        accuracy:.3f}')

        return train_losses, test_losses, train_accuracy,
            test_accuracy
```

10. Train the model over 20 epochs, reusing the functions we just implemented. Compute and store the accuracy and the loss for both the train and test sets at each epoch, for visualization purposes:

```
train_losses, test_losses, train_accuracy, test_accuracy =
train_IMDB_classification(model,
    train_dataloader, test_dataloader, criterion,
    device, epochs=20)
```

After 20 epochs, the results should be close to the following code output:

```
[epoch 20] Training: loss=0.040 accuracy=0.991 |  Test:
loss=1.155 accuracy=0.751
```

11. Plot the loss as a function of the epoch number, for both the train and test sets:

```
plt.plot(train_losses, label='train')
plt.plot(test_losses, label='test')
plt.xlabel('epoch') plt.ylabel('loss (BCE)')
plt.legend() plt.show()
```

We then get this graph:

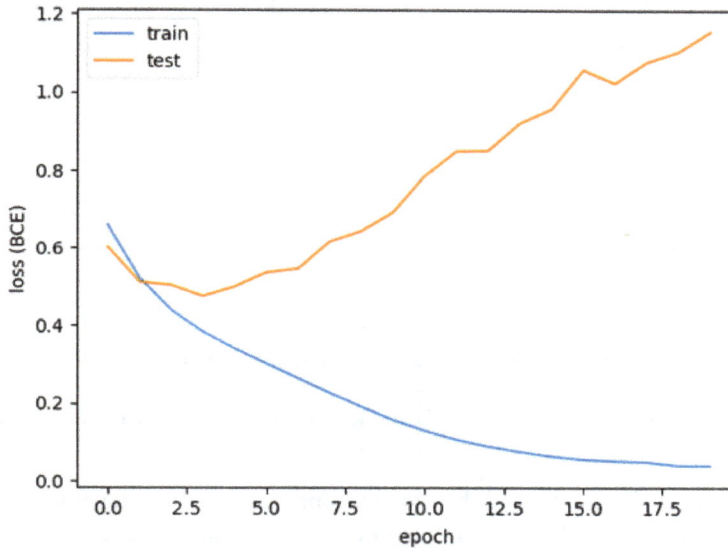

Figure 8.7 – A binary cross-entropy loss as a function of the epoch

As we can see, the loss is clearly diverging for the test set, meaning after only a few epochs, there is already overfitting.

12. Similarly, plot the accuracy as a function of the epoch number of both the train and test sets:

```
plt.plot(train_accuracy, label='train')
plt.plot(test_accuracy, label='test')
plt.xlabel('epoch') plt.ylabel('Accuracy')
plt.legend() plt.show()
```

This is the graph obtained:

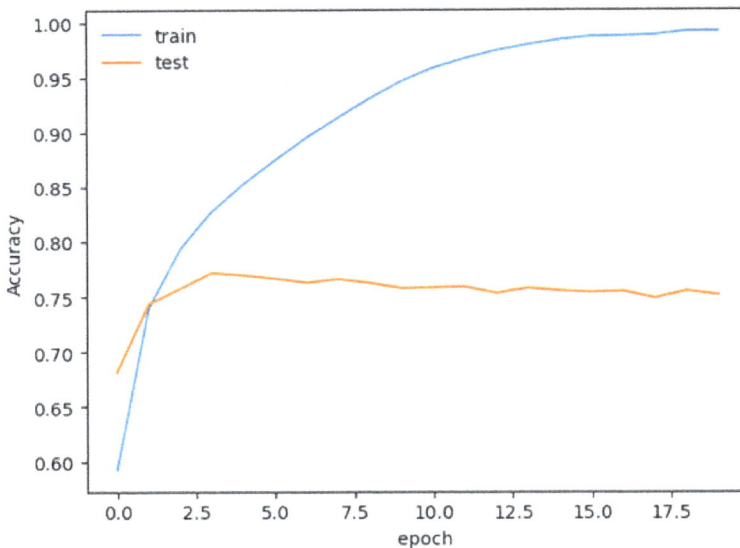

Figure 8.8 – Accuracy as a function of the epoch

As with the loss, we can see we face overfitting with an accuracy close to 100% on the train set, but it is only about a maximum of 77% on the test set.

On a side note, if you try this recipe and the previous one yourself, you may find the GRU much more stable in the results, while the RNN in the previous recipe may sometimes have a hard time properly converging.

There's more...

When working with sequential data such as text, time series, and audio, RNNs are commonly used. While simple RNNs are not so frequently used because of their limitations, GRUs are usually a better choice. Besides simple RNNs and GRUs, another type of cell is frequently used – **long short-term memory** cells, better known as **LSTMs**.

The cell of an LSTM is even more complex than the one of a GRU. While a GRU cell has a hidden state and two gates, an LSTM cell has two types of hidden states (the hidden state and the cell state) and three gates. Let's now have a quick look.

The cell state of an LSTM is described in *Figure 8.9*:

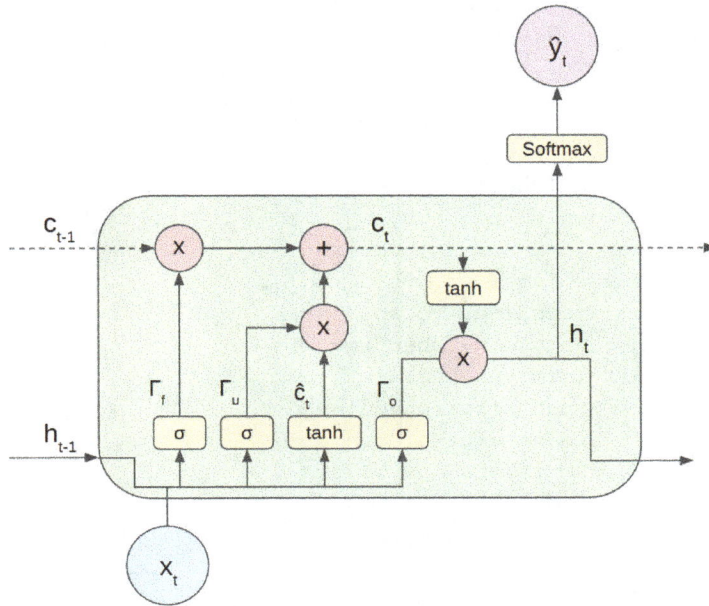

Figure 8.9 – A diagram of an LSTM cell, assuming the LSTM activation function
is a tanh and the output layer activation function is a softmax

Without getting into all the computational details of the LSTM, from *Figure 8.9* we can see there are three gates, computed with their own set of weights, based on both the previous hidden state h_{t-1} and the current features x_t, just like for a GRU, with a sigmoid activation function:

- The forget gate Γ_f

- The update gate Γ_u

- The output gate Γ_o

There are also two states, computed at each step:

- A cell state $c_t = \Gamma_f c_{t-1} + \Gamma_u \hat{c}_t$

- A hidden state $h_t = \Gamma_o c_t$

Here, the intermediary state $\hat{c}_t$ is computed with its own set of weights, just like a gate, with a free activation function.

Having more gates and states, LSTMs have more parameters than GRUs and, thus, usually need more data to be properly trained. However, they are proven to be very effective with long sequences, such as long texts.

Using PyTorch, the code for training an LSTM is not much different from the code to train a GRU. In this recipe, the only piece of code that needs to be changed would be the model implementation, replaced, for example, with the following code:

```
class LSTMClassifier(nn.Module):
    def __init__(self, vocab_size, embedding_dim,
        hidden_size, output_size, num_layers=3):
            super(LSTMClassifier, self).__init__()
            self.hidden_size = hidden_size
            self.num_layers = num_layers
            self.embedding = nn.Embedding(
                num_embeddings=vocab_size,
                embedding_dim=embedding_dim)
                self.lstm = nn.LSTM(
                    input_size=embedding_dim,
                    hidden_size=hidden_size,
                    num_layers=num_layers,
                    batch_first=True)
            self.fc = nn.Linear(hidden_size, output_size)

    def forward(self, inputs):
        batch_size = inputs.size(0)
        h_0 = torch.zeros(self.num_layers, batch_size,
            self.hidden_size)
        c_0 = torch.zeros(self.num_layers, batch_size,
            self.hidden_size)
        embedded = self.embedding(inputs)
        output,
        (final_hidden_state, final_cell_state) = self.lstm(
            embedded, (h_0, c_0))
        output = torch.softmax(self.fc(output[:, -1]),
            dim=1)
        return output
```

The main differences with GRUClassifier implemented earlier are the following:

- In init: Of course, using nn.LSTM instead of nn.GRU, since we now want an LSTM-based classifier

- In forward: We now initialize two zero vectors, h0 and c0, which are fed to the LSTM

- The output of the LSTM is now made of the output and both the hidden and cell states

Besides that, it can be trained the same way as a GRU, with the same code.

On a comparative note, let's compute the number of parameters in this LSTM, and let's compare it to the number of parameters in an "equivalent" RNN and GRU (that is, the same hidden dimension, the same number of layers, and so on).

The number of parameters in this LSTM can be computed with the following code:

```
sum(p.numel() for p in list(
    model.parameters())[1:] if p.requires_grad)
```

> **Important note**
> Note that we do not take into account the embedding part, since we omit the first layer.

Here is the number of parameters for each type of model:

- **RNN**: 6,369
- **GRU**: 19,041
- **LSTM**: 25,377

A rule of thumb to explain this is the number of gates. Compared to a simple RNN, a GRU has two additional gates requiring their own weights, hence a total number of parameters multiplied by 3. The same logic applies to the LSTM having three gates.

Overall, the more parameters a model contains, the more data it needs to be trained robustly, which is why a GRU is a good trade-off and usually a good first choice.

> **Important note**
> Up to now, we only assumed GRUs (and RNNs in general) can go from left to right – from the start of a sentence to the end of a sentence. Just because that's what we humans usually do when we read, it doesn't mean it's necessarily the most optimal way for a neural network to learn. It is possible to use RNNs in both directions, known as **bidirectional RNNs**, and this can apply to not only simple RNNs but also GRUs and LSTMs. Implementing such a model cannot be easier than with PyTorch, since the only thing to change is to add `bidirectional=True` to the model definition, such as `nn.GRU(bidirectional=True)`.

See also

- The official documentation about GRUs: `https://pytorch.org/docs/stable/generated/torch.nn.GRU.html`
- The official documentation about LSTMs: `https://pytorch.org/docs/stable/generated/torch.nn.LSTM.html`

• A somewhat out-of-date but great post about LSTMs' effectiveness by Andrej Karpathy: `https://karpathy.github.io/2015/05/21/rnn-effectiveness/`

Regularizing with dropout

In this recipe, we will add dropout to a GRU to add regularization to the IMDb classification dataset.

Getting ready

Just like fully connected neural networks, recurrent neural networks such as GRUs and LSTMs can be trained with dropout. As a reminder, dropout is just randomly setting some unit's activation to zero during training. As a result, it allows a network to have less information at once and to hopefully generalize better.

We will improve upon the results of the GRU training recipe, by using dropout on the same task – the IMDb dataset binary classification.

If not already done, the dataset can be downloaded using the Kaggle API with the following command line:

```
kaggle datasets download -d lakshmi25npathi/imdb-dataset-of-50k-
moviereviews --unzip
```

The required libraries can be installed with the following:

```
pip install pandas numpy scikit-learn matplotlib torch transformers
```

How to do it...

Here are the steps to perform this recipe:

1. We will train a GRU on the IMDb dataset, just like in the *Training a GRU* recipe. Since the five first steps of *Training a GRU* (from the imports to the `DataLoaders` instantiation) are common to this recipe, let's just assume they have been run and start directly with the model class implementation. Implement the GRU classifier model. It is made of the following elements:

    • An embedding layer (taking a zero vector as the first input), on which dropout is applied in the forward

    • Three layers of GRU, with dropout directly provided as an argument to the GRU constructor

- A fully connected layer on the last sequence step, with a sigmoid activation function, with no dropout:

```python
# Define GRU model
class GRUClassifier(nn.Module):
    def __init__(self, vocab_size, embedding_dim,
        hidden_size, output_size, num_layers=3,
        dropout=0.25):
            super(GRUClassifier, self).__init__()
            self.num_layers = num_layers
            self.hidden_size = hidden_size
            self.embedding = nn.Embedding(
                num_embeddings=vocab_size,
                embedding_dim=embedding_dim)
            self.dropout = nn.Dropout(dropout)
            self.gru = nn.GRU(
                input_size=embedding_dim,
                hidden_size=hidden_size,
                num_layers=num_layers,
                batch_first=True, dropout=dropout)
            self.fc = nn.Linear(hidden_size,
                output_size)
    def forward(self, inputs):
        batch_size = inputs.size(0)
        zero_hidden = torch.zeros(self.num_layers,
            batch_size, self.hidden_size).to(device)
        embedded = self.dropout(
            self.embedding(inputs))
        output, hidden = self.gru(embedded,
            zero_hidden)
        output = torch.sigmoid(self.fc(output[:, -1]))
        return output
```

> **Important note**
> It is not mandatory to apply dropout to the embedding, nor is it always useful. In this case, since the embedding is a large part of the model, applying dropout only to the GRU layers won't have a significant impact on performance.

2. Instantiate the GRU model, with an embedding dimension and a hidden dimension of `32`:

```
vocab_size = train_dataset.tokenizer.vocab_size
embedding_dim = 32 hidden_dim = 32 output_size = 1
# Optionally, set the device to GPU if you have one
device = torch.device(
    'cuda' if torch.cuda.is_available() else 'cpu')
model = GRUClassifier(
    vocab_size=vocab_size,
    embedding_dim=embedding_dim,
    hidden_size=hidden_dim,
     output_size=output_size, ).to(device)
```

Instantiate the optimizer as an Adam optimizer, with a learning rate of `0.001`. The loss is defined as the binary cross-entropy loss, since this is a binary classification task:

```
optimizer = optim.Adam(model.parameters(), lr=0.001)
criterion = nn.BCELoss()
```

3. Train the model over 20 epochs by reusing the helper function implemented in the previous recipe. For each epoch, we compute and store the accuracy and the loss for both the train and test sets:

```
train_losses, test_losses, train_accuracy,
test_accuracy = train_IMDB_classification(model,
    train_dataloader, test_dataloader, criterion,
    device, epochs=20)
```

The last epoch output should look like this:

```
[epoch 20] Training: loss=0.248 accuracy=0.896 |   Test:
loss=0.550 accuracy=0.785
```

4. Plot the loss as a function of the epoch number, for both the train and test sets:

```
plt.plot(train_losses, label='train')
plt.plot(test_losses, label='test')
plt.xlabel('epoch') plt.ylabel('loss (BCE)')
plt.legend() plt.show()
```

Here is the output:

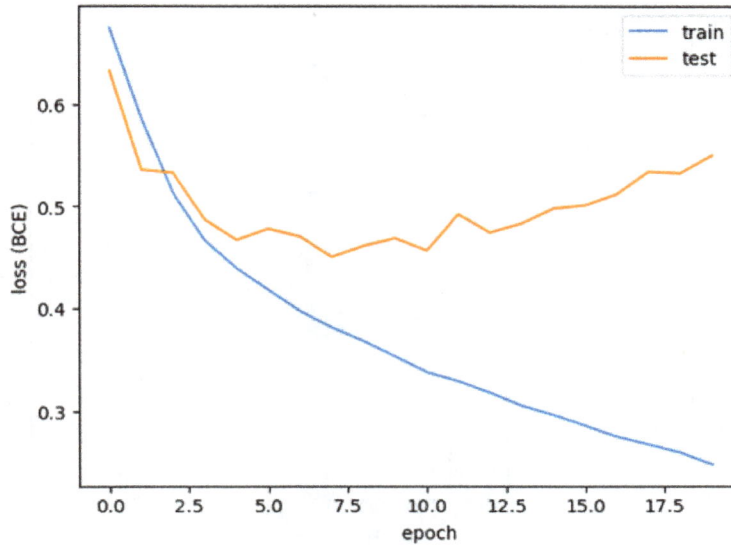

Figure 8.10 – Binary cross-entropy loss as a function of the epoch

We can see that we are still overfitting, but it's a bit less dramatic than without dropout.

5. Finally, plot the accuracy as a function of the epoch number of both the train and test sets:

```
plt.plot(train_accuracy, label='train')
plt.plot(test_accuracy, label='test')
plt.xlabel('epoch') plt.ylabel('Accuracy')
plt.legend() plt.show()
```

This is the output:

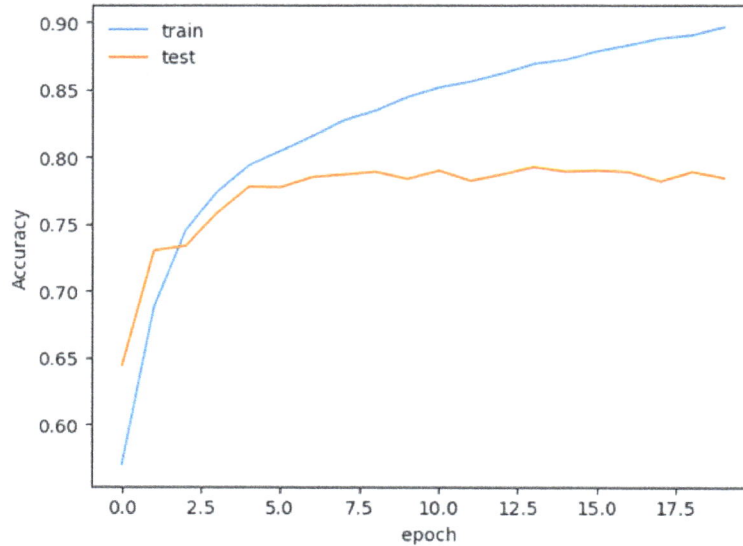

Figure 8.11 – Accuracy as a function of the epoch

Note the effect of dropout on the train accuracy.

While the accuracy did not spectacularly improve, it has increased from 77% to 79% with dropout. Also, the difference between the train and test losses is smaller than it was without dropout, allowing us to improve generalization.

There's more…

Unlike dropout, other methods that are useful to regularize fully connected neural networks can be used with GRUs and other RNN-based architectures.

For example, since we have rather substantial overfitting here, with a train loss having a steep decrease, it might be interesting to test smaller architectures, with fewer parameters to be learned.

Regularizing with the maximum sequence length

In this recipe, we will regularize by playing with the maximum sequence length, on the IMDB dataset, using a GRU-based neural network.

Getting ready

Up to now, we have not played much with the maximum length of the sequence, but it is sometimes one of the most important hyperparameters to tune.

Indeed, depending on the input dataset, the optimal maximum length can be quite different:

- A tweet is short, so having a maximum number of tokens of hundreds does not make sense most of the time
- A product or movie review can be significantly longer, and sometimes, the reviewer writes a lot of pros and cons about the product/movie, before getting to the final conclusion – in such cases, a larger maximum length may help

In this recipe, we will train a GRU on the IMDb dataset, containing movie reviews and associated labels (either positive or negative); this dataset contains some very lengthy texts. So, we will significantly increase the maximum number of words, and see how it impacts the final accuracy.

If not already done, you can download the dataset, assuming you have a Kaggle API installed, running the following command line:

```
kaggle datasets download -d lakshmi25npathi/imdb-dataset-of-50k-
moviereviews --unzip
```

The following libraries are needed: pandas, numpy, scikit-learn, matplotlib, torch, and transformers. They can be installed with the following command line:

```
pip install pandas numpy scikit-learn matplotlib torch transformers
```

How to do it...

Here are the steps to perform this recipe:

1. This recipe will be mostly the same as the *Training a GRU* recipe on the IMDb dataset; it will only change the maximum length of the sequence. Since the most significant differences are the sequence length value and the results, we will assume the four first steps of *Training a GRU* (from the imports to the dataset implementation) have been run, and we will reuse some of the code. Instantiate the `TextClassificationDataset` objects for the train and test sets (reusing the class implemented in *Training a GRU*), as well as the related data loaders.

    This time, we chose a maximum number of words of 256, significantly higher than the earlier value of 64. We will keep a batch size of 64:

    ```
    batch_size = 64 max_words = 256
    # Initialize datasets and dataloaders
    Train_dataset = TextClassificationDataset(train_data,
        max_words)
    test_dataset = TextClassificationDataset(test_data,
        max_words)
    train_dataloader = DataLoader(train_dataset,
        batch_size=batch_size, shuffle=True)
    test_dataloader = DataLoader(test_dataset,
        batch_size=batch_size, shuffle=True)
    ```

2. Instantiate the GRU model by reusing the `GRUClassifier` class implemented in *Training a GRU*, with an embedding dimension and a hidden dimension of 32:

    ```
    vocab_size = train_dataset.tokenizer.vocab_size
    embedding_dim = 32
    hidden_dim = 32
    output_size = 1
    # Optionally, set the device to GPU if you have one
    device = torch.device(
        'cuda' if torch.cuda.is_available() else 'cpu')
    model = GRUClassifier(
        vocab_size=vocab_size,
        embedding_dim=embedding_dim,
        hidden_size=hidden_dim,
        output_size=output_size, ).to(device)
    ```

Instantiate the optimizer as an Adam optimizer, with a learning rate of 0.001. The loss is defined as the binary cross-entropy loss, since this is a binary classification task:

```
optimizer = optim.Adam(model.parameters(), lr=0.001)
criterion = nn.BCELoss()
```

3. Train the model over 20 epochs reusing the `train_IMDB_classification` helper function implemented in the *Training a GRU* recipe; store the accuracy and the loss for both the train and test sets for each epoch:

```
train_losses, test_losses, train_accuracy,
test_accuracy = train_IMDB_classification(model,
    train_dataloader, test_dataloader, criterion,
    device, epochs=20)
```

After 20 epochs, the output looks like this:

```
[epoch 20] Training: loss=0.022 accuracy=0.995 |   Test:
loss=0.640 accuracy=0.859
```

4. Plot the loss as a function of the epoch number for both the train and test sets:

```
plt.plot(train_losses, label='train')
plt.plot(test_losses, label='test')
plt.xlabel('epoch') plt.ylabel('loss (BCE)')
plt.legend() plt.show()
```

Here is the output:

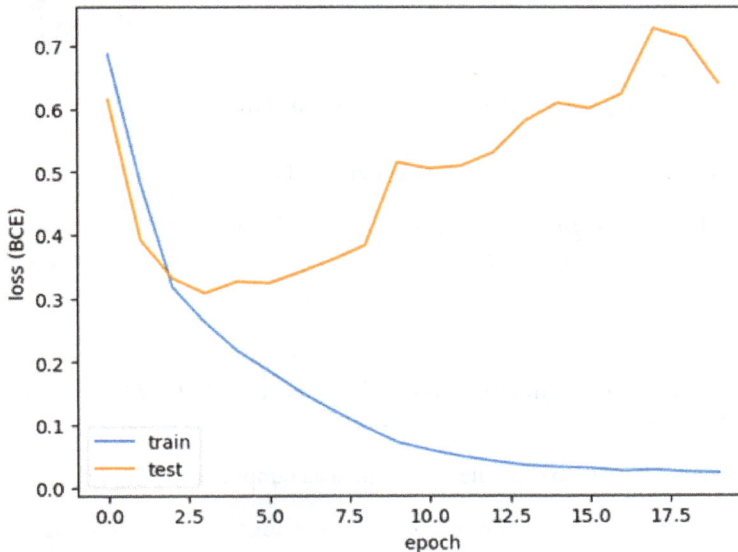

Figure 8.12 – Binary cross-entropy loss as a function of the epoch

We can see that there is overfitting after only a few epochs.

5. Finally, plot the accuracy as a function of the epoch number of both the train and test sets:

```
plt.plot(train_accuracy, label='train')
plt.plot(test_accuracy, label='test')
plt.xlabel('epoch') plt.ylabel('Accuracy')
plt.legend() plt.show()
```

This is what we get:

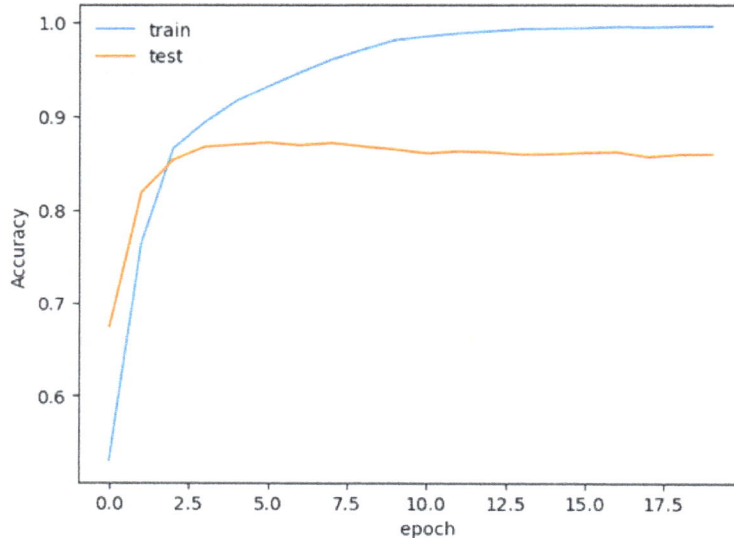

Figure 8.13 – Accuracy as a function of the epoch

The test accuracy reaches a maximum after a few epochs and then slowly decreases.

Although there is still a large overfitting effect, compared to the training with a maximum length of 64, the test accuracy went from a maximum of 77% to a maximum of 87%, a significant improvement.

There's more...

Instead of blindly choosing the maximum number of tokens, it might be interesting to first quickly analyze the length distribution.

For relatively small datasets, it's easy to compute the length of all samples; let's do it with the following code:

```
tokenizer = AutoTokenizer.from_pretrained('bert-base-uncased')
review_lengths = [len(tokens) for tokens in tokenizer(
    train_data['review'].to_list())['input_ids']]
```

We can now plot the distribution of the review lengths with a histogram, using a log scale:

```
plt.hist(review_lengths, bins=50, log=True)
plt.xlabel('Review length (#tokens)') plt.show()
```

This is the output:

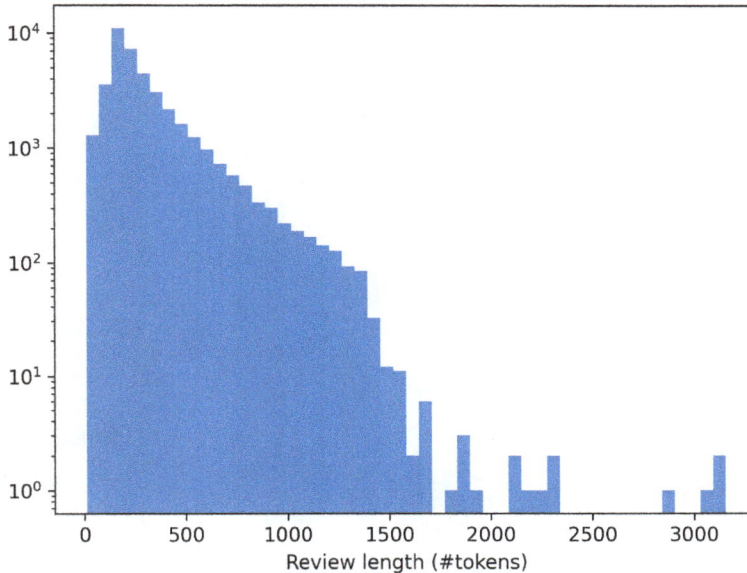

Figure 8.14 – A histogram of the review length of the IMDb dataset, in a log scale

We can see a peak of around 300 tokens in length, and almost no review has more than 1,500 tokens. As we can see from the histogram, most of the reviews seem to have a length of a couple of hundred tokens. We can also compute the average and median length:

```
print('Average length:', np.mean(review_lengths))
print('Median length:', np.median(review_lengths))
```

The computed average and median values are the following:

```
Average length: 309.757075 Median length: 231.0
```

As a result, the average length is about 309, and the median length is 231. According to this information, if the computational power allows it and depending on the task, choosing a maximum length of 256 seems to be a good first choice.

9

Advanced Regularization in Natural Language Processing

A full book could be written about regularization in **natural language processing** (**NLP**). NLP is a wide field that consists of many topics, ranging from simple classification such as review ranking to complex models and solutions such as ChatGPT. This chapter will merely scratch the surface of what can reasonably be done with simple NLP solutions such as classification.

In this chapter, we will cover the following recipes:

- Regularization using a word2vec embedding
- Data augmentation using word2vec
- Zero-shot inference with pre-trained models
- Regularization with BERT embeddings
- Data augmentation using GPT-3

By the end of this chapter, you will be able to take advantage of advanced methods for NLP tasks such as word embeddings and transformers, as well as be able to use data augmentation to generate synthetic training data.

Technical requirements

In this chapter, we will use various NLP solutions and tools, so we will require the following libraries:

- NumPy
- pandas
- scikit-learn
- Matplotlib
- Gensim
- NLTK
- PyTorch
- Transformers
- OpenAI

Regularization using a word2vec embedding

In this recipe, we will use a pre-trained word2vec embedding to improve the results of a task thanks to transfer learning. We will compare the results to the initial *Training a GRU* recipe from *Chapter 8*, on the IMDb dataset for review classification.

Getting ready

A word2vec is a rather old type of word embedding in the NLP landscape and has been widely used in many NLP tasks. While recent techniques are sometimes more powerful, the word2vec approach remains efficient and cost-effective.

Without getting into the details of word2vec, a commonly used model is a 300-dimensional embedding; each word in the vocabulary is embedded into a vector of 300 values.

word2vec is usually trained on a large corpus of texts. There are two main approaches for training a word2vec that can be roughly described as follows:

- **Continuous bag of words (CBOW)**: Uses the context of surrounding words in a sentence to predict a missing word
- **skip-gram**: Uses a word to predict its surrounding context

An example of these two approaches is proposed in *Figure 9.1*:

Source text   Training samples

                                        (context, target)

The quick brown fox jumps over the lazy dog.   (The quick fox jumps, brown)

The quick brown fox jumps over the lazy dog.   (quick brown jumps over, fox)

The quick brown fox jumps over the lazy dog.   (brown fox over the, jumps)

CBOW

Source text   Training samples

                                        (center, target)

The quick brown fox jumps over the lazy dog.   (The, quick), (the, brown)

The quick brown fox jumps over the lazy dog.   (quick, the), (quick, brown),
                                                    (quick, fox)

The quick brown fox jumps over the lazy dog.   (brown, the), (brown, quick),
                                                    (brown, fox), (brown, jumps)

The quick brown fox jumps over the lazy dog.   (fox, quick), (fox, brown),
                                                    (fox, jumps), (fox, over)

Skip-gram

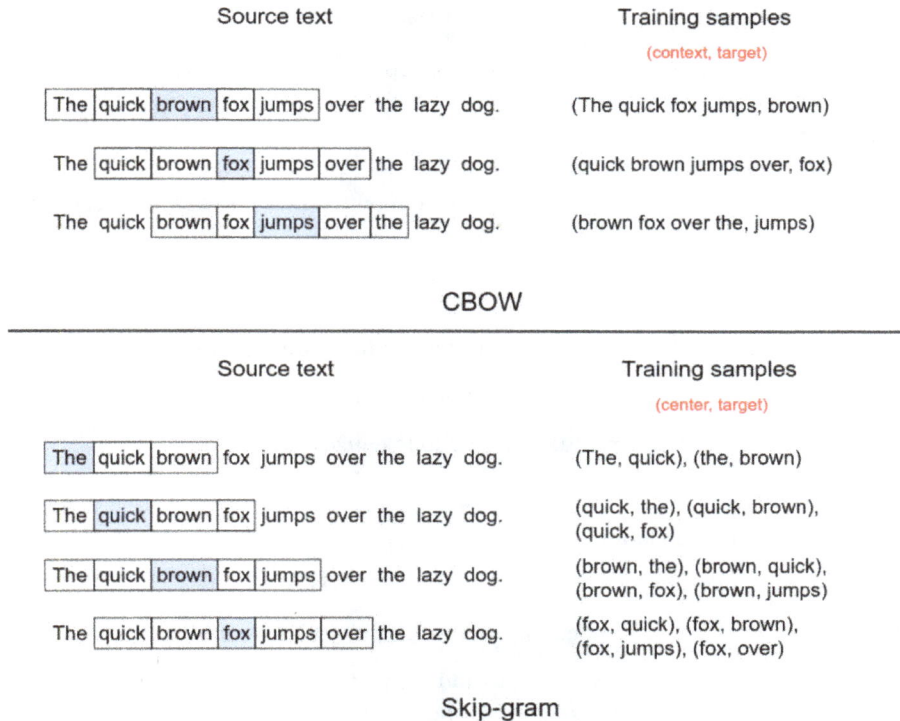

Figure 9.1 – An example of training data for both the CBOW (left) and skip-gram (right) methods

> **Note**
> In practice, CBOW is usually easier to train, while skip-gram may have better performance for rarer words.

The goal is not to train our own word2vec, but simply to reuse a trained one and take advantage of transfer learning to get a performance boost in our predictions. In this recipe, instead of training our own embedding before feeding the **gated recurrent unit** (**GRU**), we will simply reuse a pre-trained word2vec embedding, and then only train our GRU on top of these embeddings.

For that, we will again work on the IMDb dataset classification task: a dataset containing texts of movie reviews as inputs and associated binary labels, positive or negative. The dataset for this can be downloaded with the Kaggle API:

```
kaggle datasets download -d lakshmi25npathi/imdb-dataset-of-50k-
moviereviews --unzip
```

The following command will install the required libraries:

```
pip install pandas numpy scikit-learn matplotlib torch gensim nltk
```

How to do it...

In this recipe, we will train a GRU for binary classification on the IMDb review dataset. Compared to the original recipe, the main difference is in *step 5*:

1. Import the following necessary libraries:

    *   torch and some related modules and classes for the neural network
    *   train_test_split and LabelEncoder from scikit-learn for preprocessing
    *   AutoTokenizer from transformers to tokenize the reviews
    *   pandas to load the dataset
    *   numpy for data manipulation
    *   matplotlib for visualization
    *   gensim for the word2vec embedding and nltk for work tokenization

    If you haven't done so yet, you will need to add the nltk.download('punkt') line to download some required utility instances, as shown here:

    ```
    import torch
    import torch.nn as nn
    import torch.optim as optim
    from torch.utils.data import DataLoader, Dataset
    from sklearn.model_selection import train_test_split
    from sklearn.preprocessing import LabelEncoder
    import pandas as pd
    import matplotlib.pyplot as plt
    import numpy as np
    import gensim.downloader
    import nltk
    # If running for the first time nltk.download('punkt')
    ```

2. Load the pre-trained word2vec model, which contains a 300-dimension embedding. The model is about 1.6 GB and may take some time to download, depending on your available bandwidth:

    ```
    # Will take a while the first time, need to download about 1.6GB
    of the model
    word2vec_model = gensim.downloader.load('
        word2vec-google-news-300')
    ```

3. Load the data from the CSV file with `pandas`:

```
# Load data data = pd.read_csv('IMDB Dataset.csv')
```

4. Split the data into train and test sets using the `train_test_split` function, with a test size of 20% and a specified random state for reproducibility:

```
# Split data into train and test sets train_data,
    test_data = train_test_split(data, test_size=0.2,
        random_state=0)
```

5. Implement the dataset's `TextClassificationDataset` class, which handles the data. This is where the word2vec embedding is computed:

```python
# Define dataset class
class TextClassificationDataset(Dataset):
    def __init__(self, data, word2vec_model,
        max_words):
        self.data = data
        self.word2vec_model = word2vec_model
        self.max_words = max_words
        self.embeddings = data['review'].apply(
            self.embed)
        le = LabelEncoder()
        self.labels = torch.tensor(le.fit_transform(
            data['sentiment']).astype(np.float32))
    def __len__(self):
        return len(self.data)
    def __getitem__(self, index):
        return self.embeddings.iloc[index],
            self.labels[index]
    def embed(self, text):
        tokens = nltk.word_tokenize(text)
        return self.tokens_to_embeddings(tokens)
    def tokens_to_embeddings(self, tokens):
        embeddings = []
        for i, token in enumerate(tokens):
            if i >= self.max_words:
                break
            if token not in self.word2vec_model:
                continue
            embeddings.append(
                self.word2vec_model[token])
        while len(embeddings) < self.max_words:
```

```
        embeddings.append(np.zeros((300, )))
    return np.array(embeddings, dtype=np.float32)
```

At instantiation, each input movie is converted into an embedding in two ways with the embed method:

- Each movie review is tokenized with a word tokenizer (basically splitting sentences into words).

- Then, a vector of max_words length is computed, containing the word2vec embeddings of max_words first words in the review. If the review is shorter than max_words words, the vector is filled with zero padding.

6. Then, we must instantiate the TextClassificationDataset objects for the train and test sets, as well as the related data loaders. The maximum number of words is set to 64, as is the batch size:

```
batch_size = 64 max_words = 64
# Initialize datasets and dataloaders
Train_dataset = TextClassificationDataset(train_data,
    word2vec_model, max_words)
test_dataset = TextClassificationDataset(test_data,
    word2vec_model, max_words)
train_dataloader = DataLoader(train_dataset,
    batch_size=batch_size, shuffle=True)
test_dataloader = DataLoader(test_dataset,
    batch_size=batch_size, shuffle=True)
```

7. Then, we must implement the GRU classifier model. Since the embedding was computed at the data loading step, this model is directly computing a three-layer GRU, followed by a fully connected layer with a sigmoid activation function:

```
# Define RNN model
class GRUClassifier(nn.Module):
    def __init__(self, embedding_dim, hidden_size,
        output_size, num_layers=3):
            super(GRUClassifier, self).__init__()
            self.hidden_size = hidden_size
            self.num_layers = num_layers
            self.gru = nn.GRU(
                input_size=embedding_dim,
                hidden_size=hidden_size,
                num_layers=num_layers,
                batch_first=True)
        self.fc = nn.Linear(hidden_size, output_size)

    def forward(self, inputs):
```

```
batch_size = inputs.size(0)
zero_hidden = torch.zeros(self.num_layers,
    batch_size, self.hidden_size).to(device)
output, hidden = self.gru(inputs, zero_hidden)
output = torch.sigmoid(self.fc(output[:, -1]))
return output
```

8. Next, we must instantiate the GRU model. The embedding dimension, defined by the word2vec model, is 300. We have chosen a hidden dimension of 32 so that each GRU layer is made up of 32 units:

```
embedding_dim = 300
hidden_dim = 32
output_size = 1
# Optionally, set the device to GPU if you have one device =
torch.device(
    'cuda' if torch.cuda.is_available() else 'cpu')
model = GRUClassifier(
    embedding_dim=ebedding_dim,
    hidden_siz=hidden_dim,
    output_size=output_size, ).to(device)
```

9. Then, we must instantiate the optimizer as an Adam optimizer with a learning rate of 0.001; the loss is defined as the binary cross-entropy loss since this is a binary classification task:

```
optimizer = optim.Adam(model.parameters(), lr=0.001)
criterion = nn.BCELoss()
```

10. Train the model for 20 epochs using the train_model function and store the loss and accuracy for both the train and test sets at each epoch. The implementation of the train_model function can be found in this book's GitHub repository at https://github.com/PacktPublishing/The-Regularization-Cookbook/blob/main/chapter_09/chapter_09.ipynb:

```
train_losses, test_losses, train_accuracy,
test_accuracy = train_model(
    model, train_dataloader, test_dataloader,
    criterion, optimizer, device, epochs=20)
```

Here is the typical output after 20 epochs:

```
[epoch 20] Training: loss=0.207 accuracy=0.917 |  Test:
loss=0.533 accuracy=0.790
```

11. Plot the BCE loss for the train and test sets:

```
plt.plot(train_losses, label='train')
plt.plot(testlosse, label=''test'')
plt.xlabel('epoch') plt.ylabel('loss (BCE)')
plt.legend() plt.show()
```

Here is the plot for it:

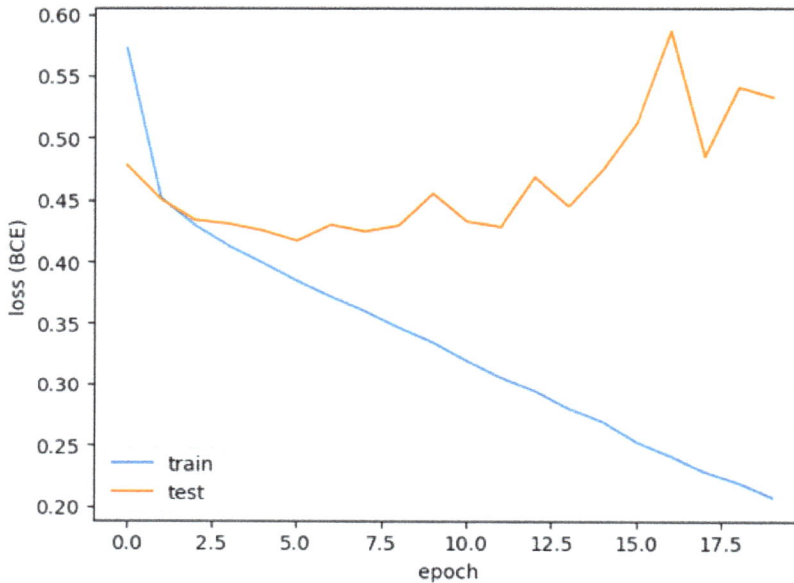

Figure 9.2 – Binary cross-entropy loss as a function of the epoch

As we can see, while the train loss keeps decreasing over the 20 epochs, the test loss soon reaches a minimum at around 5 epochs, to then increase, indicating overfitting.

12. Plot the accuracy for the train and test sets:

```
plt.plot(train_accuracy, label='train')
plt.plot(testaccurcy, label=''test'')
plt.xlabel('epoch') plt.ylabel('Accuracy')
plt.legend() plt.show()
```

Here is the plot for this:

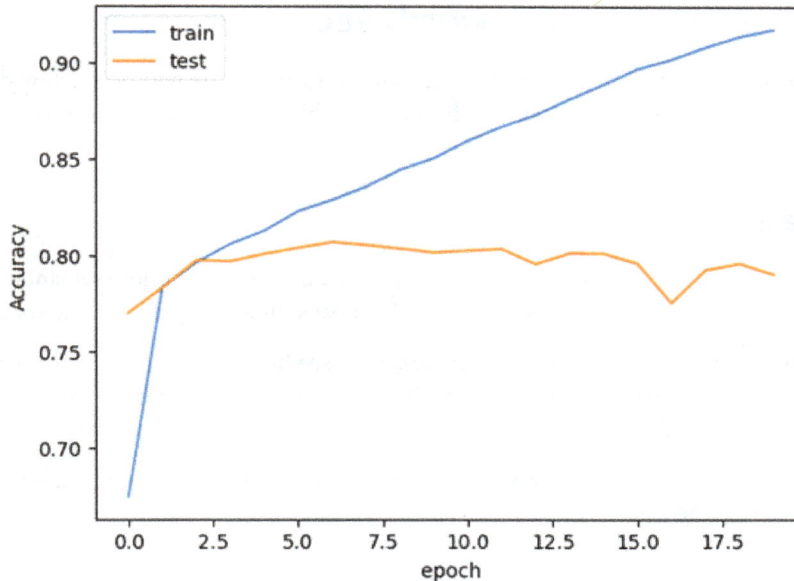

Figure 9.3 – Accuracy as a function of the epoch

As expected with the loss, the accuracy keeps increasing for the train set. For the test set, it reaches a maximum value of about 81% (against 77% without word2vec embedding, as shown in the previous chapter). Word2vec embedding allowed us to improve the results slightly, though the results may improve much more if we adjust the other hyperparameters accordingly.

There's more...

While we used the embeddings as an array in this recipe, they can be used differently; for example, we can use an average of all the embeddings in a sentence or other statistical information.

Also, while word2vec works well enough in many cases here, a few embeddings can be derived with a more specialized approach, such as doc2vec, which is sometimes more powerful for documents and longer texts.

See also

The Wikipedia article about word2vec is a valuable resource as it specifies many relevant publications: `https://en.wikipedia.org/wiki/Word2vec#cite_note-:1-3`.

This documentation from Google is also useful: `https://code.google.com/archive/p/word2vec/`.

Data augmentation using word2vec

One way to regularize a model and get better performance is to have more data. Collecting data is not always easy or possible, but synthetic data can be an affordable way to improve performance. We'll do this in this recipe.

Getting ready

Using word2vec embeddings, you can generate new, synthetic data that has a close semantic meaning. By doing this, it is fairly easy for a given word to get the most similar words in a given vocabulary.

In this recipe, using word2vec and a few parameters, we'll see how we can generate new sentences with a close semantic meaning. We will only apply it to a given sentence as an example and propose how to integrate it into a full training pipeline.

The only required libraries are `numpy` and `gensim`, both of which can be installed with `pip install numpy gensim`.

How to do it...

Here are the steps to complete this recipe:

1. The first step is to import the necessary libraries – `numpy` for random calls and `gensim` for word2vec model loading:

    ```
    import numpy as np
    import gensim.downloader
    ```

2. Load a pre-trained word2vec model. This may take some time if the model hasn't been downloaded and stored in a local cache already. Also, this is a rather large model, so it may take some time to load in memory:

    ```
    # Load the Word2Vec model
    word2vec_model = gensim.downloader.load(
        'word2vec-google-news-300')
    ```

3. Implement the `replace_words_with_similar` function so that you can randomly replace `word` with another semantically close word:

    ```
    def replace_words_with_similar(text, model,
        sim_threshold: float = 0.5,
        probability: float = 0.5,
        top_similar: int = 3,
        stop_words: list[str] = []):
        # Split in words
        words = text.split()
    ```

```
    # Create an empty list of the output words
    new_words = []
    # Loop over the words
    for word in words:
        added = False
        # If the word is in the vocab, not in stop words, and
above probability, then...
        if word in model and word not in stop_words and
np.random.uniform(0, 1) > probability:
            # Get the top_similar most similar words
            similar_words = model.most_similar(word,
                topn=top_similar)
            # Randomly pick one of those words
            idx = np.random.randint(len(similar_words))
            # Get the similar word and similarity score
            sim_word, sim_score = similar_words[idx]
            # If the similary score is above threshold, add the
word
            if sim_score > sim_threshold:
                new_words.append(sim_word)
                added = True
        if not added:
            # If no similar word is added, add the original word
            new_words.append(word)
    # Return the list as a string
    return ' '.join(new_words)
```

Hopefully, the comments are self-explanatory, but here is what this function does:

- It splits the input text into words by using a simple split (a word tokenizer can be used as well)

- For each word, it checks for the following:

  - If the word is in the word2vec vocabulary

  - If the word is not in a list of stop words (to be defined)

  - If a random probability is above the threshold probability (to draw random words)

- If a word fulfills the previous checks, the following is computed:

  - `top_similar` most similar words

  - One of those words is picked randomly

  - If the similarity score of this word is above a given threshold, add it to the output sentence

- If no updated word has been added, just add the original word so that the overall sentence remains logical

The parameters are as follows:

- `sim_threshold`: The similarity threshold

- `probability`: The probability for a word to be replaced with a similar word

- `top_similar`: The number of similar words to compute for a given word

- `stop_words`: A list of words not to be replaced in case some words are specifically important or have several meanings

4. Apply the `replace_words_with_similar` function we just implemented to a given sentence:

```
original_text = "The quick brown fox jumps over the lazy dog"
generated_text = replace_words_with_similar(
    original_text, word2vec_model, top_words=['the'])
print(""Original text: {}"".format(original_text))
print("New text: {}".format(generated_text))
```

The code output is as follows. It allows us to change a few words while keeping the overall meaning:

Original text: The quick brown fox jumps over the lazy dog New text: This quick brown squirrel jumps Over the lazy puppy

Thanks to this data augmentation technique, it is possible to generate more diverse data so that we can make the models more robust and regularized.

There's more...

One way to add such a data generation function to a classification task would be to add it at the data-loading step. This would generate synthetic data on-the-fly and could allow us to regularize the model. It could be added to the dataset class, as shown here:

```
class TextClassificationDatasetGeneration(Dataset):
    def __init__(self, data, max_length):
        self.data = data
        self.max_length = max_length
        self.tokenizer = AutoTokenizer.from_pretrained(
            'bert-base-uncased')
        self.tokens = self.tokenizer(
            data['review'].to_list(), padding=True,
            truncation=True, max_length=max_length,
            return_tensors='pt')['input_ids']
        le = LabelEncoder()
        self.labels = torch.tensor(le.fit_transform(
```

```
            data['sentiment']).astype(np.float32))
    def __len__(self):
        return len(self.data)
    def __getitem__(self, index):
        # Generate a new text
        text = replace_words_with_similar(
            self.data['review'].iloc[index])
        # Tokenize it
        tokens = self.tokenizer(text, padding=True,
            truncation=True, max_length=self.max_length,
            return_tensors='pt')['input_ids']
        return self.tokens[index], self.labels[index]
```

See also

Documentation about the `most_similar` function of the word2vec model can be found at `https://tedboy.github.io/nlps/generated/generated/gensim.models.Word2Vec.most_similar.html`.

Zero-shot inference with pre-trained models

The NLP field has faced many major advances in the last few years, which means that many pre-trained, efficient models can be reused. These pre-trained, freely available models allow us to approach some NLP tasks with zero-shot inference since we can reuse those models. We'll try this approach in this recipe.

> **Note**
> We sometimes use zero-shot inference (or zero-shot learning) and few-shot learning. Zero-shot learning means being able to perform a task without any training for this specific task; few-shot learning means performing a task while training only on a few samples.

Zero-shot inference is the act of reusing pre-trained models without doing any fine-tuning. There are many very powerful, free-to-use models available that can do just as well as a trained model of our own. Since the available models are trained on huge datasets with massive computational power, it is sometimes hard to compete with an in-house model that's been trained on much less data and with less computational power.

> **Note**
> That being said, sometimes, training on small, well-curated, task-specific data can do wonders and provide much better performance. It's all a matter of context.

Also, we sometimes have data without labels, so supervised learning is not a possibility. In such cases, labeling a small subsample of the data ourselves and evaluating a zero-shot approach against this data can be useful.

Getting ready

In this recipe, we will reuse a pre-trained model on the Tweets dataset and classify tweets as negative, neutral, or positive. Since no training is needed, we will directly evaluate the model on the test set so that it can be compared with the results we obtained with a simple RNN in the *Training an RNN* recipe from *Chapter 8*.

To do so, we need to download the dataset locally. It can be downloaded with the Kaggle API and then unzipped with the following command

```
kaggle datasets download -d crowdflower/twitter-airline-sentiment
--unzip
```

The libraries that are required to run this recipe can be installed with the following command:

```
pip install pandas scikit-learn transformers
```

How to do it...

Here are the steps to perform this recipe:

1. Import the following necessary functions and models:

   • numpy for data manipulation

   • pandas for loading the data

   • train_test_split from scikit-learn to split the dataset

   • accuracy_score from scikit-learn to compute the accuracy score

   • pipeline from transformers for instantiating the zero-shot classifier

   Here is the code for this:

   ```
   import numpy as np
   import pandas as pd
   from sklearn.model_selection import train_test_split
   from sklearn.metrics import accuracy_score
   from transformers import pipeline
   ```

2. Load the dataset. In our case, the only columns of interest are `text` for the features and `airline_sentiment` for the labels:

```
# Load dat
Data = pd.read_csv(''Tweets.csv'')
data[['airline_sentiment', 'text']].head()
```

Here is the output of this code:

	airline_sentiment	text
0	neutral	@VirginAmerica What @dhepburn said.
1	positive	@VirginAmerica plus you've added commercials t...
2	neutral	@VirginAmerica I didn't today... Must mean I n...
3	negative	@VirginAmerica it's really aggressive to blast...
4	negative	@VirginAmerica and it's a really big bad thing...

Figure 9.4 – First five rows of the dataset for the considered columns

3. Split the data into train and test sets, with the same parameters as in the *Regularization using a word2vec embedding* recipe so that the results can be compared: `test_size` set to `0.2` and `random_state` set to `0`. Since there is no training, we will only use the test set:

```
# Split data into train and test sets
Train_data, test_data = train_test_split(data,
    test_size=0.2, random_state=0)
```

4. Instantiate the classifier using a `transformers` pipeline with the following parameters:

    • `task="zero-shot-classification"`: This will instantiate a zero-shot classification pipeline

    • `model="facebook/bart-large-mnli"`: This will specify the model to be used for this pipeline

    Here is the code for this:

```
# Taking a long time first time for downloading odel...
Classifier = pipeline(task=""zero-shot-classification"",
    model="facebook/bart-large-mnli")
```

> **Note**
>
> When this is first called, it may download several files and the model itself, and it may take some time.

5. Store the candidate labels in an array. These candidate labels are needed for zero-shot classification:

```
candidate_labels = data['airline_sentiment'].unique()
```

6. Compute the predictions on the test set and store them in an array:

```
# Create an empty list to store the predictions
preds = [] # Loop over the data
for i in range(len(test_data)):
    # Compute the classifier results
    res = classifier(
        test_data['text'].iloc[i],
        candidate_labels=candidate_labels,
    )
    # Apply softmax to the results to get the predicted class
    pred = np.array(res['scores']).argmax()
    labels = res['labels']
    # Store the results in the list
    preds.append(labels[pred])
```

Refer to the *There's more…* section for a few details about what the classifier does, as well as its outputs.

7. Compute the accuracy score of the predictions:

```
print(accuracy_score(test_data['airline_sentiment'], preds))
```

The computed accuracy score is as follows:

```
0.7452725250278087
```

We got an accuracy score of 74.5%, which is equivalent to the results we had after training a simple RNN in the *Training an RNN* recipe from *Chapter 8*. Without any training costs and large, labeled datasets, we can get the same performance thanks to this zero-shot classification.

> **Note**
>
> Zero-shot learning comes with a cost since pre-trained language models are usually rather large and may require large computational power to run at scale.

There's more...

Let's look at an example of the input and output of `classifier` to get a better understanding of what it does:

```
res = classifier(
    'I love to learn about regularization',
    candidate_labels=['positive', 'negative', 'neutral'], )
print(res)
```

The output of this code is as follows:

```
{'sequence': 'I love to learn about regularization',  'labels':
['positive', 'neutral', 'negative'],  'scores': [0.6277033686637878,
0.27620458602905273, 0.09609206020832062]}
```

Here's what we can see:

- An input sentence: `I love to learn about regularization`
- Candidate labels: `positive`, `negative`, and `neutral`

The result is a dictionary with the following key values:

- `'sequence'`: The input sequence
- `'labels'`: The input candidate labels
- `'scores'`: A list of scores, one for each label, sorted in decreasing order

> **Note**
> Since the scores are always sorted in decreasing order, the labels may have a different order.

In the end, the predicted class can be computed with the following code, which will retrieve the `argmax` values of the scores and the associated label:

```
res['labels'][np.array(res['scores']).argmax()]
```

In our case, that would output `positive`.

See also

- The model card of the bart-large-mnli model: `https://huggingface.co/facebook/bart-large-mnli`

- A tutorial from Hugging Face about zero-shot classification: `https://huggingface.co/course/chapter1/3?fw=pt#zero-shot-classification`

- The documentation about the `transformers` pipelines, which allows us to do much more than zero-shot classification: `https://huggingface.co/docs/transformers/main_classes/pipelines`

Regularization with BERT embeddings

Similar to how we used a pre-trained word2vec model to compute the embeddings, it is possible to use the embeddings of a pre-trained BERT model, a transformer-based model.

In this recipe, after quickly explaining the BERT model, we will train a model using BERT embeddings.

BERT stands for **Bidirectional Encoder Representation from Transformers** and is a model that was proposed by Google in 2018. It was first deployed in late 2019 in Google Search for English queries, as well as for many other languages. The BERT model has been proven effective in several NLP tasks, including text classification and question-answering.

Before quickly explaining what BERT is, let's take a step back and look at what **attention mechanisms** and **transformers** are.

Attention mechanisms are widely used in NLP, and more and more in other fields such as computer vision, since their introduction in 2017. The high-level idea of an attention mechanism is to compute a weight for each input token concerning other tokens in the given sequence. Compared to RNNs, which process inputs as sequences, the attention mechanism considers the whole sequence at once. This allows attention-based models to handle long-range dependencies in sequences more efficiently since the attention mechanism can be considered agnostic to the sequence length.

Transformers are a type of neural network based on self-attention. They usually start with an embedding, as well as an absolute positional encoding that attention layers are trained on. Those layers usually use multi-head attention to capture various aspects of the input sequence. For more details, you can read the original paper, *Attention Is All You Need* (refer to the *See also* section for the paper).

> **Note**
> Since BERT uses absolute positional encodings, you are advised to use padding on the right, if any padding is used.

The BERT model was built on top of `transformers` and is made of 12 transformer-based encoding layers for the base model (24 layers for the large model), for about 110 million parameters. More interestingly, it was pre-trained in an unsupervised fashion, using two methods:

- **Masked language**: 15% of the tokens in a sequence are randomly masked, and the model is trained to predict the masked tokens

- **Next sentence prediction**: Given two sentences, the model is trained to predict whether they are consecutive in a given text

This pre-training approach is summarized in the following diagram, which was extracted from the BERT paper called *BERT: Pre-training of Deep Bidirectional Transformers for Language Understanding*:

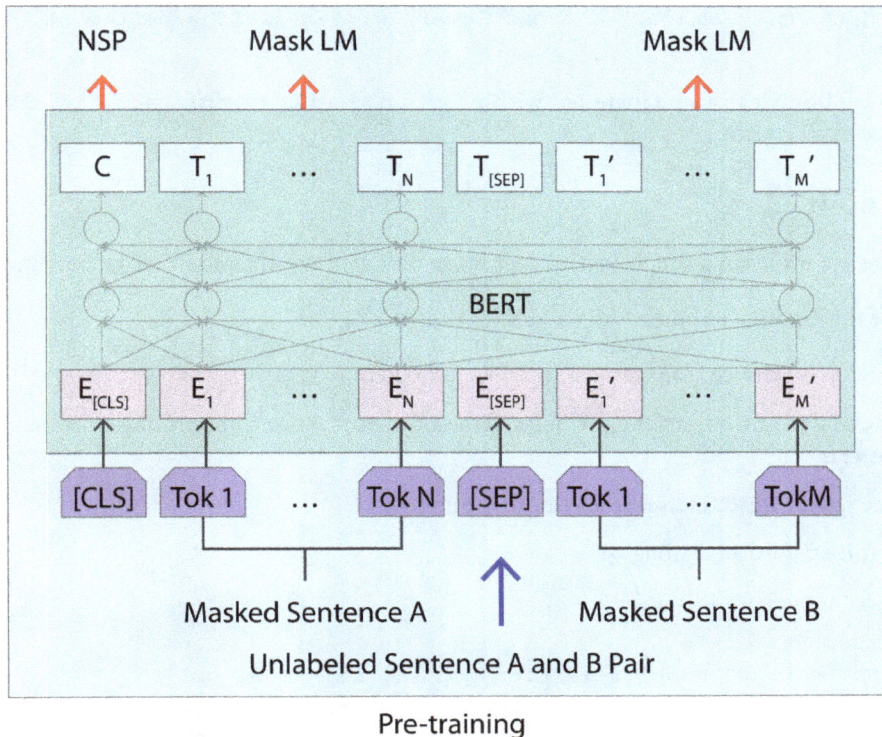

Figure 9.5 – BERT pre-training diagram proposed in the original article, BERT: Pre-training of Deep Bidirectional Transformers for Language Understanding

> **Note**
> While word2vec embedding is context-free (no matter the context, a word embedding remains the same), BERT gives a different embedding for a given word, depending on its surroundings. This makes sense since a given word may have a different meaning in two sentences (for example, apple or Apple can be either a fruit or a company, depending on the context).

Getting ready

For this recipe, we will reuse the Tweets dataset, which can be downloaded and then unzipped locally with the following command line:

```
kaggle datasets download -d crowdflower/twitter-airline-sentiment
--unzip
```

The necessary libraries can be installed with `pip install torch scikit-learn transformers pandas`.

How to do it...

In this recipe, we will train a simple logistic regression on top of pre-trained BERT embeddings:

1. Make the required imports:

    *   `torch` for device management if you have a GPU
    *   The `train_test_split` method and the `LogisticRegression` class from `scikit-learn`
    *   The related BERT classes from `transformers`
    *   `pandas` for data loading

    Here is the code for this:

    ```
    import torch
    from sklearn.model_selection import train_test_split
    from sklearn.linear_model import LogisticRegression
    from transformers import BertConfig, BertModel, BertTokenizer
    import pandas as pd
    ```

2. Load the dataset with `pandas`:

    ```
    # Load data data = pd.read_csv('Tweets.csv')
    ```

3. Split the dataset into train and test sets, keeping the same parameters as in the *Zero-shot inference* recipe with pre-trained models, so that we can compare their performance later:

```
# Split data into train and test sets train_data,
    test_data = train_test_split(data, test_size=0.2,
        random_state=0)
```

4. Instantiate the tokenizer and the BERT model. Instantiating the model is a multi-step process:

    I.   First, instantiate the model's configuration with the `BertConfig` class.

    II. Then, instantiate `BertModel` with random weights.

    III. Load the weights of the pre-trained model (this will display a warning since not all the weights will have been loaded).

    IV. Load the model on the GPU, if any, and set the model to `eval` mode.

    Here is the code for this:

```
# Instantiate the tokenizer
tokenizer = BertTokenizer.from_pretrained('bert-base-uncased')
# Initializing a BERT configuration
configuration = BertConfig()
# Initializing a BERT model with random weights
bert = BertModel(configuration)
# Loading pre-trained weights
bert = bert.from_pretrained('bert-base-uncased')
# Load the model on the GPU
if any device = torch.device(
    "cuda" if torch.cuda.is_available() else "cpu")
bert.to(device)
# Set the model to eval mode
bert.eval()
```

You may get some warning messages since some layers have no pre-trained weights.

5. Compute the embeddings for the train and test sets. This is a two-step process:

    I.   Compute the tokens with the tokenizer (and optionally load the tokens on the GPU, if any).

    II. Then, compute the embeddings.

    For more details about the inputs and outputs of the BERT model, check out the *There's more…* subsection of this recipe.

Here is the code for this:

```
max_length = 24
# Compute the embeddings for the train set
train_tokens = tokenizer(
    train_data['text'].values.tolist(),
    add_special_tokens=True,
    padding='max_length',
    truncation=True,
    max_length=max_length,
    return_tensors='pt')
    train_tokens = {k: v.to(device) for k,
        v in train_tokens.items()}
with torch.no_gad():
    train_embeddings = bert(
        **train_tokens)..pooler_output
# Compute the embeddings for the test set
test_tokens = tokenizer(
    test_data['text'].values.tolist(),
    add_special_tokens=True, padding='max_length',
    truncation=True, max_length=max_length,
    return_tensors='pt')
test_tokens = {k: v.to(device) for k,
    v in test_tokens.items()}
with torch.no_grad():
    test_embeddings = bert(
        **test_tokens).pooler_output
```

6. Then, instantiate and train a logistic regression model. It may require more iterations than the default model allows. Here, it has been set to 10,000:

```
lr = LogisticRegression(C=0.5, max_iter=10000)
lr.fit(train_embeddings.cpu(),
    train_data['airline_sentiment'])
```

7. Finally, print the accuracy on the train and test sets:

```
print('train accuracy:',
    lr.score(train_embeddings.cpu(),
    train_data['airline_sentiment']))
print('test accuracy:',
    lr.score(test_embeddings.cpu(),
    test_data['airline_sentiment']))
```

You should have output results similar to the following:

```
train accuracy: 0.8035348360655737 test accuracy:
0.7882513661202186
```

We get a final result of about 79% accuracy on the test set and 80% on the train set. As a comparison, using zero-shot inference and a simple RNN on this same dataset both provided 74% accuracy.

There's more...

To get a better understanding of what the tokenizer computes and what the BERT model outputs, let's have a look at an example.

First, let's apply the tokenizer to a sentence:

```
tokens = tokenizer('What is a tokenizer?', add_special_tokens=True,
    padding='max_length', truncation=True, max_length=max_length,
    return_tensors='pt')
print(tokens)
```

This outputs the following:

```
{'input_ids': tensor([[ 101,  2054,  2003,  1037, 19204,
17629,  1029,  2023,  2003,  1037,        2204,  3160,  1012,
   102,     0,     0,     0,     0,     0,     0,                0,
     0,     0,     0]]), 'token_type_ids': tensor([[0, 0, 0,
0, 0, 0, 0, 0, 0, 0, 0, 0, 0, 0, 0, 0, 0, 0, 0, 0,
0]]), 'attention_mask': tensor([[1, 1, 1, 1, 1, 1, 1, 1, 1, 1,
1, 1, 1, 1, 0, 0, 0, 0, 0, 0, 0, 0, 0, 0]])}
```

As we can see, the tokenizer returns three outputs:

- `input_ids`: This is the index of tokens in the vocabulary.
- `token_type_ids`: The sentence number. This is only useful for paired sentences, as in the original training of BERT.
- `attention_mask`: This is where the model will focus. As we can see, it's only been set to 1 for actual tokens, and then to 0 for padding.

These three lists are fed to the BERT model so that it can compute its output. The output is made up of the following two tensors:

- `last_hidden_state`: The values of the last hidden state, whose shape is [batch_size, max_length, 768]
- `pooler_output`: The pooled values of the outputs over the sequence steps, whose shape is [batch_size, 768]

Many other types of embeddings exist and can be more or less powerful, depending on the task at hand. For example, OpenAI also proposes embeddings that can be made available using an API. For example, the following code allows us to have embeddings for a given sentence:

```
import openai
# Give your
openai.api_key = 'xx-xxx'
# Query the API
input_text = 'This is a test sentence'
model = 'text-embedding-ada-002'
embeddings = openai.Embedding.create(input = [input_text],
    model=model)['data'][0]['embedding']
```

This would return a 1,536-dimensional embedding for the given sentence that could then be used for classification or other tasks.

Of course, to use these embeddings, you would need to do the following:

- Install the `openai` library with `pip install openai`
- Create an API key on the OpenAI website
- Provide a working payment method

See also

- The paper introducing transformers, *Attention is all you need*: `https://arxiv.org/abs/1706.03762`
- The BERT model card: `https://huggingface.co/bert-base-uncased`
- The BERT paper: `https://arxiv.org/abs/1810.04805`
- For more information about the OpenAI embeddings, here is the official documentation: `https://platform.openai.com/docs/guides/embeddings/use-cases`

Data augmentation using GPT-3

Generative models are becoming more and more powerful, especially in NLP. Using these to generate new, synthetic data sometimes allows us to significantly improve the results we get and regularize models. We'll learn how to do this in this recipe.

Getting ready

While models such as BERT are effective at tasks such as text classification, they usually do not perform very well when it comes to text generation.

Other types of models, such as **generative pre-trained transformer (GPT)** models, can be quite impressive at generating new data. In this recipe, we will use the OpenAI API and GPT-3.5 to generate synthetic yet realistic data. Having more data is key to having more regularization in our models, and data generation is one way to collect more data.

For this recipe, you will need to install the OpenAI library with `pip install openai`.

Also, since the API we will use is not free, it is necessary to create an OpenAI account with a generated API key and a working payment method.

> **Creating an API key**
> You can easily create an API key in your profile by accessing the **API keys** section.

In the *There's more…* section, we will provide a free alternative – that is, using GPT-2 to generate new data – but it will have less realistic results. For that to work, you must have Hugging Face's `transformers` library, which you can install with `pip install transformers`.

How to do it...

In this recipe, we will simply query GPT-3.5 to generate a few positive and negative movie reviews so that we have more data to train a movie review classification model. Of course, this can be derived from any classification task, as well as many other NLP tasks:

1. Import the `openai` library, as follows:

    ```
    import openai
    ```

2. Provide your OpenAI API key:

    ```
    openai.api_key = 'xxxxx'
    ```

> **Note**
> This is just a code example – never share your API key on public repositories. Use alternatives such as environment variables instead.

3. Generate three positive examples using the `ChatCompletion` API:

    ```
    positive_examples = openai.ChatCompletion.create(
        model="gpt-3.5-turbo",
        messages=[
            {"role": "system",
                "content": "You watched a movie you loved."},
            {"role": "user", "content": "Write a short,
    ```

```
                100-words review about this movie"},
    ],
    max_tokens=128,
    temperature=0.5,
    n=3, )
```

There are several parameters here:

* model: gpt-3.5-turbo, which performs well and is cost-efficient. It is based on GPT-3.5.

* messages: There can be three types of messages:

  * system: A formatting message, followed by alternating user and assistant messages

  * user: A user message

  * assistant: An assistant message; we won't use this

* max_tokens: The maximum number of tokens in the output.

* temperature: This is usually between 0 and 2. A larger value means more randomness.

* n: The number of desired outputs.

4. Now, we can display the output-generated sentences:

```
for i in range(len(positive_examples['choices'])):
    print(f'\n\nGenerated sentence {i+1}: \n')
    print(positive_examples['choices'][i]['message']['content'])
```

The following is the output of the three positive reviews generated by GPT-3.5:

Generated sentence 1:   I recently watched the movie "Inception" and was blown away by its intricate plot and stunning visuals. The film follows a team of skilled thieves who enter people's dreams to steal their secrets. The concept of dream-sharing is fascinating and the execution of the idea is flawless. The cast, led by Leonardo DiCaprio, delivers outstanding performances that add depth to the characters. The action scenes are thrilling and the special effects are mind-bending. The film's score by Hans Zimmer is also noteworthy, adding to the overall immersive experience. "Inception" is a masterpiece that will leave you pondering its themes long after the credits roll.   Generated sentence 2:   I recently watched the movie "The Shawshank Redemption" and absolutely loved it. The story follows the life of a man named Andy Dufresne, who is wrongfully convicted of murder and sent to Shawshank prison. The movie beautifully portrays the struggles and hardships faced by prisoners, and the importance of hope and friendship in such a harsh environment. The acting by Tim Robbins and Morgan Freeman is outstanding, and the plot twists keep you engaged throughout the movie. Overall, "The Shawshank Redemption" is a must-watch for anyone who loves a good drama and a heartwarming story about the power of the human spirit.   Generated sentence 3:   I recently

```
watched the movie "Parasite" and it blew me away. The story
revolves around a poor family who slowly infiltrates the lives
of a wealthy family, but things take a dark turn. The movie is
a masterclass in storytelling, with each scene building tension
and adding layers to the plot. The acting is superb, with
standout performances from the entire cast. The cinematography
is also stunning, with each shot expertly crafted to enhance the
mood and atmosphere of the film. "Parasite" is a must-watch for
anyone who loves a good thriller with a twist.
```

5. Similarly, let's generate and display three negative examples of movie reviews:

```
# Generate the generated examples
ngative_examples = openai.ChatCompletion.create(
    model="gpt-3.5-turbo",
    messages=[
        {"role": "system",
         "content": "You watched a movie you hated."},
        {"role": "user",
         "content": "Write a short,
            100-wordsreview about this movie"},
    ],
    max_tokens=128,
    temperature=0.5,
    n=3, )
# Display the generated examples
for i in range(len(
    negative_examples['choices'])):
    print(f'\n\nGenerated sentence {i+1}: \n')
    print(negative_examples[
        'choices'][i]['message']['content'])
```

The following code shows the three reviews that were generated:

```
Generated sentence 1:   I recently watched a movie that left
me feeling disappointed and frustrated. The plot was weak
and predictable, and the characters were one-dimensional and
unrelatable. The acting was subpar, with wooden performances
and lackluster chemistry between the cast. The special effects
were underwhelming and failed to add any excitement or visual
interest to the film. Overall, I found myself checking the
time and counting down the minutes until the end. I wouldn't
recommend this movie to anyone looking for a compelling and
engaging cinematic experience.   Generated sentence 2:    I
recently watched a movie that left me feeling disappointed
and underwhelmed. The plot was predictable and lacked any
real depth or complexity. The characters were one-dimensional
and unrelatable, making it hard to invest in their stories.
The pacing was slow and dragged on unnecessarily, making
the already dull plot even more tedious to sit through. The
acting was subpar, with even the most talented actors failing
```

> to bring any life to their roles. Overall, I found this
> movie to be a complete waste of time and would not recommend
> it to anyone looking for an engaging and entertaining film
> experience.   Generated sentence 3:   I recently watched a
> movie that I absolutely hated - "The Roommate". The plot was
> predictable and the acting was subpar at best. The characters
> were one-dimensional and lacked any depth or development
> throughout the film. The dialogue was cringe-worthy and the
> attempts at suspense fell flat. Overall, I found the movie to
> be a waste of time and would not recommend it to anyone. If
> you're looking for a thrilling and well-crafted thriller, "The
> Roommate" is definitely not the movie for you.

The generated examples are well written and probably good enough to be written by a human. Also, it would be possible to generate more neutral, more random, longer, or shorter examples if needed, which is quite convenient.

There's more...

Alternatively, it is possible to use the GPT-2 model for free, even though the results are less realistic. Let's learn how to do this.

First, let's instantiate a text generation pipeline based on GPT-2:

```
# Import the pipeline from transformers
from transformers import pipeline
# Instantiate a text-generation based on GPT-2 model
Generator = pipeline(''text-generation'', model=''gpt2'')
#generator = pipeline('text2text-generation')#,
# model = 'facebook/mbart-large-50')
```

This generates some text. The behavior is not the same, and it only handles text completion, so you must propose the beginning of a piece of text for the model to complete it automatically:

```
# Generate the positive examples
positive_example = generator(
    "I loved everything about this movie, ",
    max_length = 512,
    num_return_sequences=3, )
# Display the generated examples
for i in range(len(positive_example)):
    print(f'\n\nGenerated sentence {i+1}: \n')
    print(positive_example[i]['generated_text'])
```

This outputs the following three reviews:

```
Generated sentence 1:    I loved everything about this movie, the
movie itself was great and I didn't want to stop after the first movie
because I was so happy for it. But I did like the first movie and
I loved the second one and so much of what I did with it and other
things with how I handled the other films. What were your thoughts as
fans? After the opening credits, I kept thinking, "This will be my
big, special moment, I can't do this alone". I've watched the original
2 episodes, they do have a big difference in the ending. But really,
it's just my dream to put in movies you love, and I'm always looking
over every minute on how good they were and looking forward. I think
because everyone is talking about it, and so much of the movie is just
it of my dream, the movie that I am to get this, the movie that I want
to watch or maybe I'm doing my dreams. Also, it's more to a big than
your dream. I think if I get enough reviews, people will start to see
it. There will even be a few who will see it in the summer. I know I
did it in a way when I was in high school. I never read the reviews
again because I'm not a gamer and I never read the reviews because I
love that. But by watching the movies, I feel that I am better than
that. So, my dream is going to be to give up on the game I started
out with in high school and focus on the movies I like. Not just the
one I played, but the one you all enjoy watching! I really hope you
guys give it more attention and like for this one, just keep going
or stay up for the next movie for when all the things you said can
be true. Thanks in advance, and happy movie watching!    Generated
sentence 2:    I loved everything about this movie,  It was a surprise
to see. I want to say thank to the cast of the film, but don't call
me the original star. I love that I have to keep myself on top of the
world in other things. (laughs) I was excited about the ending and
I was thinking about how much fun it would be to watch that ending.
At the end of the day it was all for me. The movie was a shock to
watch. It was all about the fact that he and her father can all die.
It was so exciting. Says a fan, "I've been waiting for this movie
since childhood, and this is the first time I've seen it."    Generated
sentence 3:    I loved everything about this movie,  so I made the only
mistake I have ever made because for once it felt like this movie was
happening. It's always exciting to see a feature that gives the fans
something to feel. It's a truly beautiful world in which life isn't a
game; life is a process. But it's fun to be forced to watch something
that tells you some great things about our environment, even when only
one person actually is there, who cares about it. This film was not
just another film, it was a true movie. And while I'm still looking
forward to seeing more amazing, unique movies from the history of
cinema, I can guarantee you that there's more we'll be hearing about
from our friends at AMC and others who care about our history, the
history of film making, and the history of art-design in general...
```

As we can see, the results are less interesting and realistic than with GPT-3, but they can still be useful if we apply some manual filtering.

See also

- Chat completion documentation from OpenAI: `https://platform.openai.com/docs/guides/chat`

- Text completion documentation from OpenAI: `https://platform.openai.com/docs/guides/completion`

- Text generation documentation from HuggingFace: `https://huggingface.co/tasks/text-generation`

- GPT-2 model card: `https://huggingface.co/gpt2`

10

Regularization in Computer Vision

In this chapter, we will explore another popular field of deep learning – computer vision. **Computer vision** is a large field with many tasks, from classification through generative models to object detection. Even though we can't cover all of them, we will supply methods that can apply to all tasks.

In this chapter, we'll cover the following recipes:

- Training a **convolutional neural network (CNN)**
- Regularizing a CNN with vanilla **neural network (NN)** methods
- Regularizing a CNN with transfer learning for object detection
- Semantic segmentation using transfer learning

At the end of this chapter, you will be able to handle several computer vision tasks such as image classification, object detection, instance segmentation, and semantic segmentation. You will be able to apply several tools to regularize the trained models, such as architecture, transfer learning, and freezing weights for fine-tuning.

Technical requirements

In this section, we will train CNNs, object detection, and semantic segmentation models, requiring the following libraries:

- NumPy
- scikit-learn
- Matplotlib
- PyTorch
- torchvision

- Ultralytics
- `segmentation-models-pytorch`

Training a CNN

In this recipe, after reviewing the fundamental components of CNN, we will train one on a classification task – the CIFAR10 dataset.

Getting started

Computer vision is a special field for many reasons. The data handled in computer vision projects is usually rather large, multidimensional, and unstructured. However, its most specific aspect is arguably its spatial structure.

With its spatial structure comes a lot of potential difficulties, such as the following:

- **Aspect ratio**: Some images come with different aspect ratios depending on their source, such as 16/9, 4/3, 1/1, and 9/16
- **Occlusion**: An object can be occluded by another one
- **Deformation**: An object can be deformed, either because of perspective or physical deformation
- **Point of view**: Depending on the point of view, an object can look totally different
- **Illumination**: A picture can be taken in many light environments that may alter the image

Many of these difficulties are summarized in *Figure 10.1*.

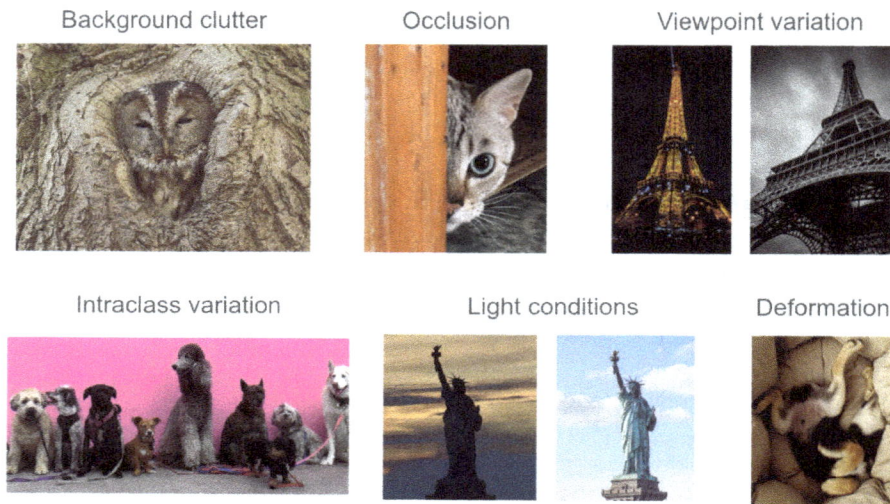

Figure 10.1 – Examples of difficulties specific to computer vision

Due to the spatial structure of data, models are needed to process it. While recurrent neural networks are well suited to sequential data, CNNs are well suited to spatially structured data.

In order to properly build a CNN, we need to introduce two new types of layers:

- Convolutional layers
- Pooling layers

Let's quickly explain them both.

Convolutional layer

A **convolutional layer** is a layer made of convolutions.

In a fully connected layer, a weighted sum of the input features (or the input activation of the previous layer) is computed, with the weights being learned while training.

In a convolutional layer, a convolution is applied to the input features (or the input activation of the previous layer), with the values of the convolution kernel being learned while training. It means the NN will learn the kernel through training to extract the most relevant features from the input images.

A CNN can be fine-tuned with several hyperparameters:

- The kernel size
- The padding size
- The stride size
- The number of output channels (i.e., the number of kernels to learn)

> Tip
>
> See the *There's more…* subsection for more information about kernels and other hyperparameters of the CNNs.

Pooling layer

A **pooling layer** allows you to reduce the dimensionality of images and is commonly used in CNNs. For example, a max pooling layer with a 2x2 kernel will reduce the dimension of an image by 4 (a factor of 2 both in width and height), as shown in *Figure 10.2*.

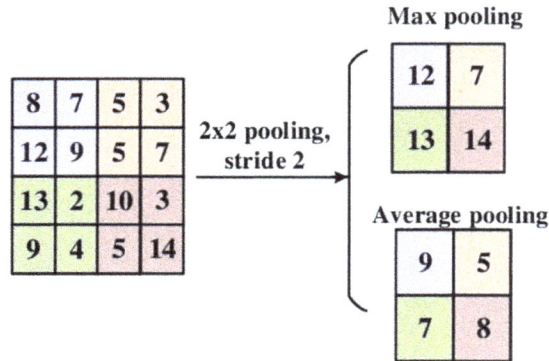

Figure 10.2 – On the left is an input image of 4x4, at the top right is the result of 2x2 max pooling, and at the bottom right is the result of a 2x2 average pooling

There are several types of pooling, such as the following:

- **Max pooling**: Computing the maximum value
- **Average pooling**: Computing the average value
- **Global average pooling**: Computing a global average value for all channels (commonly used before fully connected layers)

LeNet-5

LeNet-5 was one of the first proposed CNN architectures, by Yann Le Cun, for handwritten digit recognition. Its architecture is shown in *Figure 10.3*, taken from Yann's paper *Gradient-Based Learning Applied to Document Recognition*.

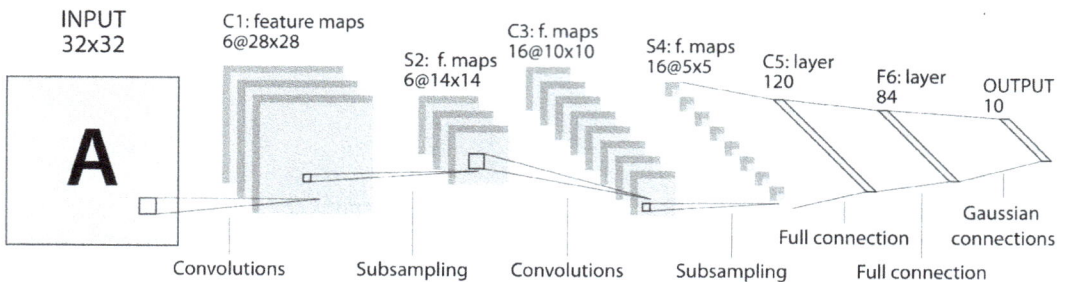

Figure 10.3 – LeNet-5's original architecture

Let's describe it in detail:

- An input image of dimension 32x32
- **C1**: A convolution layer with a 5x5 kernel and 6 output channels
- **S2**: A pooling layer with a 2x2 kernel
- **C3**: A convolution layer with a 5x5 kernel and 16 output channels
- **S4**: A pooling layer with a 2x2 kernel
- **C5**: A fully connected layer with 120 units
- **F6**: A fully connected layer with 84 units
- **Output**: An output layer with 10 units for 10 classes (0 to 9 digits)

We will implement this network in this recipe on the CIFAR-10 dataset.

To run this recipe, the needed libraries can be installed with the following command:

```
pip install numpy matplotlib torch torchvision
```

How to do it...

In this recipe, we will train a CNN for image classification on the CIFAR-10 dataset. The CIFAR-10 dataset is a dataset of 32x32 RGB images, made of 10 classes – plane, car, bird, cat, deer, dog, frog, horse, ship, and truck.

1. Import the needed modules:

   - matplotlib for visualization
   - NumPy for data manipulation
   - Several torch modules and classes
   - The dataset and transformation module from torchvision

   Here are the import statements:

   ```
   import matplotlib.pyplot as plt
   import numpy as np
   import torch
   import torch.nn as nn
   import torch.nn.functional as F
   from torch.utils.data import DataLoader
   from torchvision.utils import make_grid
   from torchvision.datasets import CIFAR10
   import torchvision.transforms as transforms
   ```

2. Instantiate the transformation to apply to images. Here, it's a simple two-step transformation:

    - Convert data to a torch tensor

    - Normalize with `0.5` mean and standard deviation values:

```
transform = transforms.Compose([
    transforms.ToTensor(),
    transforms.Normalize((0.5, 0.5, 0.5),
        (0.5, 0.5, 0.5)),
])
```

3. Load the data and instantiate the data loaders. The previously defined transformation is applied directly at loading as an argument of the `CIFAR10` constructor. The data loaders are here instantiated with a batch size of `64`:

```
# Will download the dataset at first
trainset = CIFAR10('./data', train=True,
    download=True, transform=transform)
train_dataloader = DataLoader(trainset, batch_size=64,
    shuffle=True)
testset = CIFAR10('./data', train=False,
    download=True, transform=transform)
test_dataloader = DataLoader(testset, batch_size=64,
    shuffle=True)
```

4. Optionally, we can visualize a few images to check what the inputs are:

```
# Get a batch of images and labels
images, labels = next(iter(train_dataloader))
# Denormalize the images
images = images / 2 + 0.5
# Compute a grid image for visualization
images = make_grid(images)
# Switch from channel first to channel last
images = np.transpose(images.numpy(), (1, 2, 0))
# Display the result
plt.figure(figsize=(14, 8))
plt.imshow(images)
plt.axis('off')
```

This is the output:

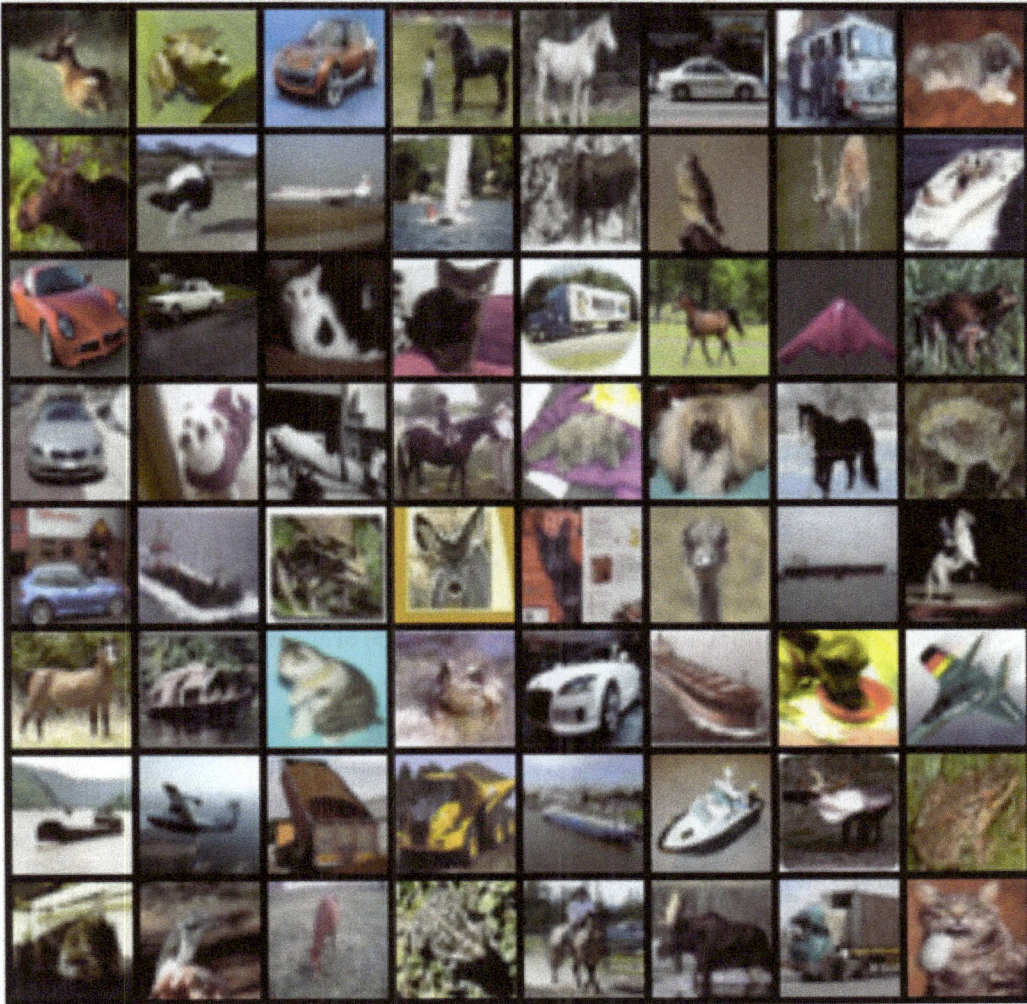

Figure 10.4 – 64 random images from the CIFAR-10 dataset. The images are
blurry, but most are clear enough for humans to classify correctly

5. Implement the `LeNet5` model:

```
class LeNet5(nn.Module):
    def __init__(self, n_classes: int):
        super(LeNet5, self).__init__()
        self.n_classes = n_classes
        self.c1 = nn.Conv2d(3, 6, kernel_size=5,
            stride=1, padding=0)
```

```
        self.s2 = nn.MaxPool2d(kernel_size=2)
        self.c3 = nn.Conv2d(6, 16, kernel_size=5,
            stride=1, padding=0)
        self.s4 = nn.MaxPool2d(kernel_size=2)
        self.c5 = nn.Linear(400, 120)
        self.f6 = nn.Linear(120, 84)
        self.output = nn.Linear(84, self.n_classes)

    def forward(self, x):
        x = F.relu(self.c1(x))
        x = self.s2(x)
        x = F.relu(self.c3(x))
        x = self.s4(x)
        # Flatten the 2D-array
        x = torch.flatten(x, 1)
        x = F.relu(self.c5(x))
        x = F.relu(self.f6(x))
        output = F.softmax(self.output(x), dim=1)
        return output
```

This implementation has almost the same layers as the original paper. Here are a few interesting points:

- nn.Conv2d is the 2D convolution layer in torch, having as hyperparameters the output dimension, kernel size, stride, and padding

- nn.MaxPool2d is the max pooling layer in torch, having as a hyperparameter the kernel size (and, optionally, the stride, defaulting to the kernel size)

- We use the ReLU activation function, even if it was not the function used in the original paper

- torch.flatten allows us to flatten a 2D tensor to a 1D tensor so that we can apply fully connected layers

- Instantiate the model, and make sure that it works well with a random input tensor:

```
# Instantiate the model
lenet5 = LeNet5(10)
# check device
device = 'cuda' if torch.cuda.is_available() else 'cpu'
lenet5 = lenet5.to(device)
# Generate randomly one random 32x32 RGB image
random_data = torch.rand((1, 3, 32, 32), device=device)
result = lenet5(random_data)
print('Resulting output tensor:', result)
print('Sum of the output tensor:', result.sum())
```

The resulting output will look like this:

```
Resulting output tensor: tensor([[0.0890, 0.1047, 0.1039,
0.1003, 0.0957, 0.0918, 0.0948, 0.1078, 0.0999,
           0.1121]], grad_fn=<SoftmaxBackward0>)
Sum of the output tensor: tensor(1.0000, grad_fn=<SumBackward0>)
```

6. Instantiate the loss and optimizer – a cross-entropy loss for multiclass classification, with an Adam optimizer:

```
criterion = nn.CrossEntropyLoss()
optimizer = torch.optim.Adam(lenet5.parameters(), lr=0.001)
```

7. Implement a helper function, epoch_step_cifar, that computes forward propagation, backpropagation (in the case of the training set), loss, and accuracy for an epoch:

```
def epoch_step_cifar(model, dataloader, device,
training_set : bool) :
    running_loss = 0.
    correct = 0.
    for i, data in enumerate(dataloader, 0):
        inputs, labels = data
        inputs = inputs.to(device)
        labels = labels.to(device)
        if training_set:
            optimizer.zero_grad()
        outputs = model(inputs)
        loss = criterion(outputs, labels)
        if training_set:
            loss.backward()
            optimizer.step()
        correct += (outputs.argmax(
            dim=1) == labels).float().sum().cpu()
        running_loss += loss.item()

    return running_loss, correct
```

8. Implement a helper function, train_cifar_classifier, that trains the model on a given number of epochs and returns the loss and accuracy:

```
def train_cifar_classifier(model, train_dataloader,
    test_dataloader, criterion, device, epochs):
        # Create empty lists to store the losses and accuracies
        train_losses = []
        test_losses = []
        train_accuracy = []
```

```
test_accuracy = []

# Loop over epochs
for epoch in range(epochs):
    ## Train the model on the training set
    running_train_loss = 0.
    correct = 0.
    lenet5.train()
    running_train_loss,
    correct = epoch_step_cifar(
        model, train_dataloader, device,
        training_set=True
    )
    # Compute and store loss and accuracy for this epoch
    train_epoch_loss = running_train_loss / len(
        train_dataloader)
    train_losses.append(train_epoch_loss)
    train_epoch_accuracy = correct / len(trainset)
    train_accuracy.append(train_epoch_accuracy)

    ## Evaluate the model on the test set
    running_test_loss = 0.
    correct = 0.
    lenet5.eval()
    with torch.no_grad():
        running_test_loss,
        correct = epoch_step_cifar(
            model, test_dataloader, device,
            training_set=False
        )

        test_epoch_loss = running_test_loss / len(
            test_dataloader)
        test_losses.append(test_epoch_loss)
        test_epoch_accuracy = correct / len(testset)
        test_accuracy.append(test_epoch_accuracy)

    # Print stats
    print(f'[epoch {epoch + 1}] Training: loss={train_epoch_
loss:.3f} accuracy={train_epoch_accuracy:.3f} |\
    \t Test: loss={test_epoch_loss:.3f} accuracy={test_epoch_
accuracy:.3f}')
```

```
        return train_losses, test_losses, train_accuracy,
            test_accuracy
```

9. Using the helper function, train the model on 50 epochs and store the loss and accuracy for both the training and test sets:

```
train_losses, test_losses, train_accuracy,
test_accuracy = train_cifar_classifier(lenet5,
    train_dataloader, test_dataloader, criterion,
    device, epochs=50)
```

The last line of the output will look like this:

```
[epoch 50] Training: loss=1.740 accuracy=0.720 |     Test:
loss=1.858 accuracy=0.600
```

10. Plot the loss for the train and test sets:

```
plt.plot(train_losses, label='train')
plt.plot(test_losses, label='test')
plt.xlabel('epoch')
plt.ylabel('loss (CE)')
plt.legend()
plt.show()
```

This is the resulting graph:

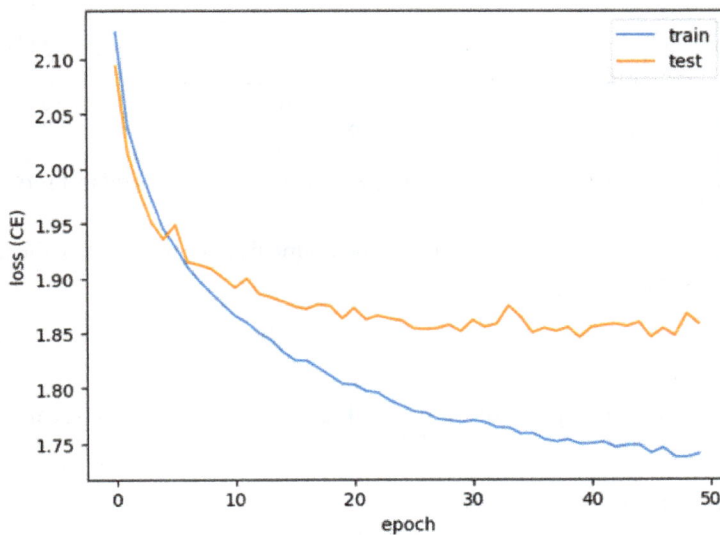

Figure 10.5 – Cross-entropy loss for the train and test sets

After less than 10 epochs, the curves start diverging.

11. Plot the accuracy as a function of the epoch:

```
plt.plot(train_accuracy, label='train')
plt.plot(test_accuracy, label='test')
plt.xlabel('epoch')
plt.ylabel('Accuracy')
plt.legend()
plt.show()
```

This is the graph that we get:

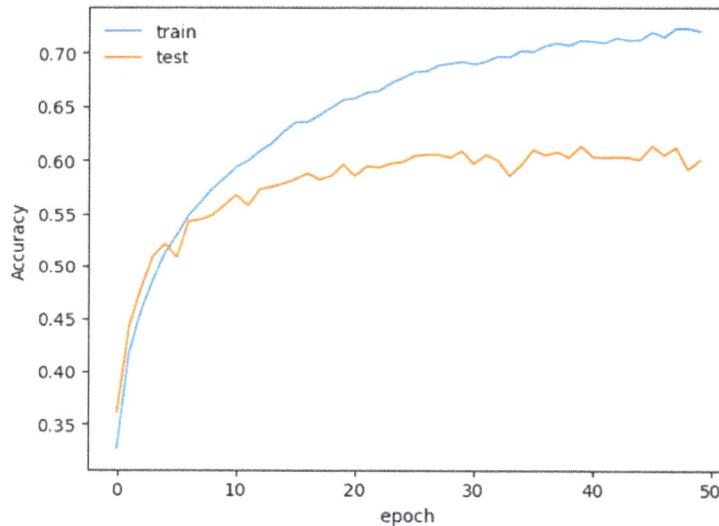

Figure 10.6 – Accuracy as a function of the epoch for both train and test sets

After about 20 epochs, the accuracy reaches a plateau of around 60% accuracy, while the train accuracy keeps growing, suggesting overfitting.

There's more...

In this subsection, let's have a quick recap of some necessary tools and concepts to properly understand CNNs:

- How to store an image
- Padding
- Kernel and convolution
- Stride

How to store an image

An image is nothing but a spatially arranged array of pixels. For example, a 1-million-pixel grayscale square image is an array of 1,000x1,000 pixels.

Each pixel is usually stored as an 8-bit value and can be represented as an unsigned integer in the range of [0, 255]. So, ultimately, such an image can be represented in Python as a NumPy array of `uint8` with a shape of (1000, 1000).

We can go one step further with color images. A color image is commonly stored with three channels – **Red, Green, and Blue (RGB)**. Each of these channels is stored as an 8-bit integer, so a squared 1M pixels color image can be stored as a NumPy array of a shape of (3, 1000, 1000), assuming the channel is stored first.

> **Tip**
> There are many other ways to describe a colored image – **hue, saturation, value** (HSV), CIELAB, transparency, and so on. However, in this book, and many computer vision cases, RGB color space is enough.

Padding

We already used padding in earlier chapters for NLP processing. It consists of adding "space" around an image, basically by adding layers of values around an image. An example of padding on a 4x4 matrix is given in *Figure 10.7*.

Figure 10.7 – An example of a 4x4 matrix padded with one layer of zeros

Padding can take several arguments, such as the following:

- The number of layers of padding
- The padding method – a given value, repetition, mirror, and so on

Most of the time, zero padding is used, but sometimes, more sophisticated padding can be useful.

Kernel and convolution

A convolution is a mathematical operation, between an image and a kernel, that outputs another image. It can be simply schematized as follows:

Convolution (input image, kernel) → Output image

A kernel is just a smaller matrix of predefined values, allowing us to get a property from an image through convolution. For example, with the right kernel, a convolution on an image allows us to blur an image, sharpen an image, detect edges, and so on.

The computation of a convolution is quite simple and can be described as follows:

1.   Spatially match the kernel and the image from the top-left corner.

2.   Compute the weights sum of all the image pixels, with the corresponding kernel value as the weight, and store this value as the top-left output image pixel.

3.   Go one pixel to the right and repeat; if you reach the rightmost edge of the image, go back to the leftmost pixel and one pixel down.

This may look complicated, but it gets much easier with a diagram, as shown in *Figure 10.8*:

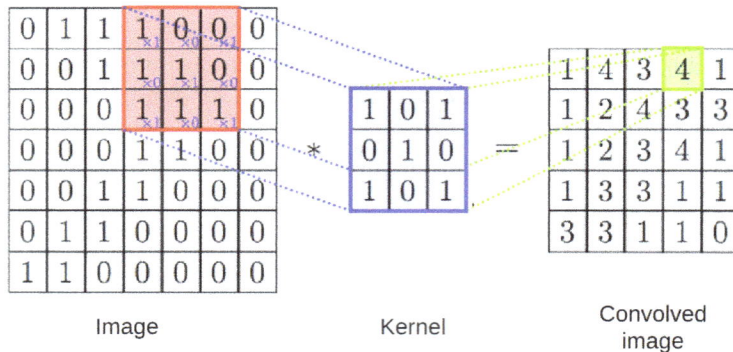

Figure 10.8 – An example of an image convolution by a kernel.
Note that the resulting image is smaller in dimension

As we can see in *Figure 10.8*, the output image is slightly smaller than the input image. Indeed, the larger the kernel, the smaller the output image.

Stride

One more useful concept about convolutions is the concept of **stride**. The stride is the number of step pixels to take between two convolutional operations. In the example in *Figure 10.8*, we implicitly considered a stride of 1 – the kernel is moved by one pixel each time.

However, it's possible to consider a larger stride – we can have a step of any number, as shown in *Figure 10.9*:

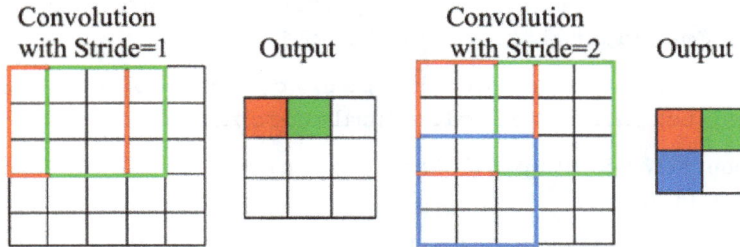

Figure 10.9 – The stride effect on convolution – the larger the stride, the smaller the output image

Having a larger stride will mainly have several, related, consequences:

- Since the convolution skips more pixels, less information remains in the output image
- The output image is smaller, allowing a reduction in dimensionality
- The computation time is lower

Depending on the needs, having a stride larger than one can be an efficient way to have lower computation time.

To summarize, we can control several aspects of a convolution with three parameters:

- The padding
- The kernel size
- The stride

They all affect the size of the output image, following this formula:

$$O = \frac{I - k + 2p}{s} + 1$$

This formula can be broken down as follows:

- I is the input image size
- k is the kernel size
- p is the padding size
- s is the stride size
- O is the output image size

Thanks to this formula, we can efficiently choose the required parameters in any case.

See also

- Documentation about CNN layers: `https://pytorch.org/docs/stable/generated/torch.nn.Conv2d.html#torch.nn.Conv2d`

- Documentation about pooling layers: `https://pytorch.org/docs/stable/generated/torch.nn.MaxPool2d.html#torch.nn.MaxPool2d`

- A paper about LeNet-5: `http://vision.stanford.edu/cs598_spring07/papers/Lecun98.pdf`

- The amazing Stanford course about deep learning for computer vision: `http://cs231n.stanford.edu/`

Regularizing a CNN with vanilla NN methods

Since CNNs are a special kind of NNs, most vanilla NN optimization methods can be applied to them. A non-exhaustive list of regularization techniques we can use with CNNs is the following:

- Kernel size

- Pooling size

- L2 regularization

- A fully connected number of units (if any)

- Dropout

- Batch normalization

In this recipe, we will apply batch normalization to add regularization, reusing the LeNet-5 model on the CIFAR-10 dataset, but any other method may work as well.

Batch normalization is a simple yet very effective method that can help NNs regularize and converge faster. The idea of batch normalization is to normalize the activation values of a hidden layer for a given batch. The method is very similar to a standard scaler for data preparation of quantitative data, but there are some differences. Let's have a look at how it works.

The first step is to compute the mean value μ and the standard deviation σ of the activation values z^i of a given layer. Assuming z^i is the activation value of the I-th unit of the layer and the layer has n units, here are the formulas:

$$\mu = \frac{1}{n}\sum_i z^i$$

$$\sigma^2 = \frac{1}{n}\sum_i \left(z^i - \mu\right)^2$$

Just like with a standard scaler, it is now possible to compute the rescaled the activation values $z^i_{rescaled}$ with the following formula:

$$z^i_{rescaled} = \frac{z^i - \mu}{\sqrt{\sigma^2 + \epsilon}}$$

Here, ϵ is just a small value to avoid division by zero.

Finally, unlike a standard scaler, there is one more step that allows the model to learn what is the best distribution with a scale and shift approach, thanks to two new learnable parameters, β and γ. They are used to compute the final batch normalization output z^i_{BN}:

$$z^i_{BN} = \gamma z^i_{rescaled} + \beta$$

Here, γ allows us to adjust the scale, while β allows us to adjust the shift. These two parameters are learned during training, like any other parameter of the NN. This allows the model to adjust the distribution if required to improve its performance.

For a more visual example, we can see in *Figure 10.10* a possible distribution of activation values on the left for a three-unit layer – the values are skewed and with a large standard deviation. After batch normalization, on the right part of *Figure 10.10*, the distributions are now close to normal distributions.

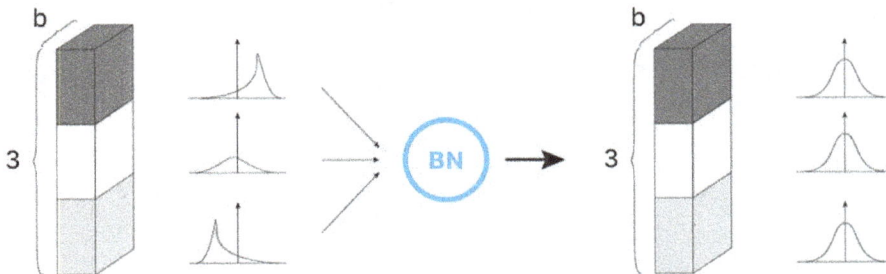

Figure 10.10 – Possible activation distributions of three units of a
layer before (left) and after (right) batch normalization

Thanks to this method, NNs tend to converge faster and generalize better, as we will see in this recipe.

Getting started

For this recipe, we will reuse torch and its integrated CIFAR-10 dataset so that all the needed libraries can be installed with the following command line (if not already installed in the previous recipe):

```
pip install numpy matplotlib torch torchvision
```

How to do it...

Since we will reuse the same data and almost the same network as in the previous recipe, we will assume the imports and instantiated classes can be reused:

1. Implement the regularized model. Here, we will mostly reuse the LeNet-5 architecture, with added batch normalization at each step:

```python
class LeNet5(nn.Module):
    def __init__(self, n_classes: int):
        super(LeNet5, self).__init__()
        self.n_classes = n_classes
        self.c1 = nn.Conv2d(3, 6, kernel_size=5,
            stride=1, padding=0, )
        self.s2 = nn.MaxPool2d(kernel_size=2)
        self.bnorm2 = nn.BatchNorm2d(6)
        self.c3 = nn.Conv2d(6, 16, kernel_size=5,
            stride=1, padding=0)
        self.s4 = nn.MaxPool2d(kernel_size=2)
        self.bnorm4 = nn.BatchNorm1d(400)
        self.c5 = nn.Linear(400, 120)
        self.bnorm5 = nn.BatchNorm1d(120)
        self.f6 = nn.Linear(120, 84)
        self.bnorm6 = nn.BatchNorm1d(84)
        self.output = nn.Linear(84, self.n_classes)

    def forward(self, x):
        x = F.relu(self.c1(x))
        x = self.bnorm2(self.s2(x))
        x = F.relu(self.c3(x))
        x = self.s4(x)
        # Flatten the 2D-array
        x = self.bnorm4(torch.flatten(x, 1))
        x = self.bnorm5(F.relu(self.c5(x)))
        x = self.bnorm6(F.relu(self.f6(x)))
        output = F.softmax(self.output(x), dim=1)
        return output
```

As shown in the code, batch normalization can be simply added as a layer with nn.BatchNorm1d (or nn.BatchNorm2d for the convolutional part), which takes as an argument the following input dimensions:

-   The number of units for fully connected layers and BatchNorm1d

-   The number of kernels for convolution layers and BatchNorm2d

> **Important note**
>
> Placing batch normalization after the activation function is arguable, and some people would rather place it before the activation function.

2. Instantiate the model, with the loss as cross-entropy and the optimizer as Adam:

```
# Instantiate the model
lenet5 = LeNet5(10)
# check device
device = 'cuda' if torch.cuda.is_available() else 'cpu'
lenet5 = lenet5.to(device)
# Instantiate loss and optimizer
criterion = nn.CrossEntropyLoss()
optimizer = torch.optim.Adam(lenet5.parameters(), lr=0.001)
```

3. Train the model over 20 epochs by reusing the `train_cifar_classifier` helper function of the previous recipe. Note that the model converges faster than without batch normalization in the previous recipe:

```
train_losses, test_losses, train_accuracy,
test_accuracy = train_cifar_classifier(lenet5,
    train_dataloader, test_dataloader, criterion,
    device, epochs=20)
```

4. Plot the loss as a function of the epoch:

```
plt.plot(train_losses, label='train')
plt.plot(test_losses, label='test')
plt.xlabel('epoch')
plt.ylabel('loss (CE)')
plt.legend()
plt.show()
```

Here is the graph:

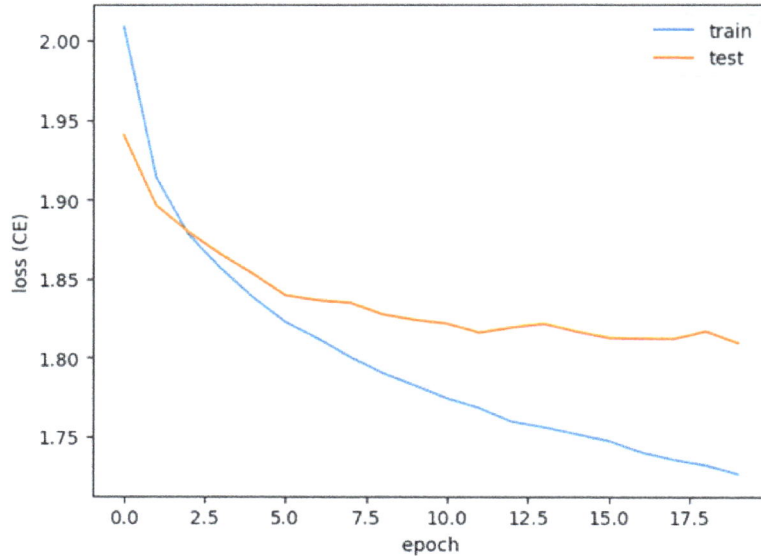

Figure 10.11 – Cross-entropy loss as a function of the epoch

Overfitting starts appearing after only a few epochs.

5. Plot the accuracy for the train and test sets:

```
plt.plot(train_accuracy, label='train')
plt.plot(test_accuracy, label='test')
plt.xlabel('epoch')
plt.ylabel('Accuracy')
plt.legend()
plt.show()
```

This is what we get:

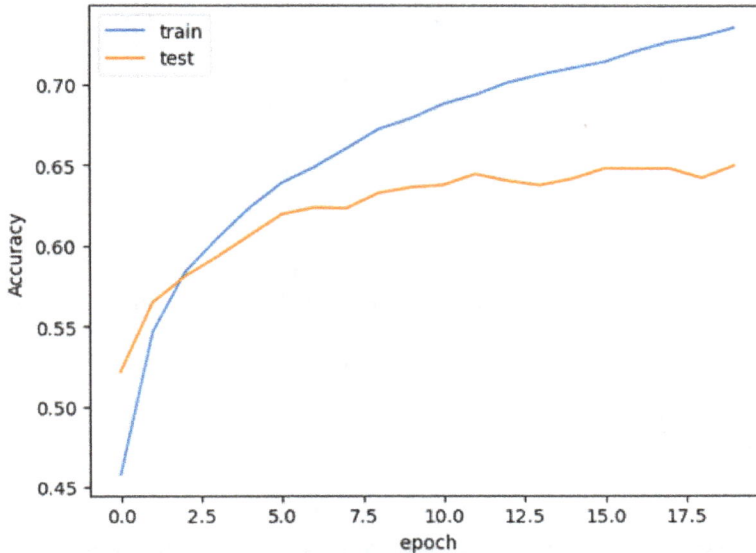

Figure 10.12 – Accuracy as a function of the epoch. The test accuracy
climbs to 66%, compared to 61% without batch normalization

As a result, we can see that while there is still some overfitting, the test accuracy improved significantly from 61% to 66%, thanks to batch normalization.

There's more...

What's interesting about CNNs is that we can have a look at what they learn from data. One way to do so is to look at the learned kernels. This can be done using the `visualize_kernels` function, as defined here:

```
from torchvision import utils

def visualize_kernels(tensor, ch=0, all_kernels=False, nrow=8,
padding=1, title=None):
    n,c,w,h = tensor.shape

    if all_kernels:
        tensor = tensor.view(n*c, -1, w, h)
    elif c != 3:
        tensor = tensor[:,ch,:,:].unsqueeze(dim=1)

    rows = np.min((tensor.shape[0] // nrow + 1, 64))
    grid = utils.make_grid(tensor, nrow=nrow,
```

```
            normalize=True, padding=padding)
    # Display
    plt.figure(figsize=(nrow, rows))
    if title is not None:
        plt.title(title)
    plt.imshow(grid.cpu().numpy().transpose((1, 2, 0)))
    plt.axis('off')
    plt.show()
```

We can now apply this function to visualize the learned kernels of the C1 and C3 layers:

```
visualize_kernels(lenet5.c1.weight.data, all_kernels=False,
    title='C1 layer')
visualize_kernels(lenet5.c3.weight.data, title='C3 layer')
```

Here is the C1 layer:

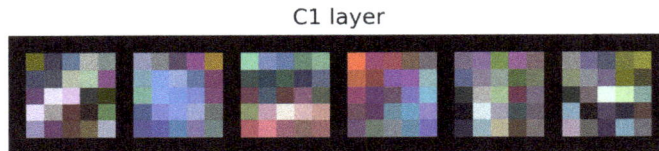

C1 layer

Here is the C3 layer:

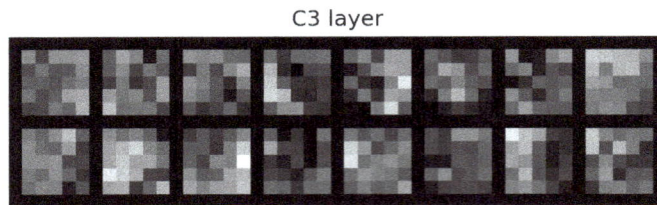

C3 layer

Figure 10.13 – Top – the learned kernels of the C1 layer, and bottom – the learned kernels of the C3 layer

Displaying kernels is not always helpful, but, depending on the task, they can give hints on what shapes a model recognizes.

See also

- The torch documentation about batch normalization: https://pytorch.org/docs/
  stable/generated/torch.nn.BatchNorm1d.html

- The batch normalization paper: https://arxiv.org/pdf/1502.03167.pdf

- A very well-written blog post about batch normalization: https://towardsdatascience.
  com/batch-normalization-in-3-levels-of-understanding-14c2da90a338

Regularizing a CNN with transfer learning for object detection

In this recipe, we will perform another typical task in computer vision – object detection. Before taking advantage of the power of transfer learning to help get better performances using a **You Only Look Once (YOLO)** model (a widely used class of models for object detection), we will give insights about what object detection is, the main methods and metrics, as well as the COCO dataset.

Object detection

Object detection is a computer vision task, involving both the identification and localization of objects of a given class (for example, a car, phone, person, or dog). As shown in *Figure 10.14*, the objects are usually localized, thanks to predicted bounding boxes, as well as predicted classes.

Figure 10.14 – An example of an image with object detection.
Objects are detected with a bounding box and a class

Researchers have proposed many methods to help solve object detection problems, some of which are heavily used in many industries. There are several groups of methods for object detection, but perhaps the two most widely used groups of methods are currently the following:

- One-stage methods, such as YOLO and SSD
- Two-stage methods, based on **Region-Based CNN (R-CNN)**

Methods based on R-CNN are powerful and usually more accurate than one-stage methods. On the other hand, one-stage methods are usually less computationally expensive and can run in real time, but they may fail at detecting small objects more often.

Mean average precision

Since this is a specific task, a specific metric is needed to assess the performances of such models – the **mean Average Precision (mAP)**. Let's get an overview of what mAP is. For that, we need to introduce several concepts, such as the **Intersection over Union (IoU)** and precision and recall in the context of object detection.

When an object is detected, it comes with three pieces of information:

- A predicted class

- A bounding box (usually four points, either center plus width and height, or top-left and bottom-right locations)

- A confidence level or probability that the box contains an object

To consider an object successfully detected, the classes must match, and the bounding box must be well localized. While knowing whether the classes match is trivial, the bounding box localization is computed using a metric called IoU.

Having an explicit name, the IoU can be computed as the intersection of the ground truth and the predicted boxes, over the union of those two same boxes, as shown in *Figure 10.15*.

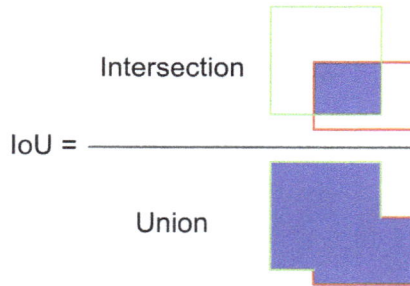

Figure 10.15 – A representation of the IoU metric

Given two bounding boxes – for example, A and B – IoU can be mathematically described with the following equation:

$$IoU = \frac{A \cap B}{A \cup B}$$

IoU has several advantages for a metric:

- The values are between 0 and 1

- A value of 0 means the two boxes don't overlap

- A value of 1 means the two boxes perfectly match

A threshold is then applied to the IoU. If the IoU is above the threshold, it is considered a **True Positive (TP)**; otherwise, it is considered a **False Positive (FP)**, allowing us to effectively compute the precision. Finally, a **False Negative (FN)** is an object that was not detected, given the IoU threshold.

Using these definitions of TP, FP, and FN, it is then possible to compute the precision and recall.

As a reminder, the precision P and recall R formulas are the following:

$$P = \frac{TP}{TP + FP}$$

$$R = \frac{TP}{TP + FN}$$

Using P and R, it is possible to plot the **Precision-Recall curve** (**PR curve**), with P as a function of R for various confidence-level thresholds, from 0 to 1. Using this PR curve, it is possible to compute the **Average Precision** (**AP**) for a given class by averaging P for different values of R (for example, averaging the interpolated P for R values in [0, 0.1, 0.2... 1]).

> **Important note**
> The average recall metric can be computed reciprocally using the same method, by reversing R and P.

Finally, the mAP is simply computed by averaging the AP over all the classes.

> **Tip**
> See the *See also* subsection for a link to a great blog post that explains in detail the mAP computation.

One drawback of this AP computation is that we considered only one IoU threshold, considering the same way almost perfect boxes with an IoU of 0.95 and not-so-good boxes with an IoU of 0.5. This is why some evaluation metrics average the AP for several IoU thresholds – for example, from 0.5 to 0.95 with a step of 0.05, usually noted as `AP@[IoU=0.5:0.95]` or `AP50-95`.

COCO dataset

The **Common Objects in Context** (**COCO**) dataset is a widely used dataset in object detection, having the following nice features:

- Hundreds of thousands of images with labels
- 80 classes of objects
- Flexible terms of use
- A wide community

This is a standard dataset when working with object detection. Thanks to that, most standard object detection models come with a set of pre-trained weights on the COCO dataset, allowing us to take advantage of transfer learning.

Getting started

In this recipe, we will use the YOLO algorithm, proposed by Ultralytics. **YOLO** stands for **You Only Look Once**, referring to the fact the method operates in a single stage, enabling real-time execution on devices with powerful enough power.

YOLO is a popular object detection algorithm that was first proposed in 2015. It has had a lot of new versions with improvements since then; version 8 is currently being developed.

It can be installed simply with the following command line:

```
pip install ultralytics
```

We will train an object detection algorithm on a vehicles dataset available on Kaggle. It can be downloaded and prepared with the following commands:

1. Download the dataset using the Kaggle API:

    ```
    kaggle datasets download -d saumyapatel/traffic-vehicles-object-
    detection  --unzip
    ```

2. Rename the folder for simplicity:

    ```
    mv 'Traffic Dataset' traffic
    ```

3. Create a datasets folder:

    ```
    mkdir datasets
    ```

4. Move the dataset to this folder:

    ```
    mv traffic datasets/
    ```

As a result, you should now have a folder dataset with the following structure:

```
traffic
├── images
│   ├── train: 738 images
│   ├── val: 185 images
│   ├── test: 278 images
├── labels
    ├── train
    ├── val
```

The dataset is split into `train`, `val`, and `test` sets, with respectively 738, 185, and 278 images. As we will see in the next subsection, these are typical road traffic images. The labels have seven classes – `Car`, `Number Plate`, `Blur Number Plate`, `Two-Wheeler`, `Auto`, `Bus`, and `Truck`. We can now proceed to train the object detection model.

How to do it...

We will first have to quickly explore the dataset and then train and evaluate a YOLO model on this data:

1. Import the required modules and functions:

    - `matplotlib` and `cv2` for image loading and visualization

    - `YOLO` for the model training

    - `glob` as `util` to list the files:

    ```
    import cv2
    from glob import glob
    import matplotlib.pyplot as plt
    from ultralytics import YOLO
    ```

2. Let's now explore the dataset. First, we will list the images in the `train` folder using `glob`, and then we will display eight of them:

    ```
    plt.figure(figsize=(14, 10))
    # Get all images paths
    images = glob('datasets/traffic/images/train/*.jpg')
    # Plot 8 of them
    for i, path in enumerate(images[:8]):
        img = plt.imread(path)
        plt.subplot(2, 4, i+1)
        plt.imshow(img)
        plt.axis('off')
    ```

    Here is the result:

Figure 10.16 – A patchwork of eight images from the train set of the traffic dataset

As we can see, these are mostly traffic-related images of different shapes and aspects.

3. If we have a look at the labels by reading a file, we get the following:

```
with open('datasets/traffic/labels/train/00 (10).txt') as file:
    print(file.read())
    file.close()
```

The output is the following:

```
2 0.543893 0.609375 0.041985 0.041667
5 0.332061 0.346354 0.129771 0.182292
5 0.568702 0.479167 0.351145 0.427083
```

The labels are an object per line, so here, we have three labeled objects in the image. Each line contains five numbers:

*   The class number
*   The box center x coordinate
*   The box center y coordinate

- The box width

- The box height

Note that all the box information is relative to the size of the image, so they are represented as floats in [0, 1].

> **Tip**
>
> There are other data formats for boxes in images such as the COCO and the Pascal VOC formats. More information about can be found in the *See also* subsection.

We can even plot this image with the boxes of the labels, using the `plot_labels` function implemented here:

```
def plot_labels(image_path, labels_path, classes):
    image = plt.imread(image_path)
    with open(labels_path, 'r') as file:
        lines = file.readlines()
        for line in lines:
            cls, xc, yc, w, h= line.strip().split(' ')

            xc = int(float(xc)*image.shape[1])
            yc = int(float(yc)*image.shape[0])
            w = int(float(w)*image.shape[1])
            h = int(float(h)*image.shape[0])
            cv2.rectangle(image, (xc - w//2,
                yc - h//2), (xc + w//2 ,yc + h//2),
                (255,0,0), 2)
            cv2.putText(image, f'{classes[int(cls)]}',
                (xc-w//2, yc - h//2 - 10),
                cv2.FONT_HERSHEY_SIMPLEX, 0.5,
                (255,0,0), 1)

    file.close()
    plt.imshow(image)
classes = ['Car', 'Number Plate', 'Blur Number Plate',
    'Two Wheeler', 'Auto', 'Bus', 'Truck']
plot_labels(
    'datasets/traffic/images/train/00 (10).jpg',
    'datasets/traffic/labels/train/00 (10).txt',
    classes
)
```

Here is the result:

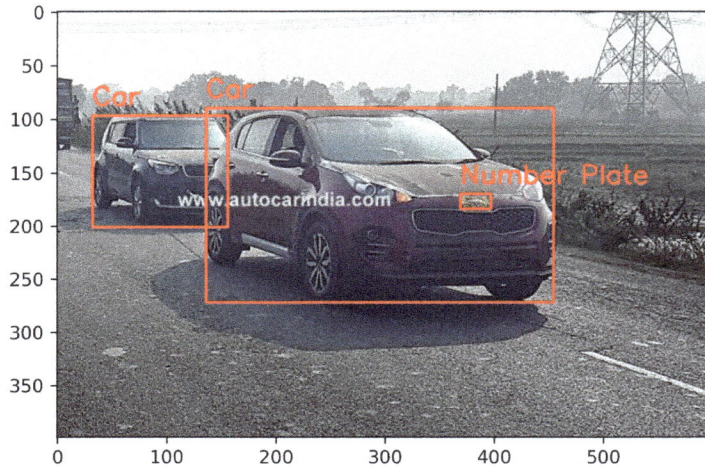

Figure 10.17 – An example of an image and its labeled bounding boxes

In this photo, we have labels for two cars and one plate. Let's go on to the next step to train a model on this data.

4.   We need to create a .yaml file, expected by the YOLO model, containing the dataset location and classes. Create and edit a file named dataset.yaml in the current directory with your favorite editor, and then fill it with the following content:

```
train: traffic/images/train
val: traffic/images/val
nc: 7
names: ['Car', 'Number Plate', 'Blur Number Plate',
    'Two Wheeler', 'Auto', 'Bus', 'Truck']
```

5.   We can now instantiate a new model. This will instantiate a YOLOv8 nano model. The YOLO model comes in five sizes:

*   'yolov8n.yaml' for the smallest model with 3.2 million parameters

*   'yolov8s.yaml' with 11.2 million parameters

*   'yolov8m.yaml' with 25.9 million parameters

*   'yolov8l.yaml' with 43.7 million parameters

*   'yolov8x.yaml' for the largest model with 68.2 million parameters:

    ```
    # Create a new YOLO model with random weights
    model = YOLO('yolov8n.yaml')
    ```

6. Train the model, providing the dataset with the previously created `dataset.yaml` file, the number of epochs, and the name (optional):

```
# Train the model for 100 epochs
model.train(data='dataset.yaml', epochs=100,
    name='untrained_traffic')
```

> **Tip**
> A lot of information is displayed when a model trains in memory, losses, and metrics. There's nothing too complicated if you want to look at it in detail.

The name is optional but allows us to easily find where the results and output are stored – in the `runs/detect/<name>` folder. If the folder already exists, it is simply incremented and not overwritten.

In this folder, several useful files can be found, including the following:

- `weights/best.pt`: The weights of the epoch that has the best validation loss
- `results.csv` with the logged results for each epoch
- Several curves and information about the data

7. Display the results. Here, we will display the automatically saved results image, `results.png`:

```
plt.figure(figsize=(14, 10))
plt.imshow(plt.imread(
    'runs/detect/untrained_traffic/results.png'))
plt.axis('off')
```

Here is the result:

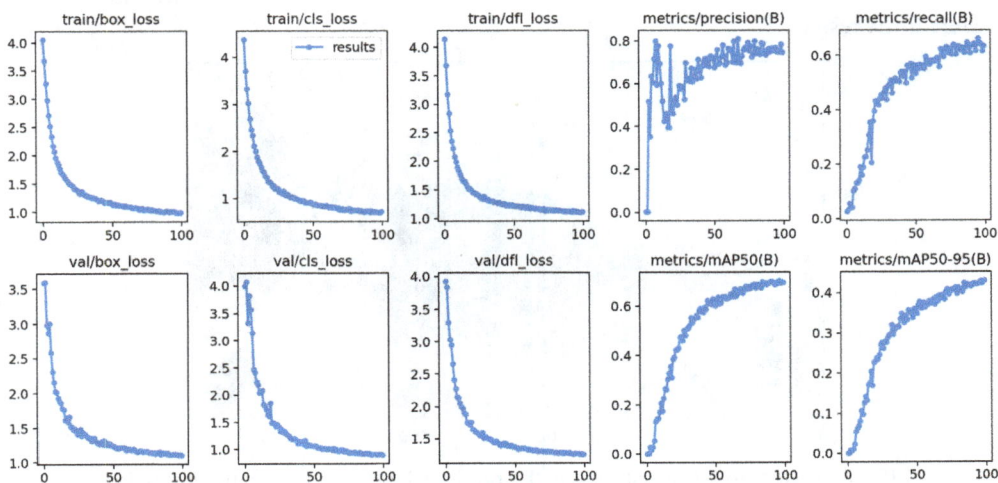

Figure 10.18 – A results summary of the YOLO model trained from scratch after 100 epochs

Several train and validation losses are displayed, as well as several losses – precision (P), recall (R), mAP50, and mAP50-95.

The results are encouraging considering the small dataset – we see a decreasing loss and a mAP50 increasing to 0.7, meaning the model is learning well.

8. Let's display the results as an example of the test set. For that, we first need to implement a function that allows us to display the image and the predicted boxes and classes, `plot_results_one_image`:

```
def plot_results_one_image(result):
    image = result[0].orig_img.copy()
    raw_res = result[0].boxes.data
    for detection in raw_res:
        x1, y1, x2, y2, p,
        cls = detection.cpu().tolist()
        cv2.rectangle(image, (int(x1), int(y1)),
            (int(x2), int(y2)), (255,0,0), 2)
        cv2.putText(image, f'{classes[int(cls)]}',
            (int(x1), int(y1) - 10),
            cv2.FONT_HERSHEY_SIMPLEX, 1, (255,0,0), 2)
    plt.imshow(image)
```

9. We can then compute the inference and display the results on an image from the test set:

```
# Compute the model inference on a test image
result = model.predict(
    'datasets/traffic/images/test/00 (100).png')
# Plot the results
plot_results_one_image(result)
```

Here is the result:

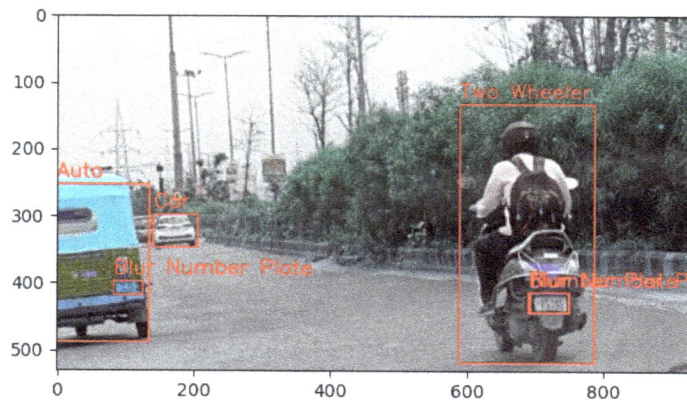

Figure 10.19 – An image from the test set and the predicted detections from the trained model

As we can see, our YOLO model has already learned to detect and correctly classify several classes. However, there is still room for improvement:

- The boxes do not perfectly match the objects; they are either too large or too small

- An object may have two classes (even if the difference between the `Blur Number Plate` and `Number Plate` classes is arguable)

Important note

It is worth mentioning that the direct output of the YOLO model usually contains many more bounding boxes. A postprocessing step, called the **non-max suppression** algorithm, has been applied here. This algorithm only keeps bounding boxes with a high enough confidence level, and small enough overlapping (computed with IoU) with other boxes of the same class.

Let's try to fix this using transfer learning.

Training with transfer learning

We will now train another model on this exact same dataset, with the same number of epochs. However, instead of using a model with random weights, we will load a model that was trained on the COCO dataset, allowing us to take advantage of transfer learning:

1. Instantiate and train a pre-trained model. Instead of instantiating the model with `yolov8n.yaml`, we only need to instantiate it with `yolov8n.pt`; this will automatically download the pretrained weights and load them:

```
# Load a pretrained YOLO model
pretrained_model = YOLO('yolov8n.pt')
# Train the model for 100 epochs
pretrained_model.train(data='dataset.yaml',
    epochs=100, name='pretrained_traffic')
```

2. Let's now display the results of this model:

```
plt.figure(figsize=(14, 10))
plt.imshow(plt.imread(
    'runs/detect/pretrained_traffic/results.png'))
plt.axis('off')
```

Here is the result:

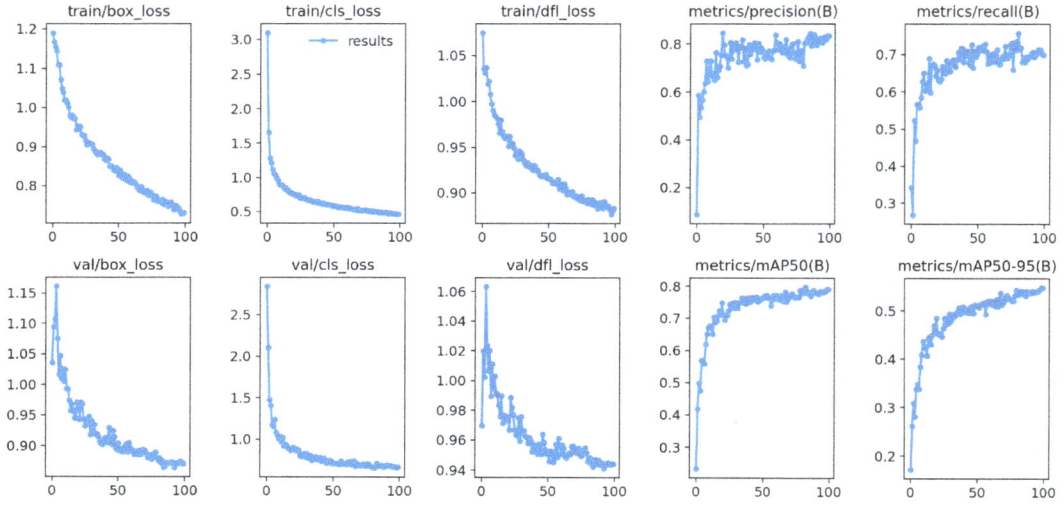

Figure 10.20 – A results summary of the YOLO model, with pretrained
weights on the COCO dataset after 100 epochs

Using transfer learning, all metrics have better performances – the mAP50 now climbs up to 0.8 against 0.7 previously, which is a significant improvement.

3.   We can now display the results in the same image as we did previously so that we can compare them:

```
result = pretrained_model.predict(
    'datasets/traffic/images/test/00 (100).png')
plot_results_one_image(result)
```

Here is the result:

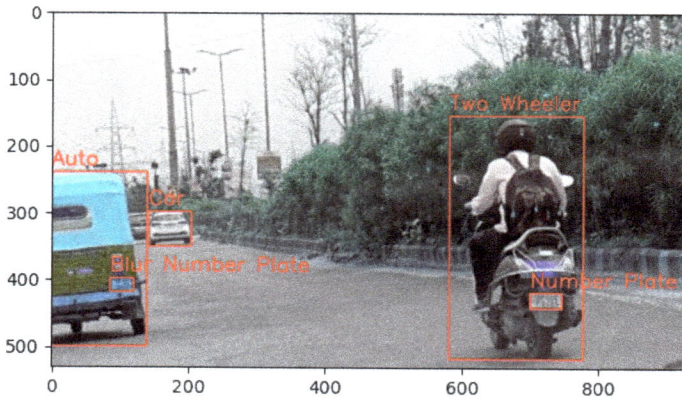

Figure 10.21 – An image from the test set and the predicted detections
from the model with the pretrained weights

This single image already shows several improvements – not only do the bounding boxes now perfectly fit the objects, but also no two objects are detected for a single number plate anymore. Thanks to transfer learning, we were able to efficiently help the model generalize.

There's more...

In this recipe, we focused on the object detection task, but YOLO models can do much more than that.

Using the same library, it is also possible to train models for the following:

- Classification
- Segmentation
- Pose

All these models also come with pretrained weights so that transfer learning can be leveraged to get good performances, even with small datasets.

See also

- A blog post explaining the mAP metric computation: `https://pyimagesearch.com/2022/05/02/mean-average-precision-map-using-the-coco-evaluator/`
- The COCO dataset website, which allows you to easily browse and display the dataset: `https://cocodataset.org/#home`
- A link to the original YOLO paper: `https://arxiv.org/abs/1506.02640`
- A link to the ultralytics documentation: `https://docs.ultralytics.com/usage/python/`
- The YOLOv8 GitHub repo: `https://github.com/ultralytics/ultralytics`
- A clear and concise post from Albumentations about the main bounding boxes formats: `https://albumentations.ai/docs/getting_started/bounding_boxes_augmentation/`

Semantic segmentation using transfer learning

In this recipe, we will take advantage of transfer learning and the fine-tuning of pretrained models to undertake a specific task of computer vision – the semantic segmentation of drone images.

Object detection and instance segmentation are about detecting objects in an image – an object is delimited by a bounding box, as well as a polygon in the case of instance segmentation. Alternatively, **semantic segmentation** is about classifying all the pixels of an image in a class.

As we can see in *Figure 10.22*, all pixels have a given color so that each one is attributed a class.

Figure 10.22 – An example of annotation of semantic segmentation. On the
left is the original image, and on the right is the labeled image – there is one
class of object per color, and each pixel is assigned to a given class

Even if it may look similar to instance segmentation, we will see in this recipe that the concepts and methods used are quite different. We will review the possible metrics, losses, architectures, and encoders to solve a semantic segmentation problem.

Metrics

Since semantic segmentation can be seen as a multiclass classification of each pixel, the most intuitive metric is the averaged accuracy score – the pixel accuracy averaged over the whole image.

Indeed, it can be used to sometimes yield solid results. However, most of the time in semantic segmentation, some classes are far less present than others – for example, in urban pictures, it is likely that there will be a lot of pixels of roads and buildings, and much less of persons or bikes. It is then likely to have good accuracy but not offer a model that accurately segments underrepresented classes.

Because of the limitation of the accuracy metric, many other metrics were proposed. One of the most used metrics in semantic segmentation is the IoU, already explained in the previous recipe. The IoU can be computed for each class independently and then averaged to compute a single metric (other averaging methods exist and are explored in more detail in *There's more…* subsection). The IoU is sometimes referred to as the **Jaccard index**.

One more frequently used metric is the **Dice coefficient**. Given two sets of pixels, A (for example, the predictions for a class) and B (for example, the ground truth for a class), the Dice coefficient can be computed with the following formula:

$$Dice = \frac{2\,|A \cap B|}{|A| + |B|}$$

Here, |A| is simply the number of pixels in A, sometimes called the cardinality of A. The Dice coefficient is usually compared to the F1 score and is mathematically equivalent. Just like the IoU, the Dice coefficient can be averaged over all the classes.

Of course, other metrics exist and can be used, but they are outside the scope of this recipe.

Losses

Several losses were developed over the years to improve the performance of semantic segmentation models. Again, if we just think of semantic segmentation as a classification task over many pixels, cross-entropy loss is an intuitive choice. However, just like the accuracy score, cross-entropy loss is not a good choice in the case of imbalanced classes.

In practice, it is common to simply use the Dice loss, which directly reuses the Dice coefficient. Dice loss is usually better in case of class imbalance, but it sometimes has bumpy training losses.

Many other losses were proposed, such as the focal loss and the Tversky loss, and all have strengths and improvements. A paper summarizing the most widely used losses is cited in the *See also* subsection.

Architectures

Semantic segmentation is a very specific task, in the sense that unlike object detection, the input and output are both images. Indeed, for a given input image of size 480x640 (purposefully omitting the RGB channels), the output image is expected to have the exact same dimension of 480x640, since each pixel must have a predicted class.

More precisely, for an *N*-class semantic segmentation task, the output dimension would be 480x640x*N*, having for each pixel a set of *N* probabilities as the output of a softmax function.

The architectures to deal with such problems are usually based on the encoder-decoder principle:

- An encoder computes describing features on the input image
- A decoder decodes those encoded features in order to have the expected outputs

One of the most famous architectures for semantic segmentation is the U-Net architecture, shown in *Figure 10.23*:

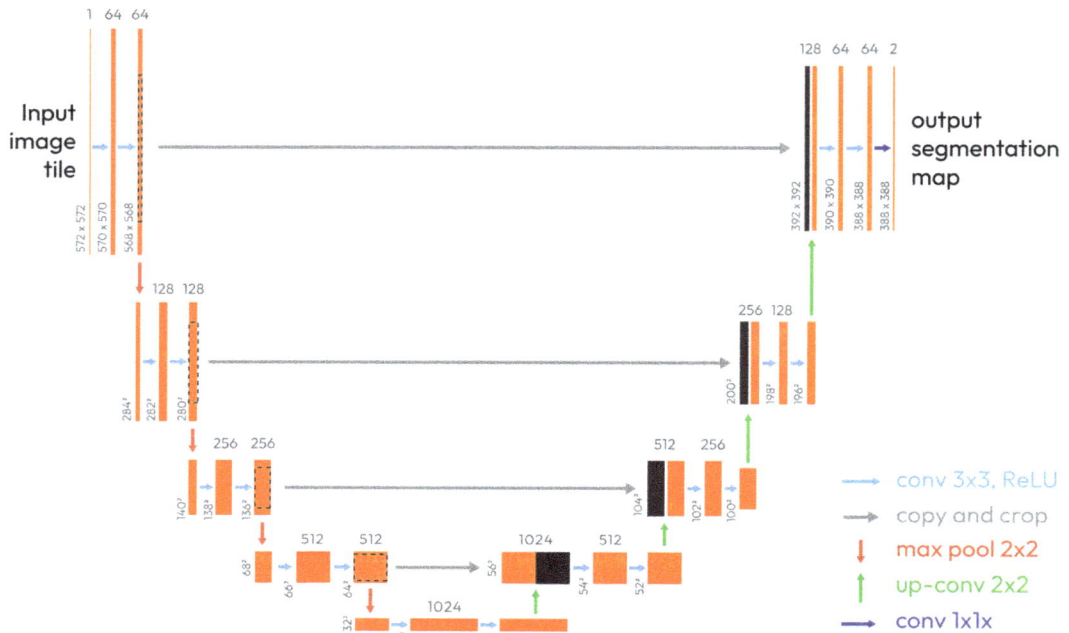

Figure 10.23 – The U-Net architecture as presented in the original paper U-Net:
Convolutional Networks for Biomedical Image Segmentation

As we can see in *Figure 10.23*, the U-Net architecture can be broken down as follows:

1. The input image is at the top left of the diagram.

2. The input image is sequentially encoded, as we can move down to the bottom of the diagram with convolutional and pooling layers.

3. As we go back up to the top right of the diagram, the output of the encoder is decoded and concatenated with the previous output of the encoder with convolutional and upscaling layers.

4. Finally, an output of the same width and height as the input image is predicted.

One strength of U-Net is that it encodes and then decodes, and it also concatenates the intermediate encodings to have efficient predictions. It is now a standard architecture when it comes to semantic segmentation.

Other architectures exist, some of which are widely used too, such as the following:

- **Feature Pyramid Networks (FPN)**
- U-Net++, a proposed improvement of U-Net

Encoders

In the original U-Net paper, as we can see in *Figure 10.23*, the encoder part is a specific one, made of convolution and pooling layers. In practice, however, it is common to use famous networks as encoders, pretrained on ImageNet or the COCO dataset, so that we can take advantage of transfer learning.

Depending on the needs and constraints, several encoders may be used, such as the following:

- MobileNet – a light encoder, developed for fast inference on the edge
- **Visual Geometry Group (VGG)** architectures
- ResNet and ResNet-based architectures
- EfficientNet architectures

The SMP library

The **Segmentation Models PyTorch (SMP)** library is an open source library allowing us to do all we need, including the following:

- Choosing architectures such as U-Net, FPN, or U-Net++
- Choosing encoders such as VGG or MobileNet
- Already implemented losses such as the Dice loss and the focal loss
- Helper functions to compute metrics such as Dice and the IoU

We will use this library in this recipe to train semantic segmentation models on a drone dataset.

Getting started

For this recipe, we will need to download a dataset containing 400 images and associated labels. It can be downloaded with the Kaggle API using the following commands:

```
kaggle datasets download -d santurini/semantic-segmentation-
dronedataset --unzip
```

We end up with three folders, containing several datasets. We will use the one in `classes_dataset`, a five-classes dataset.

We also need to install the required libraries with the following command:

```
pip install matplotlib pillow torch torchvision segmentation-models-
pytorch
```

How to do it...

We will first train a U-Net model with a MobileNet encoder with transfer learning on our task, and then we will do the same with fine-tuning techniques by freezing layers and gradually decreasing the learning rate, in order to improve the performance of the model.

Training with ImageNet weights and unfreezing all weights

We will first train a pretrained model on ImageNet in a regular fashion, with all the weights trainable:

1. We will first make the required imports for this recipe:

    ```
    from torch.utils.data import DataLoader, Dataset
    import torch
    import matplotlib.pyplot as plt
    import torchvision.transforms as transforms
    import numpy as np
    import tqdm
    from glob import glob
    from PIL import Image
    import segmentation_models_pytorch as smp
    import torch.nn as nn
    import torch.optim as optim
    ```

2. Implement the `DroneDataset` class:

    ```
    class DroneDataset(Dataset):

        def __init__(self, images_path: str,
            masks_path: str, transform, train: bool,
            num_classes: int = 5):
                self.images_path = sorted(glob(
                    f'{images_path}/*.png'))
                self.masks_path = sorted(glob(
                    f'{masks_path}/*.png'))
                self.num_classes = num_classes
                if train:
                    self.images_path = self.images_path[
    ```

```
                    :int(.8*len(self.images_path))]
            Self.masks_path = self.masks_path[
                    :int(.8*len(self.masks_path))]
        else:
            self.images_path = self.images_path[
                int(.8*len(self.images_path)):]
            self.masks_path = self.masks_path[
                int(.8*len(self.masks_path)):]
        self.transform = transform

    def __len__(self):
        return len(self.images_path)

    def __getitem__(self, idx):
        image = np.array(Image.open(
            self.images_path[idx]))
        mask = np.array(Image.open(
            self.masks_path[idx]))

        return self.transform(image), torch.tensor(
            mask, dtype=torch.long)
```

The __init__ method just reads all the available image and mask files. It also takes a Boolean variable for the train versus test dataset, allowing you to select only the first 80% or the last 20% of the files.

The __getitem__ method simply loads an image from a path and returns the transformed image as well as the mask as tensors.

3. Instantiate the transformation to apply it to images – here, it's simply a tensor conversion and a normalization:

```
transform = transforms.Compose([
    transforms.ToTensor(),
    transforms.Normalize((0.5, 0.5, 0.5),
        (0.5, 0.5, 0.5))
])
```

4. Define a few constants – the batch size, learning rate, classes, and device:

```
batch_size = 4
learning_rate = 0.005
classes = ['obstacles', 'water', 'soft-surfaces',
    'moving-objects', 'landing-zones']
device = torch.device(
    'cuda' if torch.cuda.is_available() else 'cpu')
```

5. Instantiate the datasets and data loaders:

```
train_dataset = DroneDataset(
    'classes_dataset/classes_dataset/original_images/',
    'classes_dataset/classes_dataset/label_images_semantic/',
    transform,
    train=True
)
train_dataloader = DataLoader(train_dataset,
    batch_size=batch_size, shuffle=True)

test_dataset = DroneDataset(
    'classes_dataset/classes_dataset/original_images/',
    'classes_dataset/classes_dataset/label_images_semantic/',
    transform,
    train=False
)
test_dataloader = DataLoader(test_dataset,
    batch_size=batch_size, shuffle=True)
```

6. Display an image with an overlay of the associated labels:

```
# Get a batch of images and labels
images, labels = next(iter(train_dataloader))
# Plot the image and overlay the labels
plt.figure(figsize=(12, 10))
plt.imshow(images[0].permute(
    1, 2, 0).cpu().numpy() * 0.5 + 0.5)
plt.imshow(labels[0], alpha = 0.8)
plt.axis('off')
```

Here is the result:

Figure 10.24 – An image of the Drone dataset with its mask overlay, made of five colors for five classes

As we can see, there are several colors overlayed on the image:

- Yellow for `'landing-zones'`
- Dark green for `'soft-surfaces'`
- Blue for `'water'`
- Purple for `'obstacles'`
- Light green for `'moving-objects'`

7. Instantiate the model – a U-Net architecture, with EfficientNet as an encoder (more specifically, the `'efficientnet-b5'` encoder), pretrained on `imagenet`:

```
model = smp.Unet(
    encoder_name='efficientnet-b5',
    encoder_weights='imagenet',
    in_channels=3,
```

```
        classes=len(classes),
    )
```

8. Instantiate the Adam optimizer and the loss as the Dice loss:

```
optimizer = optim.Adam(model.parameters(),
    lr=learning_rate)
criterion = smp.losses.DiceLoss(
    smp.losses.MULTICLASS_MODE, from_logits=True)
```

9. Implement a helper function, `compute_metrics`, that will help compute the IoU and F1-score (equivalent to the Dice coefficient):

```
def compute_metrics(stats):
    tp = torch.cat([x["tp"] for x in stats])
    fp = torch.cat([x["fp"] for x in stats])
    fn = torch.cat([x["fn"] for x in stats])
    tn = torch.cat([x["tn"] for x in stats])

    iou = smp.metrics.iou_score(tp, fp, fn, tn,
        reduction='micro')
    f1_score = smp.metrics.f1_score(tp, fp, fn, tn,
        reduction='micro')
    return iou, f1_score
```

10. Implement a helper function, `epoch_step_unet`, that will compute forward propagation, backpropagation if needed, the loss function, and metrics:

```
def epoch_step_unet(model, dataloader, device,
    num_classes, training_set: bool):
        stats = []
        for i, data in tqdm.tqdm(enumerate(
            dataloader, 0)):
            inputs, labels = data
            inputs = inputs.to(device)
            labels = labels.to(device)
            if training_set:
                optimizer.zero_grad()
                outputs = model(inputs)
                loss = criterion(outputs, labels)
            if training_set:
                loss.backward()
                optimizer.step()
            tp, fp, fn, tn = smp.metrics.get_stats(
                torch.argmax(outputs, dim=1), labels,
```

```
                    mode='multiclass',
                    num_classes=num_classes)
            stats.append({'tp': tp, 'fp': fp, 'fn':fn,
                'tn': tn, 'loss': loss.item()})
        return stats
```

11. Implement a `train_unet` function, allowing us to train the model:

```
def train_unet(model, train_dataloader,
    test_dataloader, criterion, device,
    epochs: int = 10, num_classes: int = 5,
    scheduler=None):
    train_metrics = {'loss': [], 'iou': [], 'f1': [],
        'lr': []}
    test_metrics = {'loss': [], 'iou': [], 'f1': []}

    model = model.to(device)

    for epoch in range(epochs):
    # loop over the dataset multiple times
        # Train
        model.train()
        #running_loss = 0.0
        train_stats = epoch_step_unet(model,
            train_dataloader, device, num_classes,
            training_set=True)
        # Eval
        model.eval()
        with torch.no_grad():
            test_stats = epoch_step_unet(model,
                test_dataloader, device, num_classes,
                training_set=False)

        if scheduler is not None:
            train_metrics['lr'].append(
                scheduler.get_last_lr())
            scheduler.step()

        train_metrics['loss'].append(sum(
            [x['loss'] for x in train_stats]) / len(
                train_dataloader))
        test_metrics['loss'].append(sum(
            [x['loss'] for x in test_stats]) / len(
                test_dataloader))
```

```
            iou, f1 = compute_metrics(train_stats)
            train_metrics['iou'].append(iou)
            train_metrics['f1'].append(f1)
            iou, f1 = compute_metrics(test_stats)
            test_metrics['iou'].append(iou)
            test_metrics['f1'].append(f1)


            print(f"[{epoch + 1}] train loss: {train_metrics['loss']
    [-1]:.3f} IoU: {train_metrics['iou'][-1]:.3f} | \
                test loss: {
                    test_metrics['loss'][-1]:.3f} IoU:
                    {test_metrics['iou'][-1]:.3f}")
        return train_metrics, test_metrics
```

The `train_unet` function does the following:

- Trains the model on the train set, and compute the evaluation metrics (the IoU and F1-score)

- Evaluates the model on the test set with the evaluation metrics

- If a learning rate scheduler is provided, applies a step (see the *There's more* subsection for more about this)

- Displays in the standard output the train and test losses and IoU

- Returns the train and test metrics

12. Train the model for 50 epochs and store the output train and test metrics:

```
train_metrics, test_metrics = train_unet(model,
    train_dataloader, test_dataloader, criterion,
    device, epochs=50, num_classes=len(classes))
```

13. Display the metrics for the train and test sets:

```
plt.figure(figsize=(10, 10))
plt.subplot(3, 1, 1)
plt.plot(train_metrics['loss'], label='train')
plt.plot(test_metrics['loss'], label='test')
plt.ylabel('Dice loss')
plt.legend()
plt.subplot(3, 1, 2)
plt.plot(train_metrics['iou'], label='train')
plt.plot(test_metrics['iou'], label='test')
plt.ylabel('IoU')
plt.legend()
plt.subplot(3, 1, 3)
plt.plot(train_metrics['f1'], label='train')
```

```
plt.plot(test_metrics['f1'], label='test')
plt.xlabel('epoch')
plt.ylabel('F1-score')
plt.legend()
plt.show()
```

Here is the result:

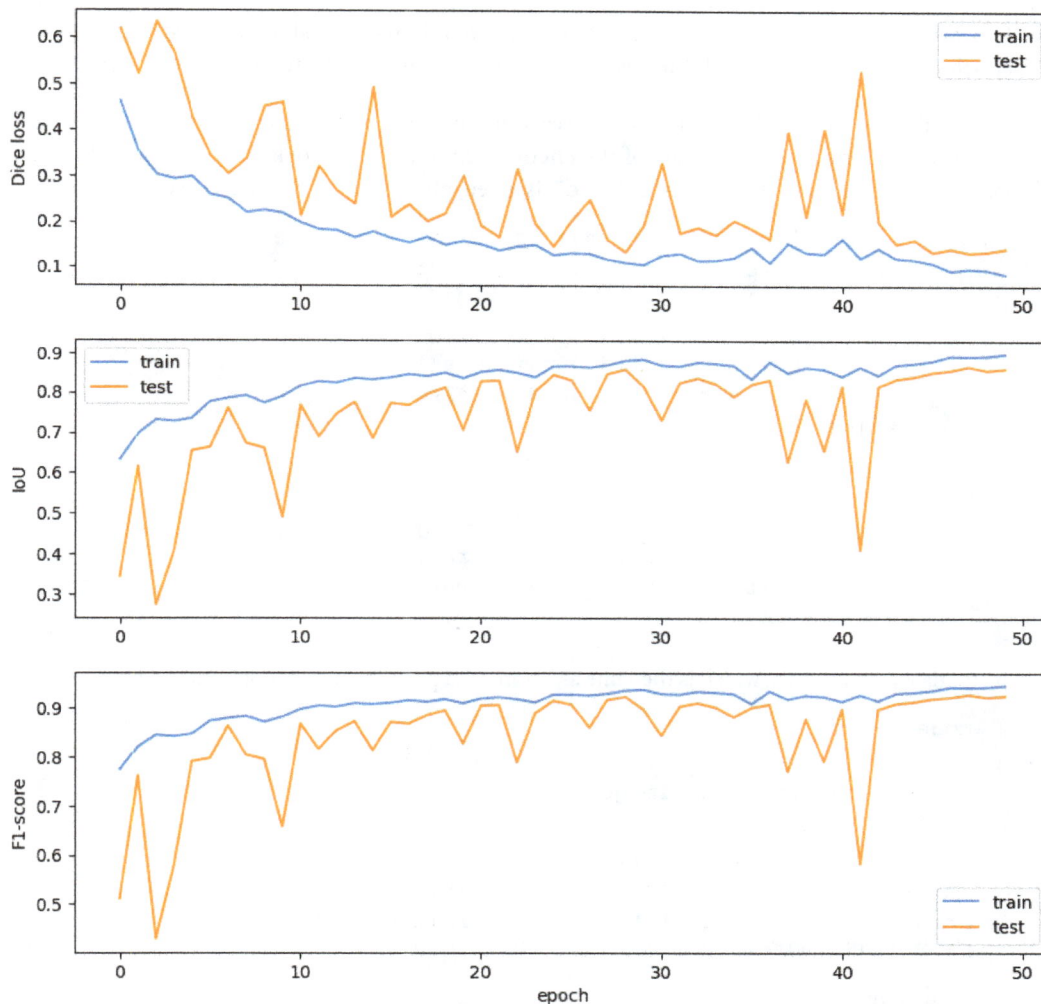

Figure 10.25 – The Dice loss (top), IoU (middle), and F1-score (bottom)
as a function of the epoch for the train and test sets

As we can see, the IoU goes up to 87% on the test set and seems to reach a plateau after about 30 epochs. Also, the test set metrics are bumpy and unstable, which can be because of a learning rate that is too high, as well as a model too large.

Let's now try to do the same with the freezing layer and gradually the decrease the learning rate.

Fine-tuning a pretrained model by freezing layers

We will now train a pretrained model in two stages – first, we will freeze most of the layers of the model for 20 epochs, then only unfreeze all the layers, and train 30 more epochs to fine-tune the model:

1. First, let's define two helper functions to freeze and unfreeze layers. The `freeze_encoder` function will freeze all the layers of the encoder up to a given block level, provided by the `max_level` argument. If no `max_level` is given, all weights of the encoder will be frozen:

```
def freeze_encoder(model, max_level: int = None):
    for I, child in enumerate(model.encoder.children()):
        if max_level is not None and i >= max_level:
            return
        for param in child.parameters():
            param.requires_grad = False
    return


def unfreeze(model):
    for child in model.children():
        for param in child.parameters():
            param.requires_grad = True
    return
```

2. Instantiate a new model, which is the same as before, and print the number of trainable parameters:

```
model = smp.Unet(
    encoder_name='efficientnet-b5',
    encoder_weights='imagenet',
    in_channels=3,
    classes=len(classes),
    )
print''Total number of trainable parameters'', sum(p.numel() for
p in model.parameters() if p.requires_grad))
```

The code output is the following:

Total number of trainable parameters: 31216581

As we can see, this model is made of ~31.2 million parameters.

3. Le"s now freeze part of the encoder – the first three blocks, which are basically most of the weights of the encoder, as we will see – and print the number of trainable parameters left over:

```
# Freeze the of the encoder
freeze_encoder(model, 3)
print('Total number of trainable parameters:', sum(p.numel() for
p in model.parameters() if p.requires_grad))
```

The output is the following:

Total number of trainable parameters: 3928469

We now have only ~3.9 million trainable parameters left. Almost 27.3 million parameters from the encoder are now frozen, out of ~28 million parameters in the encoder – the remaining parameters are from the decoder. This means we will mostly train the decoder first and use the pretrained encoder as a feature extractor.

4. Instantiate a new optimizer for training, as well as a scheduler. We will use an `ExponentialLR` scheduler here, with a gamma value of `0.95` – this means that at each epoch, the learning rate will be multiplied by 0.95:

```
optimizer = optim.Adam(model.parameters(),
    lr=learning_rate)
scheduler = optim.lr_scheduler.ExponentialLR(
    optimizer, gamma=0.95)
```

5. Train the model with frozen layers on 20 epochs:

```
train_metrics, test_metrics = train_unet(model,
    train_dataloader, test_dataloader, criterion,
    device, epochs=20, num_classes=len(classes),
    scheduler=scheduler)
```

As we can see, after 20 epochs only, the IoU on the test set already reaches 88%, slightly higher than without freezing and without any learning rate decay.

6. Now that the decoder and last layers of the encoder are warmed up against this dataset, let's unfreeze all the parameters before training for more epochs:

```
unfreeze(model)
print('Total number of trainable parameters:', sum(p.numel() for
p in model.parameters() if p.requires_grad))
```

The code output is the following:

Total number of trainable parameters: 31216581

As we can see, the trainable parameters are now back at 31.2 million, meaning that all the parameters are trainable.

7. Train the model on 30 more epochs:

```
train_metrics_unfreeze, test_metrics_unfreeze = train_unet(
model, train_dataloader, test_dataloader,
    criterion, device, epochs=30,
    num_classes=len(classes), scheduler=scheduler)
```

8. Plot the results by concatenating the results with and without freezing:

```
plt.figure(figsize=(10, 10))
plt.subplot(3, 1, 1)
plt.plot(train_metrics['loss'] + train_metrics_unfreeze['loss'],
label='train')
plt.plot(test_metrics['loss'] + test_metrics_unfreeze['loss'],
label='test')
plt.ylabel('Dice loss')
plt.legend()
plt.subplot(3, 1, 2)
plt.plot(train_metrics['iou'] + train_metrics_unfreeze['iou'],
label='train')
plt.plot(test_metrics['iou'] + test_metrics_unfreeze['iou'],
label='test')
plt.ylabel('IoU')
plt.legend()
plt.subplot(3, 1, 3)
plt.plot(train_metrics['f1'] + train_metrics_unfreeze['f1'],
label='train')
plt.plot(test_metrics['f1'] + test_metrics_unfreeze['f1'],
label='test')
plt.xlabel('epoch')
plt.ylabel('F1-score')
plt.legend()
plt.show()
```

Here is the result:

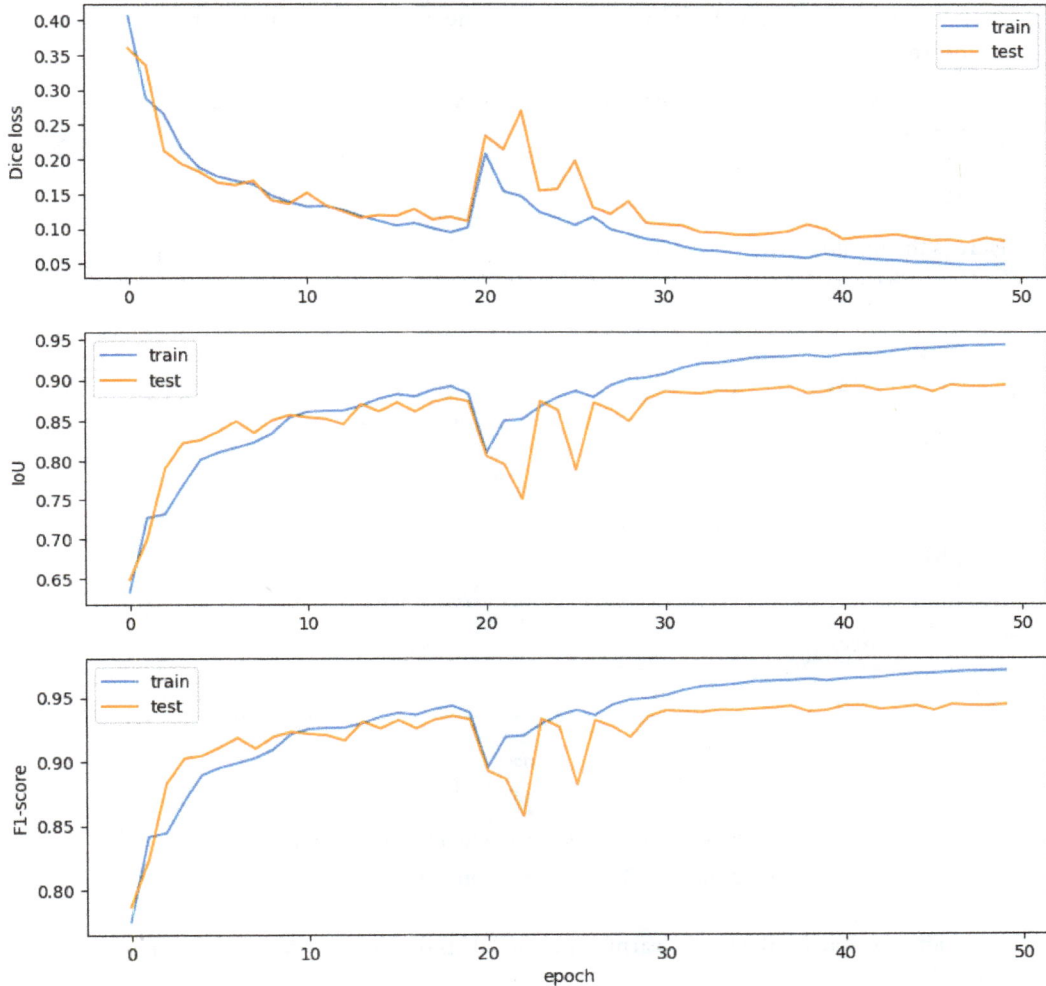

Figure 10.26 – The Dice loss (top), IoU (middle), and F1-score (bottom) as a function of the epoch for train and test sets with fine-tuning – after a drop when unfreezing the weights, the metrics improve and are stable again

We can see that as soon as we unfreeze all the parameters at epoch 20, the curves get a bit bumpy. However, after 10 more epochs at around epoch 30, the metrics become stable again.

After 50 epochs in total, the IoU reaches almost 90%, against only 87% earlier without the fine-tuning techniques (freezing and the learning rate decay).

9. Out of curiosity, we can also plot the learning rate as a function of the epoch, to look at the decrease:

```
plt.plot(train_metrics['lr'] + train_metrics_unfreeze['lr'])
plt.xlabel('epoch')
plt.ylabel('Learning rate')
plt.show()
```

Here is the result:

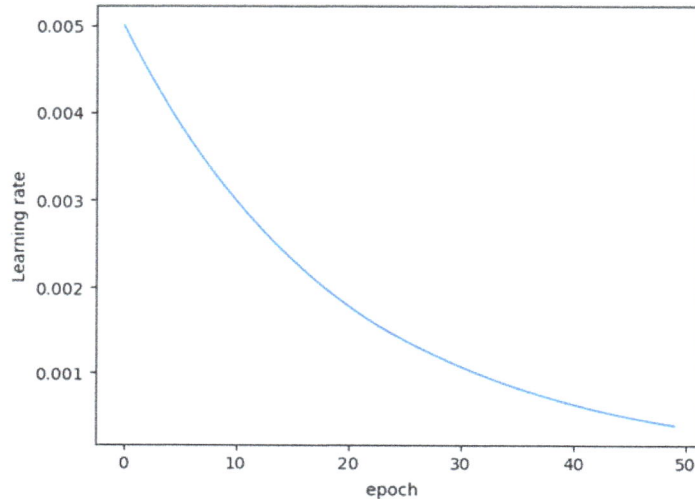

Figure 10.27 – The learning rate value as a function of the epoch for a
torch ExponentialLR class, with a gamma value of 0.95

As expected, after 50 epochs, the initial learning rate of 0.05 is divided by almost 13, down to roughly 0.0003, since $0.95^{50} \simeq 1/13$.

There's more...

There are several ways to compute the metrics such as the IoU or Dice coefficient in semantic segmentation. In this recipe, as implemented in the `compute_metrics` function, the `'micro'` option was chosen with the following code:

```
smp.metrics.iou_score(tp, fp, fn, tn, reduction='micro')
```

First, we can define the TP, FP, FN, and TN for each pixel just like in any other classification task. The metrics are then computed based on those values.

Based on that, the most common computation methods are available and well summarized in the SMP documentation:

- `'micro'`: Sum the TP, FP, FN, and TN pixels over all images and classes and then only compute the score.

- `'macro'`: Sum the TP, FP, FN, and TN pixels over all images for each label, compute the score for each label, and then average over the labels. If there is an imbalanced class (which is usually the case in semantic segmentation), this method will not take it into account and should be avoided.

- `'weighted'`: The same as `'macro'` but with a weighted average over the labels.

- `'micro-imagewise'`, `'macro-imagewise'`, and `'weighted-imagewise'`: The same as `'micro'`, `'macro'`, and `'weighted'`, respectively, but they compute the score for each image independently before averaging over the images. This can be useful when images in a dataset do not have the same dimensions, for example.

Most of the time, a `'micro'` or `'weighted'` approach works fine, but it's always useful to understand the differences and to be able to play with them.

See also

- A paper proposing a review of several losses used in semantic segmentation: `https://arxiv.org/pdf/2006.14822.pdf`

- The paper proposing the U-Net architecture: `https://arxiv.org/pdf/1505.04597.pdf`

- The GitHub repo of the SMP library: `https://github.com/qubvel/segmentation_models.pytorch`

- A link to the Kaggle dataset: `https://www.kaggle.com/datasets/santurini/semantic-segmentation-drone-dataset`

11
Regularization in Computer Vision – Synthetic Image Generation

This chapter will focus on the techniques and methods used to generate synthetic images for data augmentation. Having diverse data is often one of the most efficient ways to regularize computer vision models. Many approaches allow us to generate synthetic images; from simple tricks such as image flipping to new image creation using generative models. Several techniques will be explored in this chapter, including the following:

- Image augmentation with Albumentations
- Creating synthetic images for object detection – training an object detection model with only synthetic data
- Real-time style transfer – training a model for real-time style transfer based on Stable Diffusion, a powerful generative model

Technical requirements

In this chapter, we will train several deep learning models and generate images. We will need the following libraries:

- NumPy
- Matplotlib
- Albumentations
- Pytorch

- torchvision
- ultralytics

Applying image augmentation with Albumentations

More often than not, in **machine learning** (**ML**), data is crucial to getting better performances of models. Computer vision is no exception, and data augmentation with images can be easily taken to another level.

Indeed, it is possible to easily augment an image, for example, by mirroring it, as shown in *Figure 11.1*.

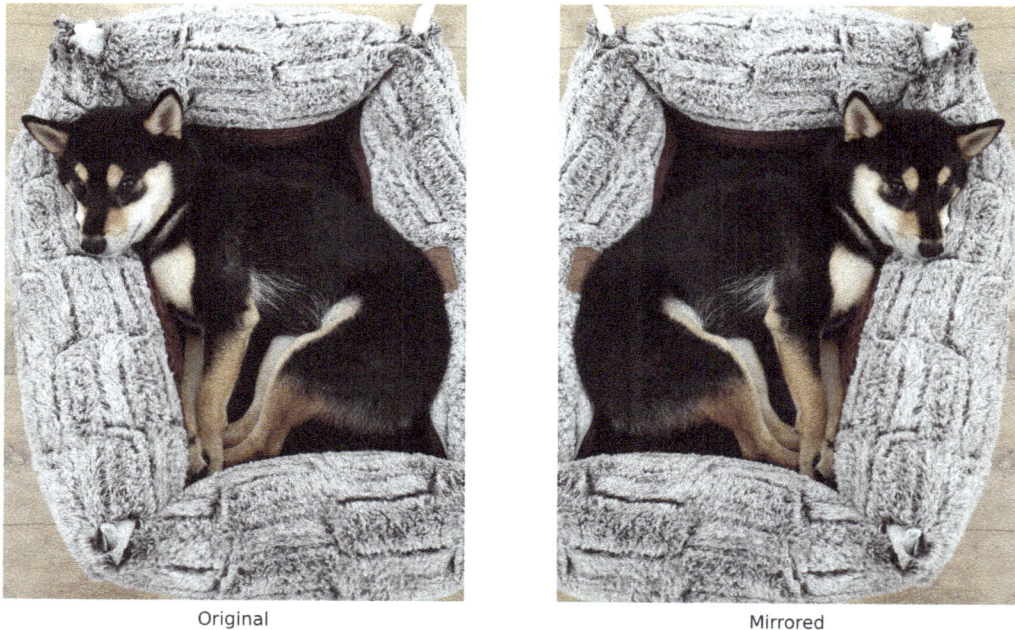

Original   Mirrored

Figure 11.1 – On the left, the original picture of my dog, and on the right, a mirrored picture of my dog

However, beyond this, many more types of augmentation are possible and can be divided into two main categories: pixel-level and spatial-level transformations.

Let's discuss both of these in the following sections.

Spatial-level augmentation

The mirroring is an example of spatial-level augmentation; however, much more than simple mirroring can be done. For example, see the following:

- **Shifting**: Shifting an image in a certain direction

- **Shearing**: Add shearing to an image

- **Cropping**: Cropping only part of an image

- **Rotating**: Applying rotation to an image

- **Transposing**: Transposing an image (in other words, applying both vertical and horizontal flipping)

- **Perspective**: Applying a 4-point perspective to an image

As we can see, there are a lot of possibilities in spatial-level augmentation. *Figure 11.2* shows some examples of these possibilities on a given image and displays some of the possible artifacts: black borders on the shifted image and mirror padding on the rotated image.

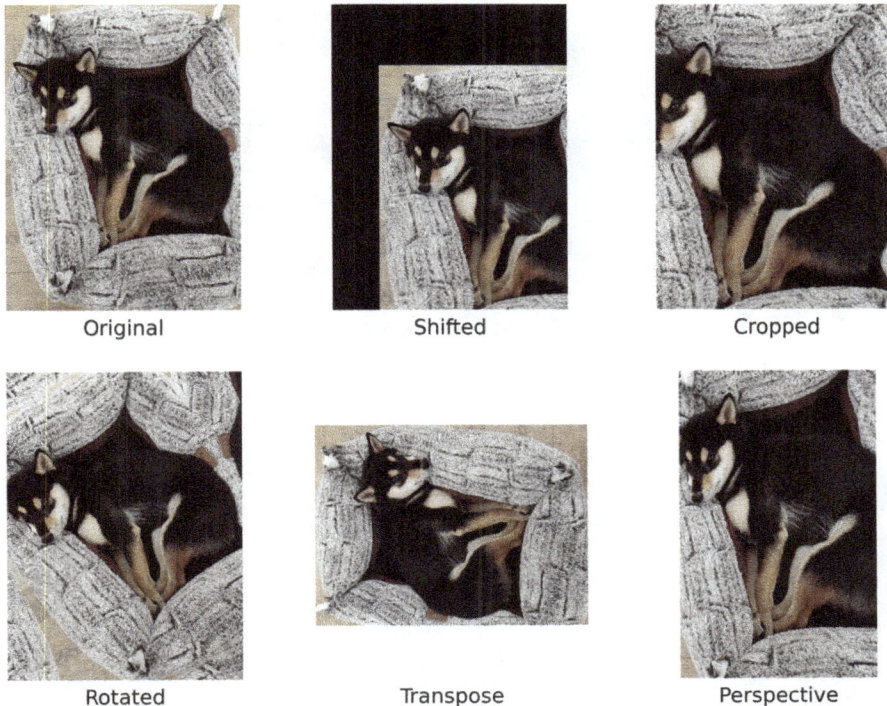

Figure 11.2 – The original image (top left) and five different augmentations
(note that some artifacts may appear, such as black borders)

Pixel-level augmentation

Another type of augmentation is at the pixel level and can be as useful as spatial-level augmentation. A simple example could be to change the brightness and contrast level of an image so that a model can be more robust in various lighting conditions.

A non-exhaustive list of pixel-level augmentation is the following:

- **Brightness**: Modify the brightness

- **Contrast**: Modify the contrast

- **Blurring**: Blur the image

- **HSV**: Randomly modify the **hue**, **saturation**, and **value** of the image

- **Color conversion**: Convert the image into black and white or sepia

- **Noise**: Add noise to the image

- There are many possibilities, and this can make a huge difference in model robustness. A few examples of the results of these augmentations are shown in the following figure:

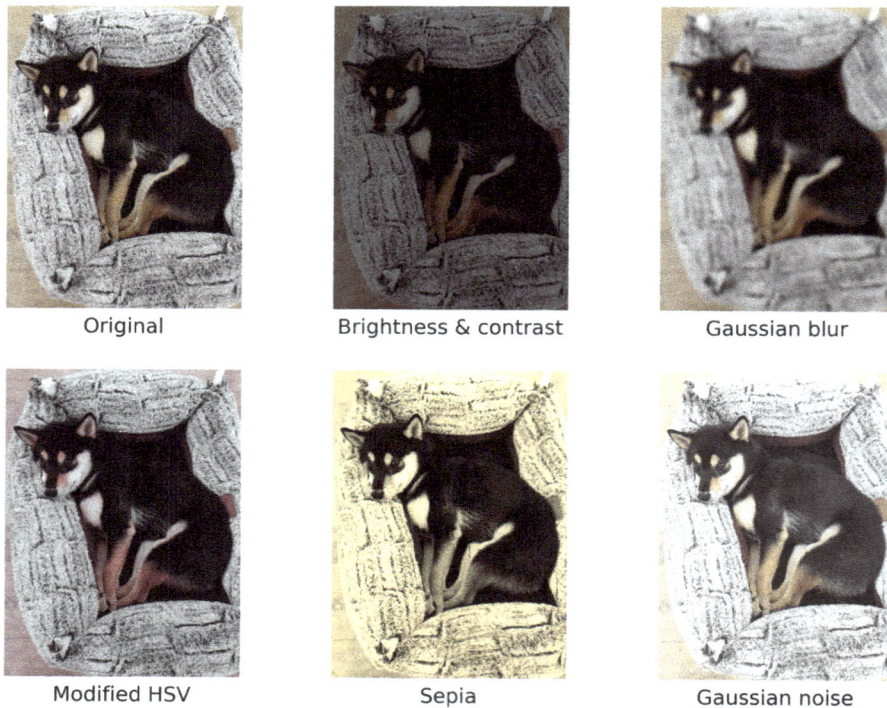

Figure 11.3 – An original image (top left) and several pixel-level augmentations

As we can see, using both pixel-level and spatial-level transformations, it is fairly easy to augment a single image into 5 or 10 images. Moreover, these augmentations can sometimes be composed together for more diversity. Of course, it does not replace a real, large, and diverse dataset, but image augmentation is usually cheaper than collecting data and allows us to get real boosts in model performance.

Albumentations

Of course, we do not have to reimplement all of these image augmentations manually. Several libraries for image augmentation exist, and Albumentations is arguably the most complete, free, and open source solution on the market. As we will see in this recipe, the Albumentations library allows powerful image augmentations with just a few lines of code.

Getting started

In this recipe, we will apply image augmentation to a simple challenge: classifying cats and dogs.

We first need to download and prepare the dataset. The dataset, originally proposed by Microsoft, is made of 12,491 cat pictures and 12,470 dog pictures. It can be downloaded with the Kaggle API with the following command-line operation:

```
kaggle datasets download -d karakaggle/kaggle-cat-vs-dog-dataset
--unzip
```

This will download a folder named `kagglecatsanddogs_3367a`.

Unfortunately, the dataset is not yet split into train and test sets. The following code will split it into 80% train and 20% test sets:

```
from glob import glob
import os
cats_folder = 'kagglecatsanddogs_3367a/PetImages/Cat/'
dogs_folder = 'kagglecatsanddogs_3367a/PetImages/Dog/'
cats_paths = sorted(glob(cats_folder + '*.jpg'))
dogs_paths = sorted(glob(dogs_folder + '*.jpg'))
train_ratio = 0.8
os.mkdir(cats_folder + 'train')
os.mkdir(cats_folder + 'test')
os.mkdir(dogs_folder + 'train')
os.mkdir(dogs_folder + 'test')
for i in range(len(cats_paths)):
    if i <= train_ratio * len(cats_paths):
        os.rename(cats_paths[i], cats_folder + 'train/' + cats_
paths[i].split('/')[-1])
    else:
        os.rename(cats_paths[i], cats_folder + 'test/' + cats_
```

```
paths[i].split('/')[-1])

for i in range(len(dogs_paths)):
    if i <= train_ratio * len(dogs_paths):
        os.rename(dogs_paths[i], dogs_folder + 'train/' + dogs_
paths[i].split('/')[-1])
    else:
        os.rename(dogs_paths[i], dogs_folder + 'test/' + dogs_
paths[i].split('/')[-1])
```

This will create subfolders of `train` and `test`, so that the `kagglecatsanddogs_3367a` folder tree now looks like the following:

```
kagglecatsanddogs_3367a
└── PetImages
    ├── Cat
    │   ├── train: 9993 images
    │   └── test: 2497 images
    └── Dog
        ├── train: 9976 images
        └── test: 2493 images
```

We will now be able to efficiently train and evaluate a model against this dataset.

The required libraries can be installed with the following command line:

```
pip install matplotlib numpy torch torchvision albumentations
```

How to do it...

We will now train a MobileNet V3 network on the train dataset and evaluate it against the test dataset. Then, we will add image augmentation using Albumentations in order to improve the performance of the model:

1. First, import the needed libraries:

   - `matplotlib` for display and visualization

   - `numpy` for data manipulation

   - `Pillow` for image loading

   - `glob` for folder parsing

   - `torch` and `torchvision` for the model training and related `util` instances

Here are the `import` statements:

```
import matplotlib.pyplot as plt
import numpy as np
import torch
import torch.nn as nn
import torch.nn.functional as F
from torch.utils.data import DataLoader, Dataset
import torchvision.transforms as transforms
from torchvision.models import mobilenet_v3_small
from glob import glob
from PIL import Image
```

2. Next, we implement the `DogsAndCats` dataset class. It takes the following arguments:

- `cats_folder`: The path to the folder containing the cat pictures
- `dogs_folder`: The path to the folder containing the dog pictures
- `transform`: The transformation to apply to images (for example, resizing, converting into tensors, and so on...)
- `augment`: To apply image augmentation, as we will do in the second part of this recipe

Here is the code for the implementation:

```
class DogsAndCats(Dataset) :

    def __init__(self, cats_folder: str,
        dogs_folder: str, transform, augment = None):
            self.cats_path = sorted(glob(
                f'{cats_folder}/*.jpg'))
            self.dogs_path = sorted(glob(
                f'{dogs_folder}/*.jpg'))
            self.images_path = self.cats_path + self.dogs_path
            self.labels = [0.]*len(
            self.cats_path) + [1.]*len(self.dogs_path)
            self.transform = transform
            self.augment = augment

    def __len__(self):
        return len(self.images_path)

    def __getitem__(self, idx):
        image = Image.open(self.images_path[
            idx]).convert('RGB')
        if self.augment is not None:
```

```
        image = self.augment(
            image=np.array(image))["image"]
    return self.transform(image),
    torch.tensor(self.labels[idx],
        dtype=torch.float32)
```

This class is rather simple: the constructor collects all the paths of the images and defines the labels accordingly. The getter simply loads an image, optionally applies image augmentation, and returns the image as tensor with its associated label.

3. Then, we instantiate the transformation class. Here, we compose three transformations:

   • Tensor conversion

   • Resizing to 224x224 images since not all images are the same size

   • Normalization of the image input

   Here is the code for it:

```
transform = transforms.Compose([
    transforms.ToTensor(),
    transforms.Resize((224, 224), antialias=True),
    transforms.Normalize((0.5, 0.5, 0.5), (0.5, 0.5,
        0.5)),
])
```

4. Then, we create a few useful variables, such as the batch size, device, and number of epochs:

```
batch_size = 64
device = torch.device(
    'cuda' if torch.cuda.is_available() else 'cpu')
epochs = 20
```

5. Instantiate the datasets and data loaders. Reusing the train and test folders prepared earlier in the *Getting ready* subsection, we can now easily create our two loaders. They both use the same transformation:

```
trainset = DogsAndCats(
    'kagglecatsanddogs_3367a/PetImages/Cat/train/',
    'kagglecatsanddogs_3367a/PetImages/Dog/train/',
    transform=transform
)
train_dataloader = DataLoader(trainset,
    batch_size=batch_size, shuffle=True)
testset = DogsAndCats(
    'kagglecatsanddogs_3367a/PetImages/Cat/test/',
    'kagglecatsanddogs_3367a/PetImages/Dog/test/',
```

```
        transform=transform
    )
    test_dataloader = DataLoader(testset,
        batch_size=batch_size, shuffle=True)
```

6. Now, we display a few images along with their labels so that we get a glimpse at the dataset using the following code:

```
def display_images(dataloader, classes = ['cat', 'dog']):
    plt.figure(figsize=(14, 10))
    images, labels = next(iter(dataloader))
    for idx in range(8):
        plt.subplot(2, 4, idx+1)
        plt.imshow(images[idx].permute(
            1, 2, 0).numpy() * 0.5 + 0.5)
        plt.title(classes[int(labels[idx].item())])
        plt.axis('off')
display_images(train_dataloader)
```

Here are the images:

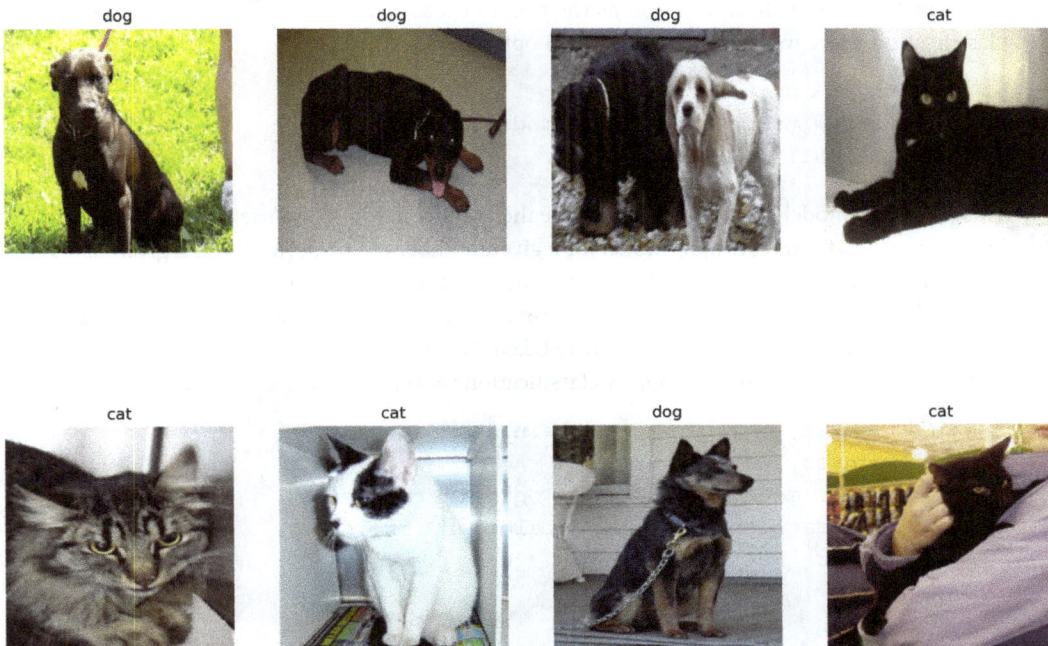

Figure 11.4 – A sample of images from the dataset

As we can see in the figure, this is a dataset made up of regular images of dogs and cats in various contexts, sometimes with humans in the pictures too.

7. Now, we implement the `Classifier` class. We will reuse the existing `mobilenet_v3_small` implementation provided in `pytorch` and simply add an output layer with one unit and a sigmoid activation function, shown as follows:

```
class Classifier(nn.Module):
    def __init__(self):
        super(Classifier, self).__init__()
        self.mobilenet = mobilenet_v3_small()
        self.output_layer = nn.Linear(1000, 1)

    def forward(self, x):
        x = self.mobilenet(x)
        x = nn.Sigmoid()(self.output_layer(x))
        return x
```

8. Next, we instantiate the model:

```
model = Classifier()
model = model.to(device)
```

9. Then, instantiate the loss function as the binary cross-entropy loss, well suited to binary classification. Here we instantiate the Adam optimizer:

```
criterion = nn.BCELoss()
optimizer = torch.optim.Adam(model.parameters(),
    lr=0.001)
```

10. Then, train the model for 20 epochs and store the outputs. To do so, we use the `train_model` function, which trains the input model for a given number of epochs and with a given dataset. It returns the loss and accuracy for the training and test set for each epoch. This function is available in the GitHub repository (`https://github.com/PacktPublishing/The-Regularization-Cookbook/blob/main/chapter_11/chapter_11.ipynb`), and is typical code for binary classification training, as we used in previous chapters:

```
train_losses, test_losses, train_accuracy,
test_accuracy = train_model(
    epochs, model, criterion, optimizer, device,
    train_dataloader, test_dataloader, trainset,
    testset
)
```

11. Finally, display the loss and accuracy as a function of the epoch:

```
plt.figure(figsize=(10, 10))
plt.subplot(2, 1, 1)
plt.plot(train_losses, label='train')
```

```
plt.plot(test_losses, label='test')
plt.ylabel('BCE Loss')
plt.legend()
plt.subplot(2, 1, 2)
plt.plot(train_accuracy, label='train')
plt.plot(test_accuracy, label='test')
plt.xlabel('Epoch')
plt.ylabel('Accuracy')
plt.legend()
plt.show()
```

Here are the plots for it:

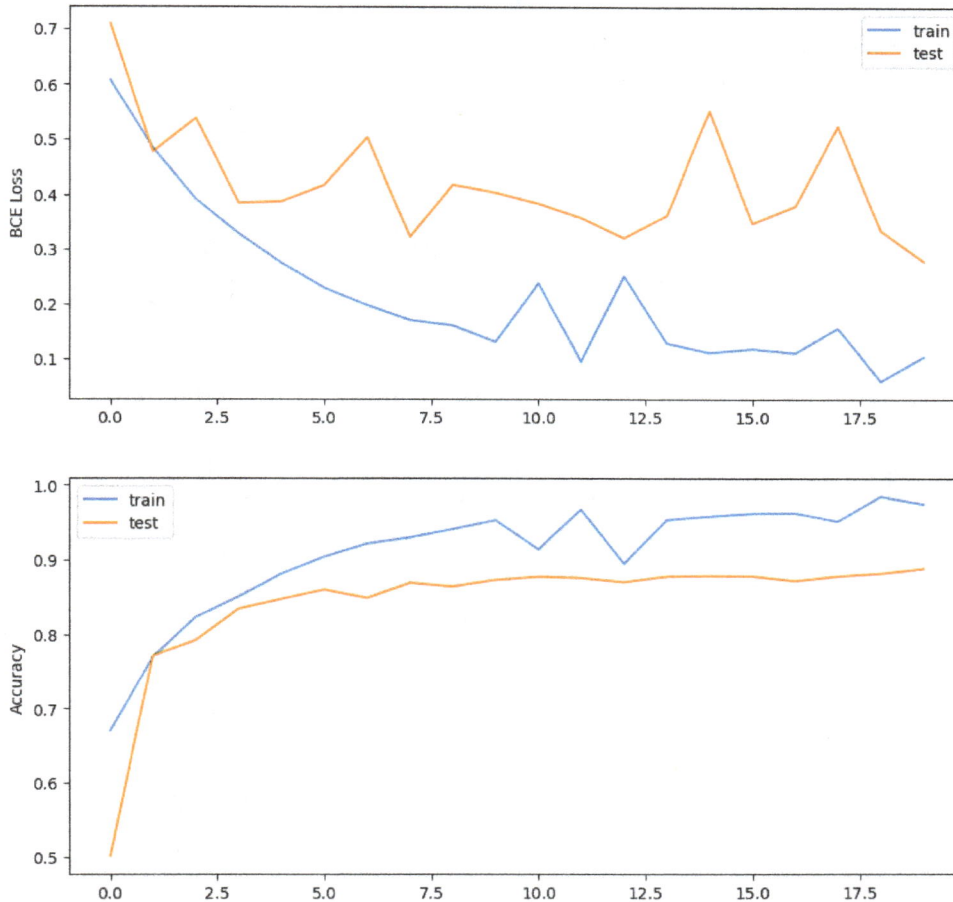

Figure 11.5 – Binary cross-entropy loss (top) and accuracy (bottom) as a function of
the epoch for both train and test sets with no augmentation (the loss and accuracy
are suggesting overfitting while the test accuracy plateaus at around 88%)

As we can see, the test accuracy seems to reach a plateau after roughly 10 epochs, with a peak accuracy of around 88%. The train accuracy gets as high as 98%, suggesting strong overfitting on the train set.

Training with image augmentation

Let's now redo the same exercise with image augmentation:

1. First, we need to implement the desired image augmentation. Using the same pattern as with the transformations in `pytorch`, using Albumentations, we can instantiate a `Compose` class with a list of augmentations. In our case, we use the following augmentations:

   - `HorizontalFlip`: This involves basic mirroring, occurring with a 50% probability, meaning 50% of the images will be randomly mirrored

   - `Rotate`: This will randomly rotate an image in the range of [-90, 90] degrees (this range can be modified) with a probability of 50%

   - `RandomBrightnessContrast`: This will randomly change the brightness and contrast of the image with a probability of 20%

   Here are the instantiations:

   ```
   import albumentations as A
   augment = A.Compose([
       A.HorizontalFlip(p=0.5),
       A.Rotate(p=0.5),
       A.RandomBrightnessContrast(p=0.2),
   ])
   ```

2. Then, instantiate a new, augmented training set and training data loader. To do so, we simply have to provide our `augment` object as an argument of the `DogsAndCats` class:

   ```
   augmented_trainset = DogsAndCats(
       'kagglecatsanddogs_3367a/PetImages/Cat/train/',
       'kagglecatsanddogs_3367a/PetImages/Dog/train/',
       transform=transform,
       augment=augment,
   )
   augmented_train_dataloader = DataLoader(
       augmented_trainset, batch_size=batch_size,
       shuffle=True)
   ```

> **Note**
>
> We do not apply augmentation to the test set since we need to be able to compare the performances to the results without augmentation. Besides that, it would be useless to augment the test set, unless you are using Test Time Augmentation (see the *There's more...* section for more about it).

3. Then, display a few images from this new, augmented dataset as follows:

```
display_images(augmented_train_dataloader)
```

Here are the images:

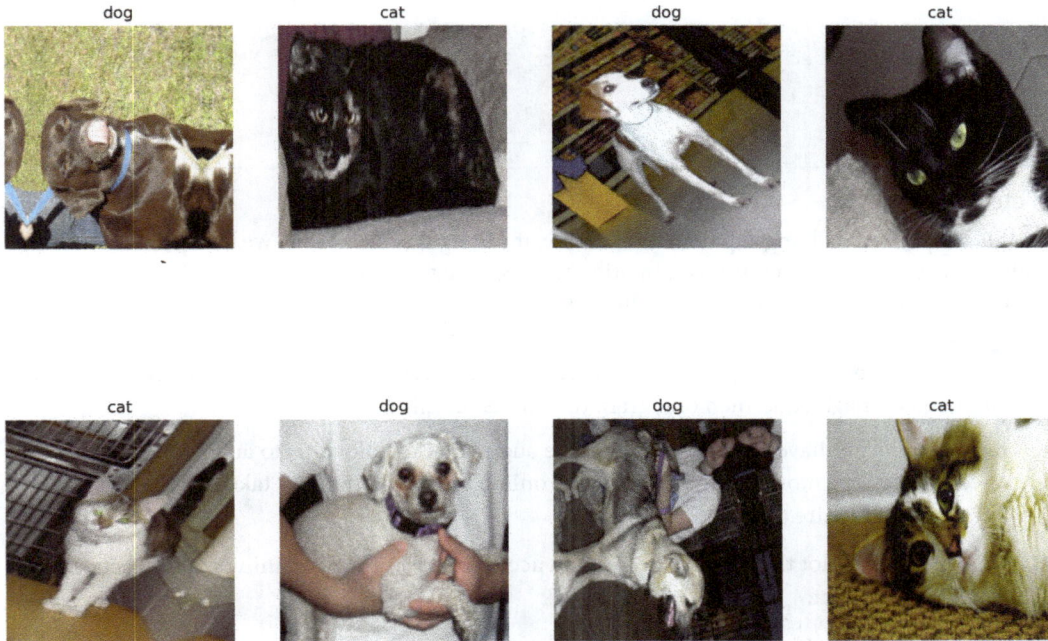

Figure 11.6 – Example of augmented images (some have been rotated, some
have been mirrored and some have modified brightness and contrast)

As we can see, some images seem now rotated. Besides, some images are also mirrored and have a modified brightness and contrast, efficiently improving the diversity of the dataset.

4. Then, we instantiate the model and the optimizer:

```
model = Classifier()
model = model.to(device)
optimizer = torch.optim.Adam(model.parameters(),
    lr=0.001)
```

5. Next, train the model on this new training set while keeping the same test set and store the output losses and metrics:

```
train_losses, test_losses, train_accuracy,
test_accuracy = train_model(
    epochs, model, criterion, optimizer, device,
    augmented_train_dataloader, test_dataloader,
    trainset, testset
)
```

> **Note**
>
> In this recipe, we are doing augmentation online, meaning that every time we load a new batch of images, we randomly apply augmentation to these images; consequently, at each epoch, we may train from differently augmented images.

Another approach is to augment data offline: we preprocess and augment the dataset, store the augmented images, and then only train the model on this data.

Both approaches have pros and cons: offline augmentation allows us to augment images only once but requires more storage space, while online preprocessing may take more time to train but does not require any extra storage.

6. Now finally, we plot the results: the loss and accuracy for both the training and test sets. Here is the code for that:

```
plt.figure(figsize=(10, 10))
plt.subplot(2, 1, 1)
plt.plot(train_losses, label='train')
plt.plot(test_losses, label='test')
plt.ylabel('BCE Loss')
plt.legend()
plt.subplot(2, 1, 2)
plt.plot(train_accuracy, label='train')
plt.plot(test_accuracy, label='test')
plt.xlabel('epoch')
plt.ylabel('Accuracy')
plt.legend()
plt.show()
```

Here are the plots:

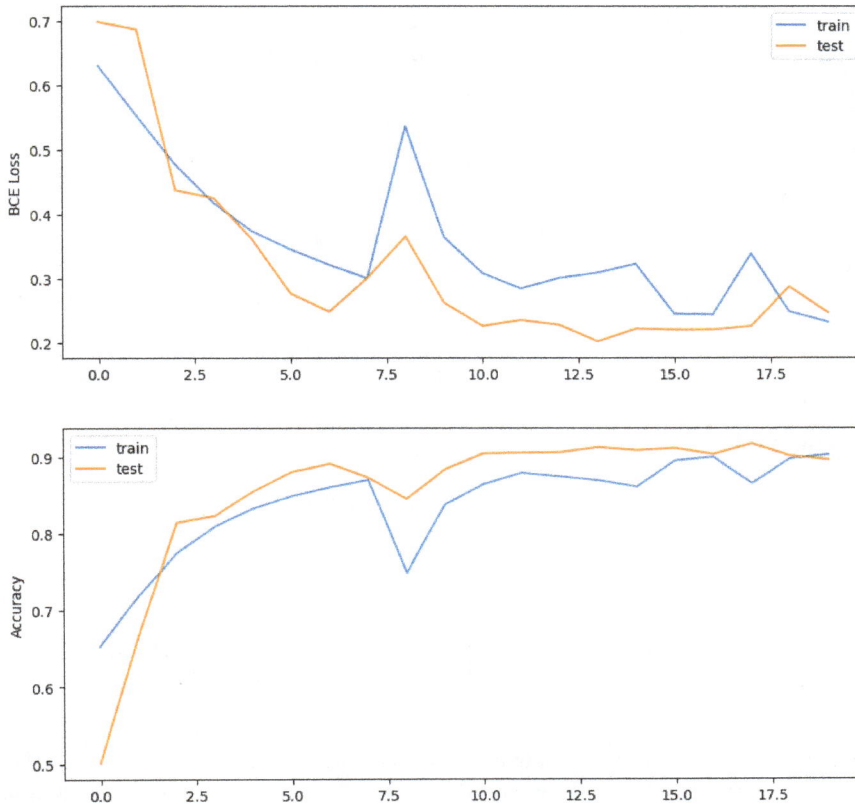

Figure 11.7 – Loss and accuracy for the augmented dataset

As you can see in the preceding figure, compared to the regular dataset, not only is the overfitting almost totally removed but the accuracy also climbs up to more than 91%, compared to 88% previously.

Thanks to this rather simple image augmentation, we could get the accuracy to climb from 88% to 91%, while reducing overfitting: the train set now has the same performances as the test set.

There's more...

While we used augmentation for training, there is a method that takes advantage of image augmentation at test time to improve the performances of models. This is sometimes called **Test Time Augmentation**.

The idea is simple: compute model inference on several, augmented images, and compute the final prediction with a majority vote.

Let's take a simple example. Assuming we have an input image that must be classified with our trained dogs and cats model, we augment this input image with mirroring and with brightness and contrast, so that we have three images:

- **Image 1**: The original image
- **Image 2**: The mirrored image
- **Image 3**: The image with modified brightness and contrast

We will now compute model inference on those three images, getting the following predictions:

- **Image 1 prediction**: Cat
- **Image 2 prediction**: Cat
- **Image 3 prediction**: Dog

We can now compute a majority vote by choosing the most represented predicted class, resulting in a cat class prediction.

> **Note**
>
> In practice, we would most likely use a soft majority vote, averaging the predicted probabilities (either for binary or multiclass classification), but the concept remains the same.

Test Time Augmentation is commonly used in competitions and can indeed improve the performance of the model for no added training cost. In a production environment though, where the inference cost is key, this method is rarely used.

See also

In this recipe, we used Albumentations for a simple classification task, but it can be used for much more than that: it allows us to perform image augmentation for object detection, instance segmentation, semantic segmentation, landmarks, and so on.

To know more about how to use it fully, have a look at the well-written documentation, with many working examples here: `https://albumentations.ai/docs/`.

Creating synthetic images for object detection

For some projects, you may have so little data that the only thing you can do is use this data in the test set. In some rare cases, it is possible to create a synthetic dataset to create a robust enough model and test it against the small, real test set.

This is what we will do in this recipe: we have a small test set of pictures of QR codes, and we want to build an object detection model for the detection of QR codes. All we have as a train set is a set of generated QR codes and downloaded images collected on open image websites such as unsplash.com.

Getting started

Download and unzip the dataset from https://www.kaggle.com/datasets/vincentv/qr-detection-yolo with the following command line:

```
kaggle datasets download -d vincentv/qr-detection-yolo --unzip
```

This dataset is made up of the following folder architecture:

```
QR-detection-yolo
├── train
│   ├── images: 9750 images
│   └── labels: 9750 text files
├── test
│   ├── images: 683 images
│   └── labels: 683 text files
└── background_images: 44 images
```

It is made up of three folders:

- **The train set**: Only generated QR codes with no context
- **The test set**: Pictures of QR codes in various contexts and environments
- **Background images**: Random images of context such as stores

The goal is to use the data in the train set and the background images to generate realistic synthetic images to train a model on and only then to evaluate the model against the test set, made of real images.

For this recipe, the needed libraries can be installed with the following command line:

```
pip install albumentations opencv-python matplotlib numpy ultralytics.
```

How to do it...

Let's divide this recipe into three parts:

1. First, we will explore the dataset and implement a few helper functions.
2. The second part is about generating synthetic data using QR codes and background images.
3. The last part is about training a YOLO model on the generated data and evaluating this model.

Let us understand each of these in the following sections.

Exploring the dataset

Let's start by creating a few helper functions and use them to display a few images of the train and test sets:

1. Import the following libraries:

    - glob for listing files

    - os for making a directory in which to store the created synthetic images

    - albumentations for data augmentation

    - cv2 for image manipulation

    - matplotlib for display

    - numpy for various data manipulation

    - YOLO for the model

    Here are the imports:

    ```
    from glob import glob
    import os
    import albumentations as A
    import cv2
    import matplotlib.pyplot as plt
    import numpy as np
    from ultralytics import YOLO
    ```

2. Let's implement a read_labels helper function, which will read the text file with the YOLO labels and return them as a list:

    ```
    def read_labels(labels_path):
        res = []
        with open(labels_path, 'r') as file:
            lines = file.readlines()
            for line in lines:
                cls,xc,yc,w,h = line.strip().split(' ')
                res.append([int(float(cls)), float(xc),
                    float(yc), float(w), float(h)])
            file.close()
        return res
    ```

3. Now let's implement a `plot_labels` helper function, which will reuse the previous `read_labels` function, read a few images and corresponding labels, and display these images with the bounding boxes:

```
def plot_labels(images_folder, labels_folder,
    classes):
        images_path = sorted(glob(
            images_folder + '/*.jpg'))
    labels_path = sorted(glob(
            labels_folder + '/*.txt'))
    plt.figure(figsize=(10, 6))
    for i in range(8):
        idx = np.random.randint(len(images_path))
        image = plt.imread(images_path[idx])
        labels = read_labels(labels_path[idx])

        for cls, xc, yc, w, h in labels:
            xc = int(xc*image.shape[1])
            yc = int(yc*image.shape[0])
            w = int(w*image.shape[1])
            h = int(h*image.shape[0])
            cv2.rectangle(image,
                (xc - w//2, yc - h//2),
                (xc + w//2 ,yc + h//2), (255,0,0), 2)
            cv2.putText(image, f'{classes[int(cls)]}',
                (xc-w//2, yc - h//2 - 10),
                cv2.FONT_HERSHEY_SIMPLEX, 0.5,
                (1.,0.,0.), 1)
        plt.subplot(2, 4, i + 1)
        plt.imshow(image)
        plt.axis('off')
```

4. Now, display a set of images from the train set and their bounding boxes with the following code:

```
plot_labels('QR-detection-yolo/train/images/',
    'QR-detection-yolo/train/labels/', 'QR Code')
```

Here are a few sample images in the form of QR codes:

Figure 11.8 – A few samples from the train set with the associated labels (this dataset is only made up of generated QR codes on a white background)

As explained, the train set is only made up of generated QR codes of various sizes on a white background with no more context.

5. Let's now display a few images from the test set with the following code:

```
plot_labels('QR-detection-yolo/test/images/',
    'QR-detection-yolo/test/labels/', 'QR Code')
```

Here are the resulting images:

Figure 11.9 – A few examples from the test set, made of real-world images of QR codes

The test set contains more complex, real examples of QR codes, and is much more challenging.

Generating a synthetic dataset from background images

In this part, we will now generate a dataset of realistic, synthetic data. To do so, we will use the images of the QR codes from the training set as well as a set of background images. Here are the steps:

1. Let's now generate a synthetic dataset using two ingredients:

    *   Real background images

    *   Numerically generated QR codes

    For that, we will use a rather long and complex function, `generate_synthetic_background_image_with_tag`, which does the following:

*   Picks a random background image in the given folder

*   Picks a random QR code image in the given folder

*   Augments the picked QR code

*   Randomly inserts the augmented QR code into the background image

*   Applies a little more augmentation to the newly created image

*   Stores the generated image and the corresponding labels in YOLO format

The code that does this is available in the GitHub repository and is too long to be displayed here, so we will only display its signature and docstring here. However, you are strongly encouraged to have a look at it and to play with it. The code can be found here:

`https://github.com/PacktPublishing/The-Regularization-Cookbook/blob/main/chapter_11/chapter_11.ipynb`

```
def generate_synthetic_background_image_with_tag(
    n_images_to_generate: int,
    output_path: str,
    raw_tags_folder: str,
    background_images_path: str,
    labels_path: str,
    background_proba: float = 0.8,
):
    """Generate images with random tag and synthetic background.
    Parameters
    ----------
    n_images_to_generate : int
        The number of images to generate.
    output_path : str
        The output directory path where to store the generated
```

```
images.
        If the path does not exist, the directory is created.
    raw_tags_folder : str
        Path to the folder containing the raw QR codes.
    background_images_path : str
        Path to the folder containing the background images.
    labels_path : str
        Path to the folder containing the labels.
        Files must be in the same order as the ones in the raw_
tags_folder.
    background_proba : float (optional, default=0.8)
        Probability to use a background image when generating a
new sample.
    """
```

> **Note**
>
> This function does this generation as many times as we want and provides a few other features;
> feel free to have a close look at it and update it.

2. We can now use this function to generate 3,000 images by calling the generate_synthetic_
   background_image_with_tag function (3,000 is a rather arbitrary choice; feel free to
   generate fewer images or more images). This may take a few minutes. The generated images
   and their associated labels will be stored in the QR-detection-yolo/generated_qr_
   code_images/ folder, which will be created if it does not exist:

```
generate_synthetic_background_image_with_tag(
    n_images_to_generate=3000,
    output_path='QR-detection-yolo/generated_qr_code_images/',
    raw_tags_folder='QR-detection-yolo/train/images/',
    background_images_path='QR-detection-yolo/background_
images/',
    labels_path='QR-detection-yolo/train/labels/'
)
```

Let's have a look at a few examples of generated images with the following code:

```
plot_labels(
    'QR-detection-yolo/generated_qr_code_images/images/',
    'QR-detection-yolo/generated_qr_code_images/labels/',
    'QR Code'
)
```

Here are the images:

Figure 11.10 – Examples of synthetically created images, made of background
images and generated QR codes with various image augmentations

As we can see, some images are simple augmented QR codes with no background context, as is possible
due to the generating function. This can be tweaked with the `background_proba` argument.

Model training

We can now start the model training part: we will train a YOLO model on the 3,000 images generated
in the previous step and evaluate this model against the test set. Here are the steps:

1. First, instantiate a YOLO model with pre-trained weights as follows:

    ```
    # Create a new YOLO model with pretrained weights
    model = YOLO('yolov8n.pt')
    ```

> **Note**
>
> You may have `FileNotFoundError` because of an incorrect dataset path. A `config` file in
> `~/.config/Ultralytics/settings.yaml` has a previous path. A quick and harmless
> fix is to simply delete this file; a new one will then be automatically generated.

2. Then, we need to create a .yaml file, data_qr_generated.yaml, with the following content:

```
train: ../../QR-detection-yolo/generated_qr_code_images/images
val: ../../QR-detection-yolo/test/images

nc: 1
names: ['QR_CODE']
```

3. This .yaml file can be used to train the model on our dataset, on 50 epochs. We also specify the initial learning rate to be 0.001 with lr0=0.001 because the default learning rate (0.01) is rather large for fine-tuning a pre-trained model in our case:

```
# Train the model for 50 epochs
model.train(data='data_qr_generated.yaml', epochs=50,
    lr0=0.001, name='generated_qrcode')
```

Results should be stored in the created folder, runs/detect/generated_qrcode.

4. Before having a look at the results, let's implement a plot_results_one_image helper function to display the output of the model, as follows:

```
def plot_results_random_images(test_images, model, classes=['QR_
code']):
    images = glob(test_images + '/*.jpg')
    plt.figure(figsize=(14, 10))
    for i in range(8):
        idx = np.random.randint(len(images))
        result = model.predict(images[idx])
        image = result[0].orig_img.copy()
        raw_res = result[0].boxes.data
        for detection in raw_res:
            x1, y1, x2, y2, p,
                cls = detection.cpu().tolist()
            cv2.rectangle(image, (int(x1), int(y1)),
                (int(x2), int(y2)), (255,0,0), 2)
            cv2.putText(image, f'{classes[int(cls)]}',
                (int(x1), int(y1) - 10),
                    cv2.FONT_HERSHEY_SIMPLEX, 1,
                    (255,0,0), 2)

        plt.subplot(2, 4, i + 1)
        plt.axis('off')
        plt.imshow(image)
```

5. We can then reload the best weights and compute the inference and display the results on an image from the test set:

```
# Load the best weights
model = YOLO(
    'runs/detect/generated_qrcode/weights/best.pt')
# Plot the results
Plot_results_random_images(
    'QR-detection-yolo/test/images/', model)
```

Here are the results:

Figure 11.11 – Examples of results of the YOLO model trained on synthetic data (even though the model is not perfect, is it capable of detecting QR codes in rather complex and various situations)

As we can see, the model is not working perfectly yet but still manages to get QR codes in several complex situations. However, in a few cases, such as with really small QR codes, bad-quality images, or highly deformed QR codes, the model does not seem to perform well.

6. Finally, we can visualize the losses and other metrics generated by the YOLO library:

```
plt.figure(figsize=(10, 8))
plt.imshow(plt.imread(
    'runs/detect/generated_qrcode/results.png'))
```

Here are the losses:

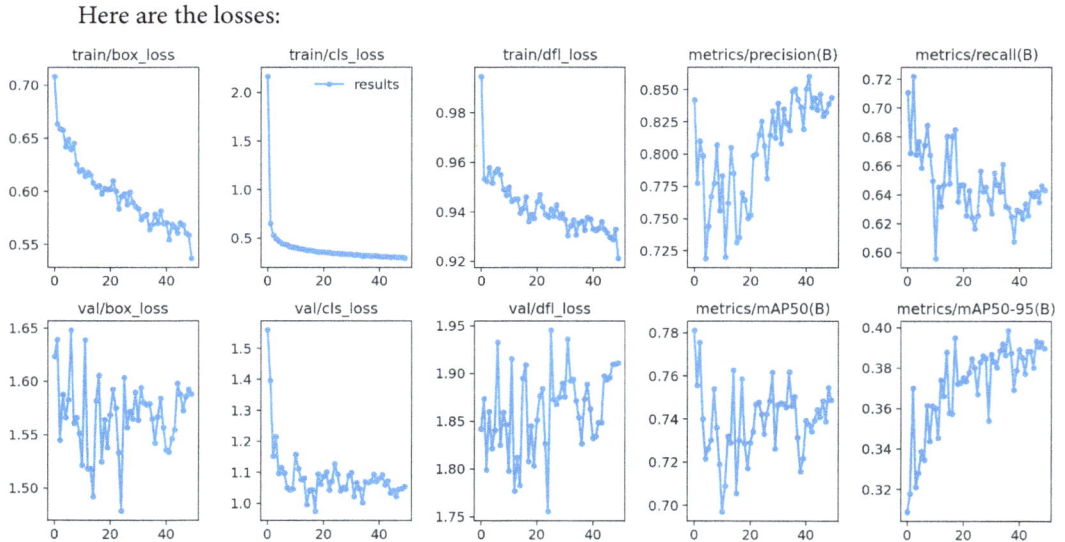

Figure 11.12 – Metrics computed by the YOLO library

In agreement with the displayed results for a few images, the metrics are not perfect, with a mAP50 around 75% only.

This could probably be improved by adding more well-chosen image augmentation.

There's more...

There are more techniques for generating images with labels even if we don't have any real data in the first place. In this recipe, we only used background images, generated QR codes, and augmentations, but it is possible to use generative models to generate even more data.

Let's see how to do this with DALL-E, a model proposed by OpenAI:

1. First, we can import the required libraries. The `openai` library can be installed with `pip install openai`:

```
import openai
import urllib
from PIL import Image
import matplotlib.pyplot as plt
openai.api_key = 'xx-xxx'
```

> **Note**
>
> You need to create your own API key by creating your own account on `openai.com`.

2. Let's now create a helper function that converts the bounding boxes and the image into a mask since we want to complete outside of the bounding box:

```python
def get_mask_to_complete(image_path, label_path, output_
filename, margin: int = 100):
    image = plt.imread(image_path)
    labels = read_labels(label_path)
    output_mask = np.zeros(image.shape[:2])
    for cls, xc, yc, w, h in labels:
        xc = int(xc*image.shape[1])
        yc = int(yc*image.shape[0])
        w = int(w*image.shape[1])
        h = int(h*image.shape[0])
        output_mask[yc-h//2-margin:yc+h//2+margin,
            xc-w//2-margin:xc+w//2+margin] = 255
    output_mask = np.concatenate([image,
        np.expand_dims(output_mask, -1)],
            axis=-1).astype(np.uint8)
    # Save the images
    output_mask_filename = output_filename.split('.')[0] + '_
mask.png'
    plt.imsave(output_filename, image)
    plt.imsave(output_mask_filename, output_mask)
    return output_mask_filename
```

3. We can now compute a mask and display the result side by side with the original image as follows:

```python
output_image_filename = 'image_edit.png'
mask_filename = get_mask_to_complete(
    'QR-detection-yolo/generated_qr_code_images/images/
synthetic_image_0.jpg',
    'QR-detection-yolo/generated_qr_code_images/labels/
synthetic_image_0.txt',
    output_image_filename
)
# Display the masked image and the original image side by side
plt.figure(figsize=(12, 10))
plt.subplot(1, 2, 1)
plt.imshow(plt.imread(output_image_filename))
plt.subplot(1, 2, 2)
plt.imshow(plt.imread(mask_filename))
```

Here are the results:

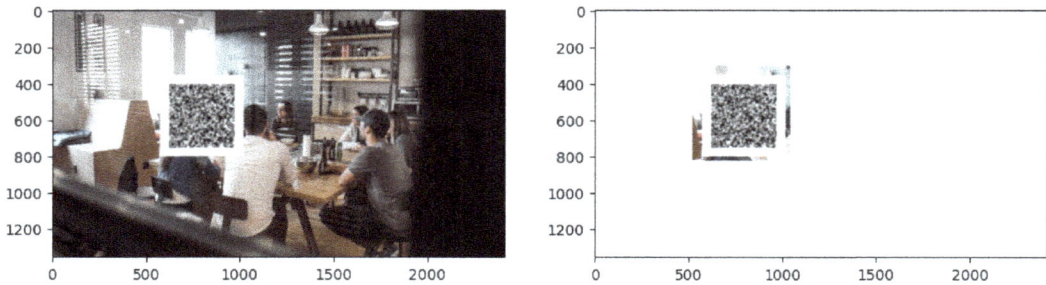

Figure 11.13 – An original image on the left, and the associated masked
image to be used for data generation on the right

> **Note**
>
> We keep a margin for the masked image so that when calling DALL-E 2, it has a sense of the surroundings. If we provide only a QR code and white surroundings in the mask, the result may not be good enough.

4. We can now query the OpenAI model DALL-E 2 to fill around this QR code and generate a new image using the `create_edit` method from the `openai` library. The function requires the following few parameters:

    - The input image (in PNG format, less than 4 MB)

    - The input mask (in PNG format and less than 4 MB too)

    - A prompt describing what the expected output image is

    - The number of images to generate

    - The output size in pixels (either 256x256, 512x512, or 1,024x1,024)

    Let's now query DALL-E on our image, and then display the original and the generated images side by side:

```
# Query openAI API to generate image
response = openai.Image.create_edit(
    image=open(output_image_filename, 'rb'),
    mask=open(mask_filename, 'rb'),
    prompt="A store in background",
    n=1,
    size="512x512"
)
```

```
# Download and display the generated image
plt.figure(figsize=(12, 10))
image_url = response['data'][0]['url']
plt.subplot(1, 2, 1)
plt.imshow(plt.imread(output_image_filename))
plt.subplot(1, 2, 2)
plt.imshow(np.array(Image.open(urllib.request.urlopen(
    image_url))))
```

Here is how the images appear:

Figure 11.14 – The original image on the left, and the generated image using DALL-E 2 on the right

As we can see in *Figure 11.14*, using this technique allows us to create more realistic images that can easily be used for training. These created images can also be augmented using Albumentations.

> **Note**
>
> There are a few drawbacks though. The generated image is of size 512x512, meaning the bounding box coordinates have to be converted (this can be done using Albumentations) and the generated image is not always good and requires a visual check.

5. We can also create variations of a given image using the `create_variation` function. This function is simpler to use and requires similar input arguments:

  - The input image (still a PNG image smaller than 4 MB)

  - The number of varied images to generate

  - The output image size in pixels (again, either 256x256, 512x512, or 1,024x1,024)

Here is the code for this:

```
# Query to create variation of a given image
response = openai.Image.create_variation(
    image=open(output_image_filename, "rb"),
    n=1,
    size="512x512"
)
# Download and display the generated image
plt.figure(figsize=(12, 10))
image_url = response['data'][0]['url']
plt.subplot(1, 2, 1)
plt.imshow(plt.imread(output_image_filename))
plt.subplot(1, 2, 2)
plt.imshow(np.array(Image.open(urllib.request.urlopen(
    image_url))))
```

Here is the output:

Figure 11.15 – The original image (left) and the generated variation using DALL-E (right)

The result presented in the preceding figure is pretty good: we can see a meeting room in the background and a QR code in the foreground, just like in the original image. However, this data would not be easy to use unless we labeled it manually since we have no certainty the QR code will be at the same location (even if we resize the bounding box coordinates). Still, using such models can be of great help for other use cases, such as classification.

See also

- The list of train parameters available with YOLOv8: `https://docs.ultralytics.com/modes/train/`

- The DALL-E API documentation: `https://platform.openai.com/docs/guides/images/usage`

Implementing real-time style transfer

In this recipe, we will build our own lightweight style transfer model based on the U-Net architecture. To do so, we will use a dataset generated using Stable Diffusion (see more next about what Stable Diffusion is). This can be seen as a kind of knowledge distillation: we will use the data generated by a large, teacher model (Stable Diffusion, which weighs several gigabytes) to train a small, student model (here, a U-Net++ of less than 30 MBs). This is a funny way to use generative models to create data, but the concepts developed here can be used in many other applications: some will be proposed in the *There's more…* section, along with guidance on creating your own style transfer dataset using Stable Diffusion. But before that, let's give some context about style transfer.

Style transfer is a famous and fun use of deep learning, allowing us to change the style of a given image into another style. Many examples exist, such as Mona Lisa in Van Gogh's Starry Night style, as represented in the following figure:

Figure 11.16 – Mona Lisa in the Starry Night style

Until recently, style transfer was mostly performed using **Generative Adversarial Networks (GANs)**, which are quite hard to train properly.

It is now simpler than ever to apply style transfer to images, using pre-trained models based on Stable Diffusion. Unfortunately, Stable Diffusion is a large and complex model that sometimes requires several seconds to generate a single image on a recent graphics card.

In this recipe, we will train a U-Net-like model allowing for real-time transfer learning on any device. To accomplish this, we will employ a form of knowledge distillation. Specifically, we will train the U-net model using Stable Diffusion data and incorporate a VGG perceptual loss for that purpose.

> **Note**
>
> **VGG** stands for **Visual Geometry Group**, the name of the Oxford team who proposed this deep learning model architecture. It is a standard deep learning model in Computer Vision.

Before moving on to the recipe, let's have a look at two important concepts for this recipe:

- Stable Diffusion
- Perceptual loss

Stable Diffusion

Stable Diffusion is a complex and powerful model, allowing us to use image and text prompts to generate new images.

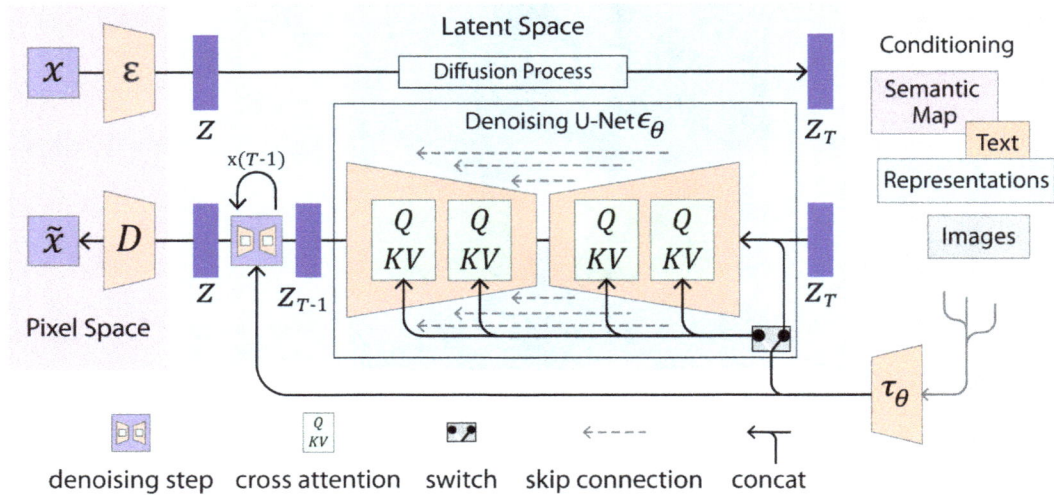

Figure 11.17 – Architecture diagram of Stable Diffusion

As we can see from the preceding figure, the way Stable Diffusion is trained can be summarized, with simplifications, as follows:

- Diffusion is gradually applied **T** times to an input image **Z**: this is like adding random noise to the input image

- These diffused images are passed through a denoising U-Net model

- A condition is optionally added, such as a descriptive text or another image prompt, as an embedding

- The model is trained to output the input image

Once a model is well trained, it can be used for inference by skipping the first part as follows:

- Given a seed, a random image is generated and given as input to the denoising U-Net

- An input prompt is added as condition: it can be text or an input image, for example

- An output image is then generated: this is the final result

Although this is a simplistic explanation of how it works, it allows us to get a general understanding of what it does and what are the expected inputs to generate a new image.

Perceptual loss

Perceptual loss has been proposed to train a model to learn about perceptual features in an image. It was developed specifically for style transfer and allows you to focus not only on the pixel-to-pixel content itself but also on the style of the image.

It takes two images as input: the model prediction and the label image, and it is commonly based on a VGG neural network pre-trained on the ImageNet dataset or any similar generic image dataset.

More specifically, for both images (for example, the model prediction and the label), the following computations are made:

- The feedforward computation of the VGG model is applied to each image

- The outputs after each block of the VGG model are stored, allowing us to get more and more specific features with deeper blocks

---

Note

In a deep neural network, the first layers are commonly about learning generic features and shapes, while the deepest layers are about learning more specific features. The perceptual loss takes advantage of this property.

Using these stored computations, perceptual loss is finally computed as the sum of the following values:

- The differences (for example, L1 or L2 norm) between the computations for each block output: These can be represented as the feature reconstruction loss and will focus on image features.

- The differences between the Gram matrices of the computations for each block output: These can be represented as the style reconstruction loss and will focus on the image's style.

> **Note**
>
> A Gram matrix of a given set of vectors is made by computing the dot product of each pair of vectors and then arranging the results into a matrix. It can be seen as a similarity or a correlation between the given vectors.

At the end, minimizing this perceptual loss should allow us to apply a style from a given image to another, as we will see in this recipe.

Getting started

You can download the full dataset that I created for this recipe on Kaggle with the following command line:

```
kaggle datasets download -d vincentv/qr-detection-yolo --unzip
```

You then have the following folder architecture:

```
anime-style-transfer
├── train
│   ├── images: 820 images
│   └── labels: 820 images
└── test
    ├── images: 93 images
    └── labels: 93 images
```

This is a rather small dataset, but it should allow us to get good enough performances to show the potential of this technique.

For more about how to create such a dataset using ControlNet yourself, have a look at the *There's more…* subsection.

The libraries needed for this recipe can be installed with the following command line:

```
pip install matplotlib numpy torch torchvision segmentation-models-
pytorch albumentations tqdm
```

How to do it...

Here are the steps for this recipe:

1. Import the required libraries:

    - `matplotlib` for visualization

    - numpy for manipulation

    - Several `torch` and `torchvision` modules

    - `segmentation models pytorch` for the model

    - `albumentations` for image augmentation

    Here is the code for it:

    ```
    import matplotlib.pyplot as plt
    import numpy as np
    import torch
    import torch.nn as nn
    import torch.nn.functional as F
    from torch.utils.data import DataLoader, Dataset
    import torchvision.transforms as transforms
    from torchvision.models import vgg16, VGG16_Weights
    from glob import glob
    import segmentation_models_pytorch as smp
    from torch.optim.lr_scheduler import ExponentialLR
    import albumentations as A
    import tqdm
    ```

2. Implement `AnimeStyleDataset`, allowing us to load the dataset. Note that we use the `ReplayCompose` tool from `Albumentations`, allowing us to apply the exact same image augmentation to the image and the associated label:

    ```
    class AnimeStyleDataset(Dataset):
        def __init__(self, input_path: str,
            output_path: str, transform, augment = None):
                self.input_paths = sorted(glob(
                    f'{input_path}/*.png'))
                self.output_paths = sorted(glob(
                    f'{output_path}/*.png'))
                self.transform = transform
                self.augment = augment

        def __len__(self):
            return len(self.input_paths)
    ```

```
def __getitem__(self, idx):
    input_img = plt.imread(self.input_paths[idx])
    output_img = plt.imread(
        self.output_paths[idx])

    if self.augment:
        augmented = self.augment(image=input_img)
        input_img = augmented['image']
        output_img = A.ReplayCompose.replay(
            augmented['replay'],
            image=output_img)['image']

    return self.transform(input_img),
        self.transform(output_img)
```

3. Instantiate the augmentation, which is a composition of the following transformations:

    • `Resize`

    • A horizontal flip with a probability of 50%

    • `ShiftScaleRotate`, allowing us to randomly add geometrical variety

    • `RandomBrightnessContrast`, allowing us to add variety to the light

    • `RandomCropFromBorders`, which will randomly crop the borders of the images

    Here is the code for it:

```
augment = A.ReplayCompose([
    A.Resize(512, 512),
    A.HorizontalFlip(p=0.5),
    A.ShiftScaleRotate(shift_limit=0.05,
        scale_limit=0.05, rotate_limit=15, p=0.5),
    A.RandomBrightnessContrast(p=0.5),
    A.RandomCropFromBorders(0.2, 0.2, 0.2, 0.2, p=0.5)
])
```

4. Instantiate the transformation, allowing us to convert the torch tensors and rescale the pixel values. Also, define the batch size and the device as follows:

```
batch_size = 12
device = torch.device(
    'cuda' if torch.cuda.is_available() else 'cpu')
mean = (0.485, 0.456, 0.406)
std = (0.229, 0.224, 0.225)

transform = transforms.Compose([
```

```
    transforms.ToTensor(),
    transforms.Resize((512, 512), antialias=True),
    transforms.Normalize(mean, std),
])
```

> **Note**
>
> Here, we use the mean and standard deviation rescaling specific to the ImageNet dataset because the VGG perceptual loss (see ahead) is trained on this specific set of values. Also, the batch size may need to be adjusted depending on your hardware specifications, especially the memory of your graphics processing unit.

5. Instantiate the datasets and data loaders, providing the train and test folders. Note that we apply augmentation to the train set only:

```
trainset = AnimeStyleDataset(
    'anime-style-transfer/train/images/',
    'anime-style-transfer/train/labels/',
    transform=transform,
    augment=augment,
)
train_dataloader = DataLoader(trainset,
    batch_size=batch_size, shuffle=True)

testset = AnimeStyleDataset(
    'anime-style-transfer/test/images/',
    'anime-style-transfer/test/labels/',
    transform=transform,
)
test_dataloader = DataLoader(testset,
    batch_size=batch_size, shuffle=True)
```

6. Display a few images along with their labels so that we have a glimpse at the dataset. For that, we first need a helper unnormalize function to rescale the images' values to the range [0, 1]:

```
def unnormalize(x, mean, std):
    x = np.asarray(x, dtype=np.float32)
    for dim in range(3):
        x[:, :, dim] = (x[:, :, dim] * std[dim]) + mean[dim]

    return x
plt.figure(figsize=(12, 6))

images, labels = next(iter(train_dataloader))
```

```
for idx in range(4):
    plt.subplot(2, 4, idx*2+1)
    plt.imshow(unnormalize(images[idx].permute(
        1, 2, 0).numpy(), mean, std))
    plt.axis('off')
    plt.subplot(2, 4, idx*2+2)
    plt.imshow(unnormalize(labels[idx].permute(
        1, 2, 0).numpy(), mean, std))
    plt.axis('off')
```

Here are the results:

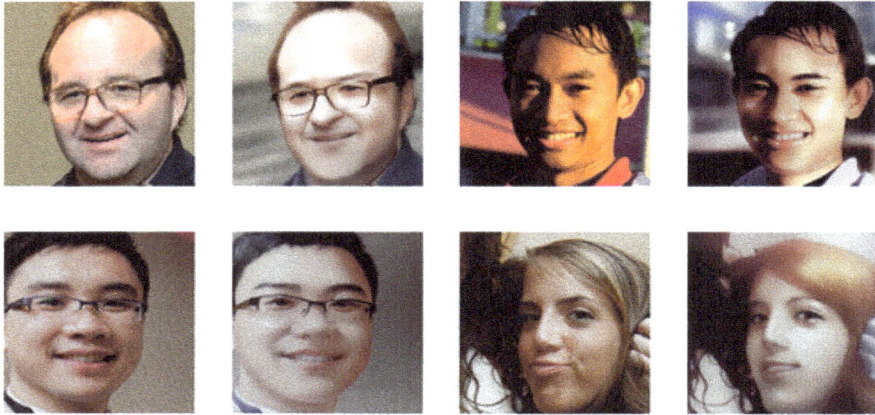

Figure 11.18 – A set of four images with their associated anime labels

As we can see in the preceding figure, the dataset is made up of images of faces, and the labels are the equivalent pictures with some drawing and anime style applied. These images were generated using Stable Diffusion and ControlNet; see how to do that yourself in the *There's more…* section.

7. Now we instantiate the model class. Here, we reuse the existing `mobilenetv3_large_100` implementation provided in the SMP library with U-Net++ architecture. We specify the input and output channels to be 3, using the `in_channels` and `n_classes` parameters respectively. We also reuse `imagenet` weights for the encoder. Here is the code:

```
model = smp.UnetPlusPlus(
    encoder_name='timm-mobilenetv3_large_100',
    encoder_weights='imagenet',
    in_channels=3,
```

```
        classes=3,
        )
model = model.to(device)
```

> **Note**
>
> For more about the SMP library, refer to the *Semantic segmentation using transfer learning recipe* of *Chapter 10*.

8. Now, we implement the VGG perceptual loss as follows:

```python
class VGGPerceptualLoss(torch.nn.Module):
    def __init__(self):
        super(VGGPerceptualLoss, self).__init__()
        blocks = []
        blocks.append(vgg16(weights=VGG16_Weights.DEFAULT).
features[:4].eval())
        blocks.append(vgg16(weights=VGG16_Weights.DEFAULT).
features[4:9].eval())
        blocks.append(vgg16(weights=VGG16_Weights.DEFAULT).
features[9:16].eval())
        blocks.append(vgg16(weights=VGG16_Weights.DEFAULT).
features[16:23].eval())
        for block in blocks:
            block = block.to(device)
            for param in block.parameters():
                param.requires_grad = False
            self.blocks = torch.nn.ModuleList(blocks)
            self.transform = torch.nn.functional.interpolate

    def forward(self, input, target):
        input = self.transform(input, mode='bilinear',
            size=(224, 224), align_corners=False)
        target = self.transform(target,
            mode='bilinear', size=(224, 224),
            align_corners=False)
        loss = 0.0
        x = input
        y = target
        for i, block in enumerate(self.blocks):
            x = block(x)
            y = block(y)
```

```
                        loss += torch.nn.functional.l1_loss(
                            x, y)
                        act_x = x.reshape(x.shape[0],
                            x.shape[1], -1)
                        act_y = y.reshape(y.shape[0],
                            y.shape[1], -1)
                        gram_x = act_x @ act_x.permute(
                            0, 2, 1)
                        gram_y = act_y @ act_y.permute(
                            0, 2, 1)
                        loss += torch.nn.functional.l1_loss(
                            gram_x, gram_y)
                return loss
```

In this implementation, we have two methods:

- The `init` function, defining all the blocks and setting them as non-trainable

- The `forward` function, resizing the image to 224x224 (the original VGG input shape) and computing the loss for each block

9. Next, we define the optimizer, an exponential learning rate scheduler, and the VGG loss, as well as the weights of the style and content loss:

```
optimizer = torch.optim.Adam(model.parameters(),
    lr=0.001)
scheduler = ExponentialLR(optimizer, gamma=0.995)
vgg_loss = VGGPerceptualLoss()

content_loss_weight=1.
style_loss_weight=5e-4
```

> **Note**
>
> The style loss (that is, the VGG perceptual loss) is by default much larger than the content loss (that is, the L1 loss). So, here we counterbalance that by applying a low weight to the style loss.

10. Train the model over 50 epochs and store the losses for the train and test sets. To do so, let's use the `train_style_transfer` function available in the GitHub repository (https://github.com/PacktPublishing/The-Regularization-Cookbook/blob/main/chapter_11/chapter_11.ipynb):

```
train_losses, test_losses = train_style_transfer(
    model,
    train_dataloader,
    test_dataloader,
    vgg_loss,
    content_loss_weight,
    style_loss_weight,
    device,
    epochs=50,
)
```

> **Note**
>
> This function is a typical training loop as we implemented many times already. The only difference is the loss computation, which is computed as follows:
>
> `style_loss = vgg_loss(outputs, labels)`
>
> `content_loss = torch.nn.functional.l1_loss(outputs, labels)`
>
> `loss = style_loss_weight*style_loss + content_loss_weight*content_loss`

11. Plot the loss as a function of the epoch for the train and test sets:

```
plt.plot(train_losses, label='train')
plt.plot(test_losses, label='test')
plt.ylabel('Loss')
plt.xlabel('Epoch')
plt.legend()
plt.show()
```

Here are the results:

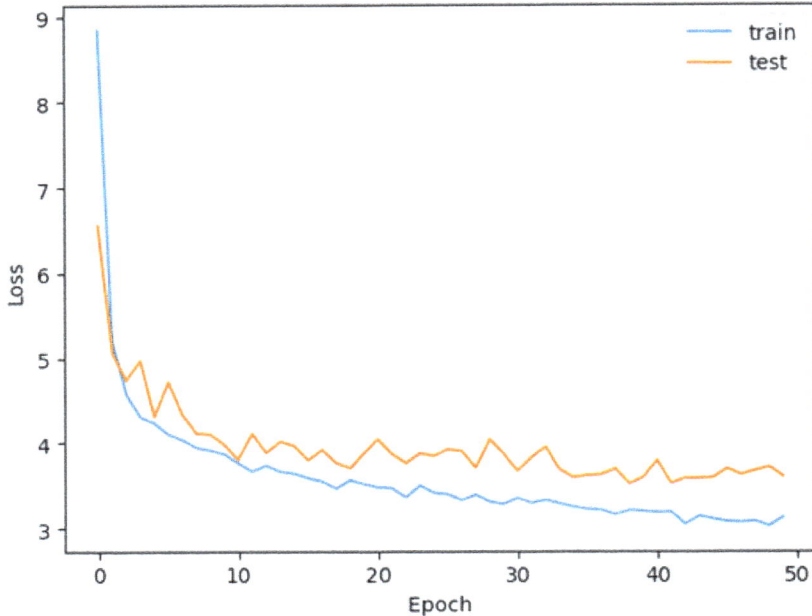

Figure 11.19 – Train and test losses as a function of the epoch for the style transfer network

As we can see in the preceding figure, the model is learning but tends to overfit slightly as the test loss does not decrease significantly anymore after about 40 epochs.

12. Finally, test the trained model on a bunch of images in the test set and display the results:

```
images, labels = next(iter(test_dataloader))

with torch.no_grad():
    outputs = model(images.to(device)).cpu()

plt.figure(figsize=(12, 6))
for idx in range(4):
    plt.subplot(2, 4, idx*2+1)
    plt.imshow(unnormalize(images[idx].permute(
        1, 2, 0).numpy(), mean, std))
    plt.axis('off')
    plt.subplot(2, 4, idx*2+2)
    plt.imshow(unnormalize(outputs[idx].permute(
        1, 2, 0).numpy(), mean, std).clip(0, 1))
    plt.axis('off')
```

Here are the results displayed:

Figure 11.20 – A few images and their predicted transferred style

As we can see in the preceding figure, the model manages to transfer some of the style, by giving a smooth skin and sometimes coloring the hair. It is not perfect though, as the output images seem to be too bright.

It is likely that by fine-tuning the loss weights and other hyperparameters, it would be possible to achieve better results.

There's more...

While this recipe showed how to use generative models such as Stable Diffusion to create new data in a fun way, it is possible to use it in many other applications. Let's see here how to use Stable Diffusion to create your own style transfer dataset, as well as a few other possible applications.

As mentioned earlier, Stable Diffusion allows us to create realistic and creative images based on input prompts. Unfortunately, on its own, it cannot effectively apply a style to a given image without compromising the original image's details (for example, the face shape, etc.). To do so, we can use another model based on Stable Diffusion: ControlNet.

ControlNet works like Stable Diffusion: it takes input prompts and generates output images. However, unlike Stable Diffusion, ControlNet will take control information as input, allowing us to specifically generate data based on a control image: this is exactly what was done to create the dataset of this recipe, efficiently adding a drawing style to faces while keeping the overall facial features.

The control information can take many forms, such as the following:

- Image contours with Canny edges or Hough lines, allowing us to perform realistic and limitless image augmentation for image classification

- Depth estimation, allowing us to efficiently generate background images

- Semantic segmentation, allowing image augmentation for semantic segmentation tasks

- Pose estimation, generating more images with people in a given pose, which can be useful for object detection, semantic segmentation, and more

- Much more, such as scribbles and normal maps

> **Note**
>
> The Canny edge detector and Hough line transform are typical image processing algorithms, allowing us to detect edges and straight lines in images, respectively.

As a concrete example, in the following figure, using an input image and the computed Canny edges as input, as well as a text prompt such as *A realistic cute shiba inu in a fluffy basket*, ControlNet allows us to generate a new image really close to the first one. Refer to the following figure:

Original image | Canny edges | Generated image

Figure 11.21 – On the left, the input image; at the center, the computed Canny edges; and on the right, the generated image with ControlNet and the prompt "A realistic cute shiba inu in a fluffy basket"

There are several ways to install and use ControlNet, but the official repository can be installed with the following commands:

```
git clone git@github.com:lllyasviel/ControlNet.git

cd ControlNet
```

```
conda env create -f environment.yaml

conda activate control
```

From there, you have to download the models specifically for your needs, available on HuggingFace. For example, you can download the Canny model with the following command:

```
wget https://huggingface.co/lllyasviel/ControlNet/resolve/main/models/
control_sd15_canny.pth -P models/
```

> **Note**
> The downloaded file is more than 6 GB, so this might take time.

Finally, you can launch ControlNet UI with the following command line, `python gradio_canny2image.py`, and then follow the instructions by going to the created localhost, `http://0.0.0.0:7860`.

Using ControlNet and Stable Diffusion, given a powerful enough computer, you can now generate almost limitless new images, allowing you to train really robust and well-regularized models for computer vision.

See also

- Paper on the U-Net++ architecture: `https://arxiv.org/abs/1807.10165`
- More about neural style transfer: `https://en.wikipedia.org/wiki/Neural_style_transfer`
- The Wikipedia page about Stable Diffusion: `https://en.wikipedia.org/wiki/Stable_Diffusion`
- The paper on perceptual loss: `https://arxiv.org/pdf/1603.08155.pdf`
- This recipe was inspired by `https://medium.com/@JMangia/optimize-a-face-to-cartoon-style-transfer-model-trained-quickly-on-small-style-dataset-and-50594126e792`
- The official ControlNet repository: `https://github.com/lllyasviel/ControlNet`
- An advanced way of using ControlNet, allowing us to compose several models: `https://github.com/Mikubill/sd-webui-controlnet`
- The Wikipedia page about the Canny edge detector: `https://en.wikipedia.org/wiki/Canny_edge_detector`

Index

G

gate 235
gated recurrent unit (GRU) 261
  gates 235
  parameters 247
  reference link 247
  training 234-247
**Generative Adversarial Networks
    (GANs)** 374
**Generative Pre-trained Transformer
    3 (GPT-3)** 2
  using, for data augmentation 282-287
**generative pre-trained transformer
    (GPT) models** 283
Gini impurity 78, 79
GPT-2 model card
  reference link 288
Gradient boosting
  principles 107, 108
greedy algorithm 76

H

hashing 114-119
high cardinality features
  aggregating 120-124
  hashing 114-119
hue, saturation, value (HSV) 301
HuggingFace, text generation
  reference link 288
HuggingFace tokenizer 232, 233
human-level error 17
hyperparameter 43
  strategies, implementing to optimize 44
hyperparameter optimization 70
  performing 43-46

hyperparameters
  reference link 97
  used, for regularizing decision tree 95-97

I

IBM's case 3
image augmentation
  applying, with Albumentations 344-354
  training 354-357
ImageNet weights
  pretrained model, training with 328-336
images, with labels
  generating, techniques 368-372
imbalanced data
  resampling, with SMOTE 134-139
imbalanced dataset
  cons 130
  oversampling 130-134
  pros 130
  undersampling 125-128
imbalanced data, with SMOTE
  pros and cons 135
inference 86
Intersection over Union (IoU) 312
  advantages 312

J

Jaccard index 325

K

Kaggle API
  installing 115, 116
kernel 302

‹packt›

Packtpub.com

Subscribe to our online digital library for full access to over 7,000 books and videos, as well as industry leading tools to help you plan your personal development and advance your career. For more information, please visit our website.

Why subscribe?

- Spend less time learning and more time coding with practical eBooks and Videos from over 4,000 industry professionals

- Improve your learning with Skill Plans built especially for you

- Get a free eBook or video every month

- Fully searchable for easy access to vital information

- Copy and paste, print, and bookmark content

Did you know that Packt offers eBook versions of every book published, with PDF and ePub files available? You can upgrade to the eBook version at packtpub.com and as a print book customer, you are entitled to a discount on the eBook copy. Get in touch with us at customercare@packtpub.com for more details.

At www.packtpub.com, you can also read a collection of free technical articles, sign up for a range of free newsletters, and receive exclusive discounts and offers on Packt books and eBooks.

Other Books You May Enjoy

If you enjoyed this book, you may be interested in these other books by Packt:

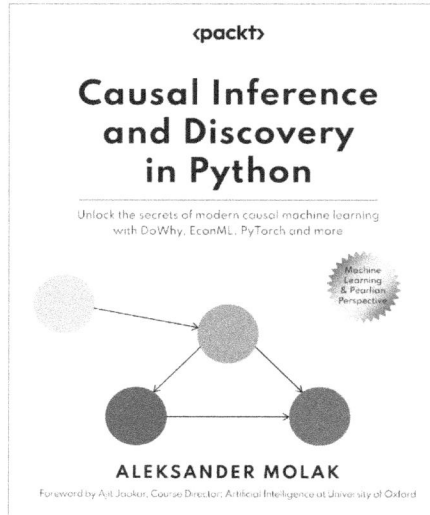

Causal Inference and Discovery in Python

Aleksander Molak

ISBN: 978-1-80461-298-9

- Master the fundamental concepts of causal inference
- Decipher the mysteries of structural causal models
- Unleash the power of the 4-step causal inference process in Python
- Explore advanced uplift modeling techniques
- Unlock the secrets of modern causal discovery using Python
- Use causal inference for social impact and community benefit

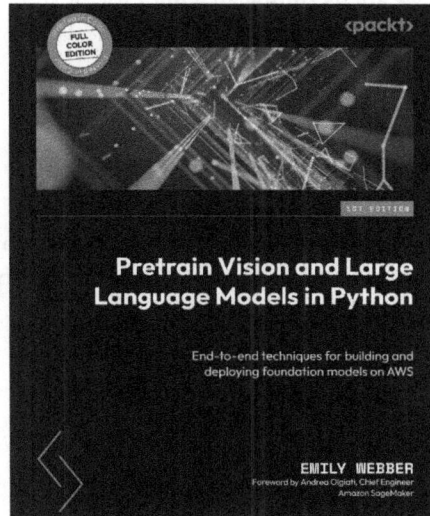

Pretrain Vision and Large Language Models in Python

Emily Webber

ISBN: 978-1-80461-825-7

- Find the right use cases and datasets for pretraining and fine-tuning
- Prepare for large-scale training with custom accelerators and GPUs
- Configure environments on AWS and SageMaker to maximize performance
- Select hyperparameters based on your model and constraints
- Distribute your model and dataset using many types of parallelism
- Avoid pitfalls with job restarts, intermittent health checks, and more
- Evaluate your model with quantitative and qualitative insights
- Deploy your models with runtime improvements and monitoring pipelines

Packt is searching for authors like you

If you're interested in becoming an author for Packt, please visit `authors.packtpub.com` and apply today. We have worked with thousands of developers and tech professionals, just like you, to help them share their insight with the global tech community. You can make a general application, apply for a specific hot topic that we are recruiting an author for, or submit your own idea.

Share Your Thoughts

Now you've finished *The Regularization Cookbook*, we'd love to hear your thoughts! Scan the QR code below to go straight to the Amazon review page for this book and share your feedback or leave a review on the site that you purchased it from.

`https://packt.link/r/1837634084`

Your review is important to us and the tech community and will help us make sure we're delivering excellent quality content.

Download a free PDF copy of this book

Thanks for purchasing this book!

Do you like to read on the go but are unable to carry your print books everywhere?

Is your eBook purchase not compatible with the device of your choice?

Don't worry, now with every Packt book you get a DRM-free PDF version of that book at no cost.

Read anywhere, any place, on any device. Search, copy, and paste code from your favorite technical books directly into your application.

The perks don't stop there, you can get exclusive access to discounts, newsletters, and great free content in your inbox daily

Follow these simple steps to get the benefits:

1. Scan the QR code or visit the link below

https://packt.link/free-ebook/9781837634088

2. Submit your proof of purchase
3. That's it! We'll send your free PDF and other benefits to your email directly

www.ingramcontent.com/pod-product-compliance
Lightning Source LLC
Chambersburg PA
CBHW081038220326
41598CB00038B/6919